Solutions manual to accompany

Organic Chemistry

by Clayden, Greeves, Warren, and Wothers

STUART WARREN

University of Cambridge

OXFORD

UNIVERSITY PRESS

OXFORD
UNIVERSITY PRESS

Great Clarendon Street, Oxford OX2 6DP

Oxford New York

Athens Auckland Bangkok Bogotá Buenos Aires Calcutta
Cape Town Chennai Dar es Salaam Delhi Florence Hong Kong Istanbul
Karachi Kuala Lumpur Madrid Melbourne Mexico City Mumbai
Nairobi Paris São Paulo Singapore Taipei Tokyo Toronto Warsaw

with associated companies in
Berlin Ibadan

Oxford is a trade mark of Oxford University Press

Published in the United States
by Oxford University Press Inc., New York

British Library Cataloguing in Publication Data
Data available

Library of Congress Cataloging in Publication Data

3 5 7 9 10 8 6 4

ISBN 0 19 870038 5 (Pbk)

Typeset by EXPO Holdings, Malaysia
Printed in Great Britain on acid-free paper by
Bath Press Ltd., Bath, Avon

Suggested solutions for Chapter 2

2

Problem 1

Draw good diagrams of saturated hydrocarbons with seven carbon atoms having (a) linear, (b) branched, and (c) cyclic frameworks. Draw molecules based on each framework having both ketone and carboxylic acid functional groups.

Purpose of the problem

To get you drawing simple structures well and to steer you away from rules and names towards creative structural ideas.

Suggested solution

There is only one linear saturated hydrocarbon with seven carbon atoms but there is a wide choice for the rest. We offer some possibilities but you may well have thought of others.

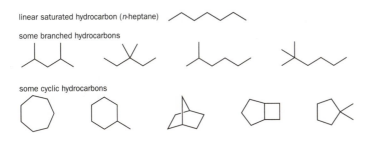

We give just a few examples for the ketones and carboxylic acids. You will notice that no C_7 carbocyclic acid is possible based on, say, cycloheptane without adding an extra carbon atom.

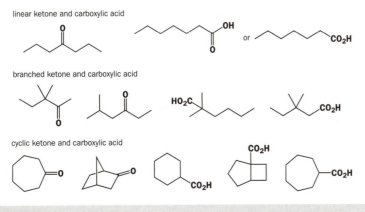

Problem 2

Study the structure of brevetoxin on p. 33. Make a list of the different types of functional group (you already know that there are many ethers) and of the numbers of rings of different sizes. Finally, study the carbon framework – is it linear, cyclic, or branched?

Purpose of the problem

To persuade you that functional groups are easily recognized even in complicated molecules and that, say, an ester is an ester whatever company it may keep. You were not expected to see the full implications of the carbon framework part of the question. That was to amuse and surprise you.

Suggested solution

The ethers are all the unmarked oxygen atoms in the rings: all are cyclic, seven in six-membered rings, two in seven-membered rings, and one in an eight-membered ring. There are two carbonyl groups, one an ester and one an aldehyde, and three alkenes.

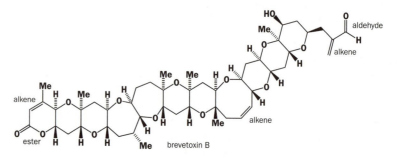

The carbon chain is branched because it has seven methyl groups branching off it and the aldehyde is also a branch. Amazingly, under this disguise, you can detect a basically linear carbon chain, shown with a thick black line, although it twists and turns throughout the entire molecule!

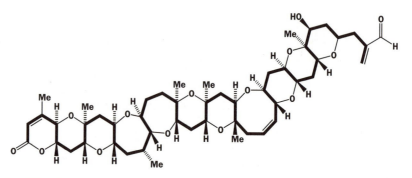

Problem 3

What is wrong with these structures? Suggest better ways of representing these molecules.

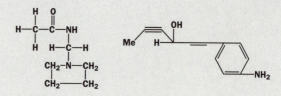

Purpose of the problem

To shock you with two dreadful structures and to try and convince you that well drawn realistic structures are more attractive to the eye as well as easier to understand.

Suggested solution

The bond angles are grotesque with square planar saturated carbon, alkynes at 120°, alkenes at 180°, bonds coming off benzene rings at the wrong angle, and so on. The left-hand structure would be clearer if most of the hydrogens were omitted. Here are two possible better structures for each molecule. There are many other correct possibilities.

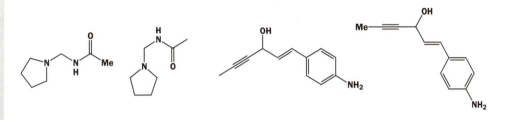

Problem 4

Draw structures corresponding to these names. In each case suggest alternative names that might convey the structure more clearly to someone who is listening to you speak.
(a) 1,4-di-(1,1-dimethylethyl)benzene
(b) 3-(prop-2-enyloxy)prop-1-ene
(c) cyclohexa-1,3,5-triene

Purpose of the problem

To help you appreciate the limitations of names, the usefulness of names for part structures, and, in the case of (c), to amuse.

Suggested solution

(a) 1,4-di-(1,1-dimethylethyl)benzene. More helpful name *para*-di-*t*-butyl benzene. It is sold as 1,4-di-*tert*-butyl benzene, an equally helpful name.

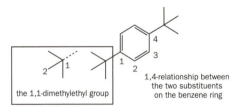

the 1,1-dimethylethyl group

1,4-relationship between the two substituents on the benzene ring

(b) 3-(prop-2-enyloxy)prop-1-ene. This name does not convey the simple symmetrical structure nor that it contains two allyl groups. Most chemists would call this 'diallyl ether' though it is sold as 'allyl ether'.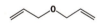

(c) cyclohexa-1,3,5-triene. This is, of course, benzene, but even IUPAC has not tried to impose this 'correct' name for such an important compound.

Problem 5

Draw one possible structure for each of these molecules, selecting any group of your choice for the 'wild card' substituents.

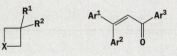

Purpose of the problem

To help you appreciate the wide range and versatility of general structures with X, R^1, Ar^1, etc. These become more important when you start a database search for a part structure.

Suggested solution

There are, of course, many possible solutions. X could be a heteroatom or a structural fragment while Ar could be any of a very large number of substituted benzene rings or even other types of aromatic rings. Our first two solutions in each case are ones you might have found and the rest are more inventive. The four-membered ring could have X = NH or CO (a ketone) while the substituents R^1 and R^2 could be the same (both methyl groups) or different (a benzene ring and an ether). In the last two structures X itself carries extra groups – the two oxygen atoms in the SO_2 group or an alkene – while R^1 and R^2 could be in a ring or be different highly functionalized groups. Of course, there are also some things that R^1, R^2, and X cannot be. R^1 and R^2 cannot be N or CH_2 while X cannot be Ph or Cl.

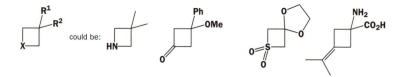

The three aryl groups in the second example might all be different or some might be the same. In the last two structures we show some unusual aromatic rings including some linked together and one with a nitrogen atom in the ring.

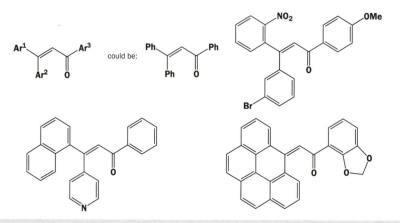

Problem 6

Translate these very poor 'diagrams' of molecules into more realistic structures. Try to get the angles about right and, whatever you do, don't include any square planar carbon atoms or other bond angles of 90°!

$C_6H_5CH(OH).(CH_2)_4COC_2H_5$ $O(CH_2CH_2)_2O$ $(CH_3O)_2CHCH=CHCH(OMe)_2$

Purpose of the problem

An exercise in interpretation and composition – this sort of 'structure', which is used when structures must be represented by ordinary printing, gives no clue to the shape of the molecule and you must decide that for yourself.

Suggested solution

You probably needed a few trial and error drawings first, but simply drawing out the carbon chains gives you the answers. The first is straightforward though you may not previously have seen the dot on the middle of the formula. This does not represent any atom but simply shows that the atom immediately before the dot is not joined to that immediately after it. The (OH) group is a substituent off the chain, not part of the chain itself. The second has no ends (Me groups, etc.) and so must be an unbroken ring. The third gives no clue as to the shape of the alkene and we have chosen *trans*. It also uses two ways to represent MeO. Either is correct but it is best to stick to one representation in any given molecule.

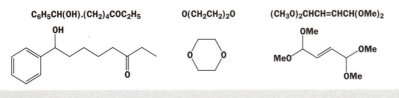

$C_6H_5CH(OH).(CH_2)_4COC_2H_5$ $O(CH_2CH_2)_2O$ $(CH_3O)_2CHCH=CHCH(OMe)_2$

Problem 7

Suggest at least six different structures that would fit the formula C_4H_7NO. Make good realistic diagrams of each one and say which functional group(s) are present.

Purpose of the problem

The identification and naming of functional groups is more important than the naming of compounds. This was your chance to experiment with different functional groups as well as different carbon skeletons.

Suggested solution

You will have found the carbonyl and amino groups very useful, but did you also use alkenes and alkynes, rings, ethers, alcohols, and cyanides? Here are twelve possibilities but there are many more. The functional group names in brackets are alternatives: some you will not have known. You need not have classified the alcohols and amines.

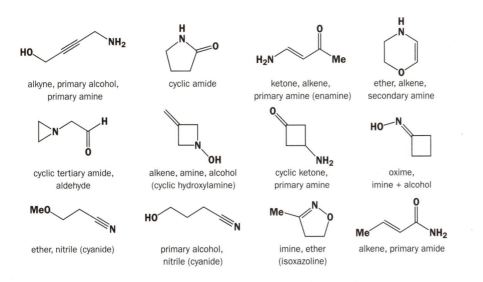

alkyne, primary alcohol, primary amine

cyclic amide

ketone, alkene, primary amine (enamine)

ether, alkene, secondary amine

cyclic tertiary amide, aldehyde

alkene, amine, alcohol (cyclic hydroxylamine)

cyclic ketone, primary amine

oxime, imine + alcohol

ether, nitrile (cyanide)

primary alcohol, nitrile (cyanide)

imine, ether (isoxazoline)

alkene, primary amide

Problem 8

Draw and name a structure corresponding to each of these descriptions.

(a) An aromatic compound containing one benzene ring with the following substituents: two chlorine atoms having a *para* relationship, a nitro group having an *ortho* relationship to one of the chlorine atoms, and an acetyl group having a *meta* relationship to the nitro group.

(b) An alkyne having a trifluoromethyl substituent at one end and a chain of three carbon atoms at the other with a hydroxyl group on the first atom, an amino group on the second, and the third being a carboxyl group.

Purpose of the problem

Just a little practice in converting verbal descriptions into structures and then naming them is justified. We hope you agree that the diagrams are more compact than the description in the question and more easily understood than the name.

Suggested solution

The structures are uniquely described by the rather verbose descriptions.

Naming the compounds requires (1) identifying the boss functional group, (2) numbering the skeleton, and (3) locating the functional group by name and number. The aromatic compound is a ketone with two carbon atoms – an ethanone (even though there cannot be a two-carbon ketone!). The carbonyl group is C1. The aromatic ring is joined on by its C1 at C1 and is numbered so as to give the smallest possible numbers to the substituents. It is (2,5-dichloro-3-nitrophenyl)ethanone.

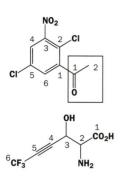

The aliphatic compound is a carboxylic acid which becomes C1. The rest is straightforward. Groups are usually named in alphabetical order. It is 2-amino-3-hydroxy-6,6,6-trifluorohex-4-ynoic acid. If you don't agree with our names, don't worry. It doesn't really matter. What does matter is: did you draw clear comprehensible structures?

Problem 9

Draw full structures for these compounds, displaying the hydrocarbon framework clearly and showing all the bonds present in the functional groups. Name the functional groups.

AcO(CH$_2$)$_3$NO$_2$ MeO$_2$C.CH$_2$.OCOEt   CH$_2$=CH.CO.NH(CH$_2$)$_2$CN

Purpose of the problem

This is rather like Problem 6 except that more thought is needed for the details of the functional groups and you may have needed to check the 'organic elements' Ac, Et, and so on in the chapter.

Suggested solution

For once the solution can be simply stated as no variation is really possible The tricks for the first one are to see that 'AcO' represents an ester and to have only four bonds to nitrogen in NO$_2$. The second has two ester groups on the central CH$_2$ group but one is joined to it by a CO bond and the other by a CC bond. The last is straightforward except for the dots used to separate the substituents (explained in the answer to Problem 6).

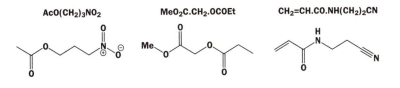

Problem 10

Identify the oxidation level of each of the carbon atoms in these structures with some sort of justification.

Purpose of the problem

This important exercise is one you will get used to very quickly and, before long, do without thinking. If you do, it will save you from many trivial errors. Remember that the oxidation *state* of carbon is +4, or C(IV), in all these compounds. The oxidation *level* of a functional group tells you with which oxygen-based functional group it is interchangeable without oxidation or reduction.

Suggested solution

Just count the number of bonds to heteroatoms. These can range from none to four. The only tricky ones are alkenes and alkynes, which have no heteroatoms but are formed by dehydration of alcohols, aldehydes, or ketones. There is a summary chart on p. 36 of the textbook, but briefly:

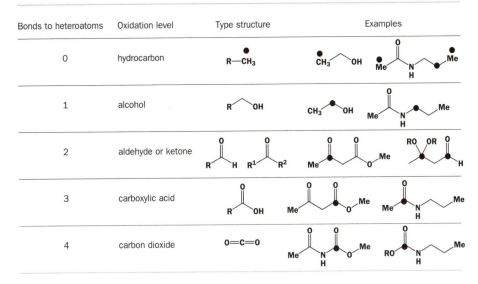

Bonds to heteroatoms	Oxidation level	Type structure	Examples
0	hydrocarbon		
1	alcohol		
2	aldehyde or ketone		
3	carboxylic acid		
4	carbon dioxide		

In these cases we have examples of all oxidation levels. Check the answer against yours and the table. In the case of the alkene, formally a dehydration product from an alcohol, either but not both of the C atoms is at the alcohol oxidation level.

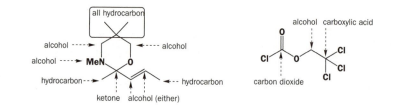

Problem 11

If you have not already done so, complete the exercises on pp. 23 (drawing amino acids) and 44 (giving structures for the 10 common compounds and three common solvents).

Purpose of the problem

These compounds are important so drawing them again is not just a useful exercise; it also helps reinforce your knowledge of these structures.

Suggested solution

Here are the drawings we suggested for the amino acids on p. 23. There are other ways.

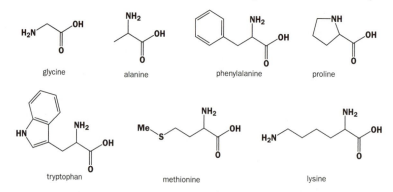

The 'ten common compounds' and 'three common solvents' are specially important and we recommend that you learn these structures. We generally pour scorn on the idea of memorizing things, so take note when we recommend it!

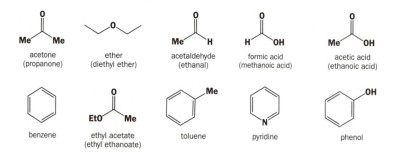

The three common solvents below join ether, acetone, ethyl acetate (EtOAc), toluene, and pyridine as commonly used solvents for organic reactions. One or other of these will dissolve almost any organic compound.

tetrahydrofuran (THF)   dimethylformamide (DMF)   dimethylsulfoxide (DMSO)

Suggested solutions for Chapter 3

3

Problem 1

How does the mass spectrum gives evidence of isotopes in the compounds of bromine, chlorine, and carbon? Assuming the molecular ion of each of these compounds is of 100% abundance, what peaks (and in what intensity) would appear around that mass number? (a) C_2H_5BrO, (b) C_{60}, (c) C_6H_4BrCl? Give in cases (a) and (c) a possible structure for the compound. What compound is (b)?

Purpose of the problem

To give you practice in spotting the important atoms that have isotopes and in interpreting multiple molecular ions. The molecular ion is the most important peak in the mass spectrum and is often the only peak to interest us.

Suggested solution

Bromine has two isotopes, ^{79}Br and ^{81}Br, in about a 1:1 ratio, chlorine has two, ^{35}Cl and ^{37}Cl, in about a 3:1 ratio and there is about 1.1% ^{13}C in normal compounds. Hence the molecular ions of the three compounds will be as follows.

(a) C_2H_5BrO, MW 124/126. There will also be weak peaks (2.2% of each main peak) at 125 and 127 for the same ions with ^{13}C.

(b) C_{60} has a molecular ion at 720 with a strong peak at 721. The chance that one C atom is ^{13}C is simply $60 \times 1.1\% = 66\%$ so the M + 1 peak is more than half as strong as M$^+$ itself. The chance of two ^{13}C atoms is small.

■ Compounds containing Br or Cl should have their molecular ions specified like this. The first figure gives M$^+$ with ^{79}Br and the second with ^{81}Br.

(c) C_6H_4BrCl is more complicated as the molecular ion will contain a 1:1 ratio of ^{79}Br and ^{81}Br, and a 3:1 ratio of ^{35}Cl and ^{37}Cl. The molecular ion consists of four peaks (ratios in brackets): $C_6H_4^{79}Br^{35}Cl$ (3), $C_6H_4^{81}Br^{35}Cl$ (3), $C_6H_4^{79}Br^{37}Cl$ (1), and $C_6H_4^{81}Br^{37}Cl$ (1). The masses of these peaks are 190, 192, 192, and 194. The molecular ion will therefore have three peaks at 190, 192, and 194 in a ratio of 3:4:1 with peaks at 191, 193, and 195 at $6 \times 1.1\% = 6.6\%$ of the peak before it. So the complete picture is 190 (75%), 191 (5%), 192 (100%), 193 (6.6%), 194 (25%), and 195 (1.7%).

Compound (b) is, of course, buckminsterfullerene and the other compounds might be isomers such as these.

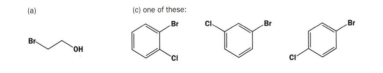

Problem 2

The ^{13}C NMR spectrum for ethyl benzoate contains these peaks: 17.3, 61.1, 100–150 p.p.m. (four peaks), and 166.8 p.p.m. Which peak belongs to which carbon atom?

Purpose of the problem

To familiarize you with the four main regions of the ^{13}C NMR spectrum: saturated carbons at 0–50 p.p.m., saturated carbons next to oxygen at 50–100 p.p.m., alkenes and aromatic compounds with unsaturated carbon at 100–150 p.p.m., and unsaturated carbon next to oxygen (usually C=O) at 150–200 p.p.m.

Suggested solution

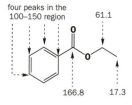

four peaks in the
100–150 region   61.1

There are four different types of carbon atom in the benzene ring (don't forget the 'ipso' carbon joined to the substituent) and it is not easy to say which is which. The rest can be deduced from the four main regions.

Problem 3

The thinner used in typists' correction fluids is a single compound, $C_2H_3Cl_3$, having ^{13}C NMR peaks at 45.1 and 95.0 p.p.m. What is its structure? A commercial paint thinner gives two spots on chromatography and has ^{13}C NMR peaks at 7.0, 27.5, 35.2, 45.3, 95.6, and 206.3 p.p.m. Suggest what compounds might be used to make up this thinner.

Purpose of the problem

To start you on the road of structure identification with one very simple problem and then some deductive reasoning. It is necessary to think about the size of chemical shifts in this case.

Suggested solution

There are only two possible structures for a compound $C_2H_3Cl_3$ – it must be trichloroethane and the chlorines can be distributed in these two ways.

■ We have ignored the effects of the Cl atoms on the other C atom, but they will also increase the chemical shift to a lesser extent.

The first isomer would have a peak for the methyl group in the region 0–50 p.p.m. and one for the CCl_3 group at a much larger chemical shift because of the three chlorine atoms. The second isomer would have a peak for CH_2Cl in the 50–100 p.p.m. region and one for $CHCl_2$ at a larger shift but the two shifts would not be so far apart. The observed shifts are 50 p.p.m. apart and, in fact, the compound is 1,1,1-trichloroethane. The NMR spectrum of the 1,1,2-trichloro isomer has peaks at 50.5 and 70.7 p.p.m. – much closer together.

The mixture probably contains the same trichloroethane as the peaks at 45.3 and 95.6 p.p.m. suggest (the chemical shifts of the mixture need not be quite the same as those of the pure compound as each compound is effectively dissolved in the other). The remaining peaks at 7.0, 27.3, 35.2, and 206.3 p.p.m. definitely belong to a carbonyl compound, probably a ketone since 206.3 p.p.m. is so large. Butanone fits the bill as it has one methyl and one CH_2 group on the carbonyl group and one methyl away from any electronegative atoms. There are other possibilities.

27.3   35.2

206.3   7.0

Problem 4

The 'normal' O–H stretch (i.e. without hydrogen bonding) comes at about 3600 cm^{-1}. What is the reduced mass (μ) for O–H? What happens to the reduced mass when you double the atomic weight of each atom in turn, that is, what is μ for O–D and what is μ for S–H? In fact, both O–D and S–H stretches come at about 2500 cm^{-1}. Why?

Purpose of the problem

To get you thinking about the position of IR bands in terms of the two main influences – reduced mass and bond strength. The relationship between the frequency (v) of the vibration, the force constant f (more or less equals bond strength), and the reduced mass (μ) is given by this equation.

$$v = \frac{1}{2\pi c}\sqrt{\frac{f}{\mu}}$$

Suggested solution

The reduced mass (μ) of a vibrating bond is given by this equation.

$$\mu = \frac{m_1 m_2}{m_1 + m_2}$$

So the reduced mass of OH is 16/17 or about 0.94. When you double the mass of H, the reduced mass of OD becomes 32/18 or about 1.78 which is roughly double that for OH. But when you double the mass of O, the reduced mass for SH is 32/33 or about 0.97 – hardly changed from OH! The change in reduced mass is enough to account for the changed frequency of the OD bond – it changes by about $\sqrt{2}$ – but cannot account for the change from OH to SH as the two reduced masses are about the same. The only explanation of this can be that the SH bond is weaker than the OH bond by a factor of about 2.

There is an important principle to be deduced from this problem. Very roughly, the reduced masses of all bonds involving heavier atoms (C, N, O, S, etc.) are about the same and the differences in IR stretching frequency are mostly due to changes in bond strength. This is most dramatic in comparing single, double, and triple bonds. Only with bonds involving hydrogen does the reduced mass become the more important factor, though it is also significant in comparing, say, C–O with C–Cl.

Problem 5

Three compounds, each having the formula C_3H_5NO, have the IR spectra summarized here. What are their structures? Without ^{13}C NMR data, it may be easier to tackle this problem by first writing down all the possible structures for C_3H_5NO. In what specific ways would ^{13}C NMR data help?

(a) One sharp band above 3000 cm^{-1}; one strong band at about 1700 cm^{-1}
(b) Two sharp bands above 3000 cm^{-1}; two bands between 1600 and 1700 cm^{-1}
(c) One strong broad band above 3000 cm^{-1}; a band at about 2200 cm^{-1}

Purpose of the problem

To show that IR alone has some usefulness in the identification of molecules but that NMR is necessary even with very simple molecules. In answers to examination questions of this type it is essential to show how you interpret the data as well as to give a structure. If you get the answer right, the interpretation is not so important, but, if you get the answer wrong, you should still get some credit for your interpretation.

Suggested solution

(a) One sharp band above 3000 cm^{-1} must be an N–H and one strong band at about 1700 cm^{-1} is probably a C=O. That leaves C_2H_4 and so we might have one of these (there are other less likely structures). ^{13}C NMR would help because a(i) would have a carbonyl group and two signals for saturated C while a(ii) would also have a C=O but only one signal for saturated carbon as the compound is symmetrical.

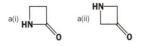

(b) Two sharp bands above 3000 cm^{-1} must be an NH$_2$ group and two bands between 1600 and 1700 cm^1 suggests two double bonds, presumably a C=O and a C=C. This leaves just three hydrogen atoms and gives us structures such as b(i) or b(ii). ^{13}C NMR would help because the C=O shift would show whether the compound was an amide or an aldehyde and the alkene shifts would reveal the presence of the NH$_2$ group in b(ii).

(c) One strong broad band above 3000 cm^{-1} must be an OH and a band at about 2200 cm^{-1} must be a triple bond, presumably CN as there would otherwise be NHs as well. That leaves C$_2$H$_4$ again but we do not need a ring and we have structures like these.

Problem 6

Four compounds having the molecular formula C$_4$H$_6$O$_2$ have the IR and ^{13}C NMR spectra given below. How many DBEs (Double Bond Equivalents) are there in C$_4$H$_6$O$_2$? What are the structures of the four compounds? You might again find it helpful to draw out some or all possibilities before you start.
(a) IR: 1745 cm^{-1}; ^{13}C NMR: 214, 82, 58, and 41 p.p.m.
(b) IR: 3300 (broad) cm^{-1}; ^{13}C NMR: 62 and 79 p.p.m.
(c) IR: 1770 cm^{-1}; ^{13}C NMR: 178, 86, 40, and 27 p.p.m.
(d) IR: 1720 and 1650 (strong) cm^{-1}; ^{13}C NMR: 165, 131, 133, and 54 p.p.m.

Purpose of the problem

First steps in real identification using two different methods. Because the molecules are so small (only four carbon atoms) drawing out a few trial structures gives you some ideas as to the types of compounds likely to be found.

Suggested solution

Here are some possible structures for C$_4$H$_6$O$_2$. It is clear that two double bonds or one double bond and a ring are likely to feature. The double bonds have got to be C=O or C=C (or both if there is no ring). Functional groups are likely to include alcohol, aldehyde, acid, and ketone.

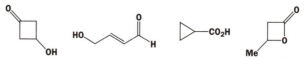

(a) IR: 1745 cm^{-1} must be a C=O group; ^{13}C NMR: 214 p.p.m. (aldehyde or ketone), 82 and 58 p.p.m. (two saturated carbons next to oxygen?), and 41 p.p.m. (one saturated carbon not next to oxygen but near some electron-withdrawing group). The second oxygen does not show up in the IR so it must be an ether. As there is only one double bond, there must be a ring. This suggests one structure.

(b) IR: 3300 (broad) cm^{-1} must be an OH; ^{13}C NMR: 62 and 79 p.p.m. must be a symmetrical molecule with no alkenes and no C=O so it must be a triple bond (usually 70–80 p.p.m.) and a saturated carbon next to oxygen. This again gives one structure. Note that the alkyne does not show up in the IR because it is symmetrical.

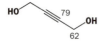

(c) IR: 1770 cm^{-1} must be some sort of C=O; ^{13}C NMR: 178 p.p.m. (C=O of acid derivative), 86 p.p.m. (saturated C next to O), 40 and 27 p.p.m. (saturated Cs not next to O). Again only one double bond so it must have a ring too. Looks like a close relative of (a)!

(d) IR: 1720 and 1650 (strong) cm^{-1} must be conjugated C=C and C=O; ^{13}C NMR: 165 p.p.m. (C=O of acid derivative), 131 and 133 p.p.m. (alkene), and 54 p.p.m. (saturated C next to O). This defines all the carbon atoms and it must be a simple unsaturated ester. Notice that we cannot tell which signal corresponds to which alkene carbon but that this does not affect our conclusion.

Problem 7

Three compounds of molecular formula C_4H_8O have the IR and ^{13}C NMR spectra given below. Suggest a structure for each compound, explaining how you make your deductions.

compound A   IR: 1730 cm^{-1}; ^{13}C NMR: 13.3, 15.7, 45.7, and 201.6 p.p.m.
compound B   IR: 3200 (broad) cm^{-1}; ^{13}C NMR: 36.9, 61.3, 117.2, and 134.7 p.p.m.
compound C   IR: no peaks except CH and fingerprint; ^{13}C NMR: 25.8 and 67.9 p.p.m.
(The rest of the problem as stated in the text is given below in the 'Suggested solution'.)

Purpose of the problem

More practice in the same essential skill. Notice that we have two more H atoms in this formula so we have either a ring or a double bond but not both. In addition, a bit of chemistry is added.

Suggested solution

compound A   IR: 1730 cm^{-1} (C=O of some sort); ^{13}C NMR: 13.3, 15.7, and 45.7 p.p.m. (three saturated carbon atoms with one next to some electron-withdrawing group but not oxygen), and 201.6 p.p.m. (aldehyde or ketone). No symmetry. A ketone would have two Cs next to C=O so this is just butanal.

compound B   IR: 3200 (broad) cm^{-1} (OH); ^{13}C NMR: 36.9 p.p.m. (saturated C), 61.3 p.p.m. (saturated C next to O), 117.2 and 134.7 p.p.m. (alkene). There are two possibilities here.

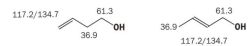

compound C   IR: no peaks except CH and fingerprint (oxygen must be ether); ^{13}C NMR: 25.8 p.p.m. (saturated C not next to O) and 67.9 p.p.m. (saturated C next to O). Note symmetry. As there are no double bonds, there must be a ring and it can only be THF.

Now the extra bit of the original problem, which may resolve our doubts on the structure of compound B:

Compound A reacts with NaBH$_4$ to give compound D. Compound B reacts with hydrogen gas over a palladium catalyst to give the same compound D. Compound C reacts with neither reagent. Suggest a structure for compound D from the data given and explain the reactions. (*Note*: H$_2$ reduces alkenes to alkanes in the presence of a palladium catalyst.)

compound D   IR: 3200 (broad) cm^{-1} (OH); ^{13}C NMR: 15.2, 20.3, and 36.0 p.p.m. (three saturated Cs not next to O) and 62.9 p.p.m. (saturated C next to O). Must be *n*-butanol.

NaBH$_4$ reduces the aldehyde to the alcohol and H$_2$/Pd would reduce the alkene in either of our candidates for compound B so we are none the wiser. Compound C does not react because it is an ether with no unsaturation. In fact, the data given are for the first of these possibilities. The data for compound B are for the first possibility and those for the second are given in the margin. You might perhaps have suspected that the difference between the two alkene carbons (117.2 and 134.7 p.p.m.)

suggests unequal substitution. You should still be content if you got either or both of these suggestions.

Problem 8

The situation is as follows. You have dissolved *t*-BuOH (Me$_3$COH) in MeCN with an acid catalyst, left the solution overnight, and found crystals with the following characteristics there in the morning. What are they?

IR: 3435 and 1686 cm^{-1}; ^{13}C NMR: 169, 50, 29, and 25 p.p.m.; mass spectrum (%): 115 (7), 100 (10), 64 (5), 60 (21), 59 (17), 58 (100), and 56 (7). (Don't try to assign all of these!)

Purpose of the problem

Structure determination in real life. This is a common situation – you carry out a reaction and you find a product that is not starting material – what is it? You need to use all the information and some logic. What you must *not* do is decide what the product is from your (limited) knowledge of chemistry and then make the data fit this structure.

Suggested solution

The mass spectrum shows that the compound has a molecular ion at 115 and is presumably C$_6$H$_{13}$NO – the sum of the two reagents *t*-BuOH and MeCN. It appears that they have added together but the infrared tells us that neither OH nor CN has survived. So what do we know?

- The IR spectrum shows that we have an NH group and a C=O group, which accounts for both heteroatoms.
- The NMR spectrum shows a carbonyl group (169 p.p.m.) and three types of saturated carbon atom. There must be a lot of symmetry and one explanation might be that the *t*-Bu group has survived. This would give us NH, C=O, and Me$_3$C as three fragments and would leave CH$_3$ left over. We can join these fragments up in two ways.

Each of these structures has four types of carbon atom and could fit the spectra. We might prefer 8b as the Me–C–N group is still present but a better decision is taken using the mass spectrum fragmentation pattern. The peak at 100 is the loss of 15 (Me) which could occur from either structure, though more typical of 8b. The base peak (100%) at 58 suggests Me$_2$C=NH$_2^+$, which can easily come from 8b but could come from 8a only with extensive reorganization of the skeleton. The correct structure is 8b, though to be certain we really need ^{1}H NMR too (Chapter 11).

Problem 9

How many isomers of trichlorobenzene are there? The 1,2,3-trichloro isomer is illustrated. Could they be distinguished by ^{13}C NMR?

Purpose of the problem

To get you thinking about the possible different structures answering one formula as detemined, say, by the mass spectrum. It is important not to stop trying when you have deduced one possible structure for a set of data but to continue thinking about alternatives. This exercise should help and should also make you think about symmetry.

Suggested solution

If two of the Cl atoms are next to each other ('*ortho*') there are two possible places for the third giving the 1,2,3- and 1,2,4-isomers. If the first two are 1,3-related ('*meta*') there are again two structures but we have just used one of them (the 1,2,4-isomer). If the first two are placed 1,4 ('*para*') there is only one place for the third and we have already found that (the 1,2,4-isomer). There are three isomers in all.

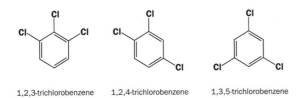

1,2,3-trichlorobenzene   1,2,4-trichlorobenzene   1,3,5-trichlorobenzene

The ^{13}C NMR spectrum will be different for each simply on account of the number of different carbon atoms. The 1,3,5-isomer is most symmetrical, having just two different signals (135.5 and 127.2 p.p.m.). The 1,2,3-isomer has four different signals (134.3, 131.6, 128.7, and 127.5 p.p.m.) but the 1,2,4-isomer has no symmetry and has six signals (133.4, 133.0, 131.1, 131.0, 130.2, and 127.9 p.p.m.). All these signals are in the 100–150 p.p.m. region.

Problem 10

How many signals would you expect in the ^{13}C NMR of the following compounds?

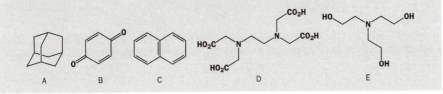

A   B   C   D   E

Purpose of the problem

To get you thinking about symmetry.

Suggested solution

Compound A has tetrahedral symmetry and there are only two types of C atom: all the CHs are the same (a) and all the CH_2s are the same (b). This is the famous compound adamantane – a crystalline solid in spite of its being a hydrocarbon with only ten carbon atoms. If you do not see the symmetry, make a molecular model – it is a beautiful structure.

Compound B is symmetrical too: the two C=O groups are the same (a) and so are the other atoms in the ring (b). This compound is called quinone and is an orange solid.

Compound C is naphthalene (moth balls) and has high symmetry as shown by the labelling a, b, and c.

Compound D is 'EDTA' (EthyleneDiamineTetracetic Acid – an important metal chelator). This time there are three types of carbon atom – two of type a and four each of b and c.

Compound E is 'triethanolamine' used a lot by biochemists. It is also symmetrical with only two types of carbon atom.

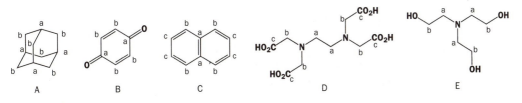

A   B   C   D   E

If you also said roughly where these types of carbon atom appeared in the spectrum, you should be pleased as that enhances your answer. Here are the actual numbers for interest only:

A (adamantane): 37.8 and 28.4 p.p.m.
B (quinone): 187.0 and 136.4 p.p.m.
C (naphthalene): 133.4, 127.8, and 125.7 p.p.m.
D (EDTA): 173.8, 60.9, and 54.5 p.p.m.
E (triethanolamine): 59.5 and 57.1 p.p.m.

Problem 11

How would mass spectra help you distinguish these structures?

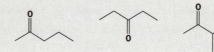

Purpose of the problem

To show you that fragmentation in the mass spectrum is sometimes useful.

Suggested solution

The three compounds are isomers, $C_5H_{10}O$, and have the same molecular ion at 86.

■ *Reminder.* Only positively charged fragments are detected in the mass spectrum.

Fragmentation of ketones usually occurs next to the C=O group to give first an acylium ion and then, by loss of CO, a carbocation. The fragment ions are different in the three cases. The first compound might lose the methyl or propyl side chains and give two series of fragment ions. Two of the fragments happen to have the same mass (43) so there will be peaks at 86, 71, 43, and 15 as well as others.

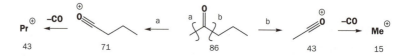

The second ketone is symmetrical and we need show only one fragmentation pattern. There will be peaks at 86, 57, and 29 and the two fragments have different masses from those above.

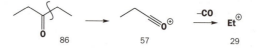

The third ketone is branched but will give the same basic pattern of fragments as the first compound: 86, 71, 43, 15. How can we tell the difference? The first compound has a long enough chain to do the McLafferty rearrangement losing a molecule (ethylene or ethene) and giving a fragment with an even mass.

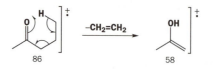

So, to summarize, the first ketone shows peaks at 86, 71, 58, 43, and 15; the second at 86, 57, and 29; and the third at 86, 71, 43, and 15 but with no peak at 58.

Suggested solutions for Chapter 4

4

Problem 1

In the (notional and best avoided in practice) formation of NaCl from a sodium atom and a chlorine atom, descriptions like this abound in textbooks: 'an electron is transferred from the valency shell of the sodium atom to the valency shell of the chlorine atom'. What is meant, in quantum number terms, by 'valency shell'? Give a complete description in terms of all four quantum numbers of that transferred electron: (a) while it is in the sodium atom and (b) after it has been transferred to the chlorine atom. Why is the formation of NaCl by this process to be discouraged?

Purpose of the problem

To encourage you to use the periodic table in working out the electronic configuration of atoms and to familiarize you with the four quantum numbers.

Suggested solution

Sodium and chlorine are at either end of the second row of the periodic table: Na, Mg, Al, Si, P, S, Cl, Ar. Sodium is in group 1 and has the configuration $1s^2$, $2s^2$, $2p^6$, $3s^1$. Chlorine is in group VII (17) and has the configuration $1s^2$, $2s^2$, $2p^6$, $3s^2$, $3p^5$. The 'valency shell' refers to the outer shell of the atom where all the chemistry occurs and is known by the value of the principal quantum number (n) for that shell. This is the number at the front of the electronic description: in this case (Na or Cl) the electrons are in 3s or 3p orbitals so $n = 3$.

While the electron is on the sodium atom it is a 3s electron, that is, $n = 3$; $l = 0$. There is only one type of s orbital so m_l must also be 0. The spin is arbitrarily up or down so it doesn't matter whether you say $m_s = -\frac{1}{2}$ or $+\frac{1}{2}$. When the electron is transferred to the chlorine atom it goes into a $3p_z$ orbital so we have $n = 3$ again but now $l = 1$ (p orbital), m_l is 1 (p_z orbital), and again $m_s = -\frac{1}{2}$ or $+\frac{1}{2}$. If any of this gave you trouble, you should refer to the chapter.

This method of making sodium chloride is to be discouraged because both elements are extremely reactive and would combine with an enormous release of energy. This is, of course, because the odd electron on sodium is much more stable if it is transferred to chlorine. We are more likely to make sodium and chlorine from abundant NaCl – there's plenty in the sea, for example.

Problem 2

What is the electronic structure of these species? You should consult a periodic table before answering.

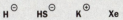

Purpose of the problem

To encourage you to use the periodic table in working out the electronic configuration of atoms. We really mean it!

Suggested solution

H⁻   Hydrogen is the first element in the periodic table and has just one proton in the nucleus and just one orbital (1s) available to be filled. If we fill it, we get a species with one proton and two electrons and that is H⁻. The answer is $1s^2$.

■ If you said that this species is hybridized and put electrons into $3sp^3$ orbitals you have done well. This answer is right too.

HS⁻   Sulfur is in group VI (16) in the same row as Na and Cl. It makes a σ bond to hydrogen and uses the remainder of the electrons as lone pairs to give a full noble gas configuration (same as Ar) of: $1s^2, 2s^2, 2p^6, 3s^2, 3p^6$.

K⁺   Potassium is below sodium in the periodic table so this cation has lost its 4s electron and is: $1s^2, 2s^2, 2p^6, 3s^2, 3p^6$. This configuration is the same as that of Cl⁻.

Xe   Xenon is a noble gas in group VIII (18). It comes some way down the table after He, Ne, Ar, and Kr and after two rows of transition elements that fill the 3d and 4d orbitals so it has $1s^2, 2s^2, 2p^6, 3s^2, 3p^6, 4s^2, 3d^{10}, 4p^6, 5s^2, 4d^{10}, 5p^6$.

Problem 3

What sort of bonds can be formed between s orbitals and p orbitals? Which will provide better overlap, 1s + 2p or 1s + 3p? Which bonds will be stronger, those between hydrogen and C, N, O, and F on the one hand or those between hydrogen and Si, P, S, and Cl on the other? Within the first group, bond strength goes in this order: HF > OH > NH > CH. Why?

Purpose of the problem

To encourage you to think about the energies of orbitals as well as just about their quantum description.

Suggested solution

s orbitals and p orbitals can combine to form σ bonds. In the chapter (p. 108) we discussed the structure of PH_3 which has bond angles of about 90° and is made of σ bonds between 1s(H) and 3p(P) orbitals. There is no special problem in overlapping either 2p or 3p orbitals with 1s orbitals though the 3p orbitals are larger.

1s   2p   1s   3p

The difference comes in the energy of the orbitals. The 2p orbitals are much closer in energy to the 1s orbital than the 3p orbitals and so the energy gain is greater on σ bond formation.

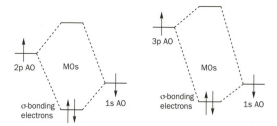

Bonds between H and C, N, O, and F are all stronger than bonds between H and Si, P, S, and Cl. This is partly because 2p AOs are used for the first but 3p AOs for the second group. The full story includes the fact that CH_4, NH_3, and H_2O are hybridized (so the lower energy 2s orbitals are used as well) while PH_3 and H_2S are not hybridized. SiH_4 is, of course, tetrahedral while it is difficult to say whether linear HF and HCl are hybridized or not!

■ This factor was used in Chapter 3 (p. 68) to explain the relative positions of the stretching frequencies of OH, NH, and CH in the infrared spectrum.

Within the group C, N, O, and F, the energy of the 1s orbital stays the same but the energy of the 2p (or of the sp^3 hybrid orbitals) drops as the elements get more electronegative. These orbitals get closer in energy to the 1s orbital and the gain in bond formation is correspondingly greater.

Problem 4

Though no helium 'molecule' He_2 exists, an ion He_2^+ does exist. Explain.

Purpose of the problem

To encourage you to think about the filling of atomic orbitals and to accept surprising conclusions.

Suggested solution

The orbitals of the He atom and the fact that no He_2 molecule exists are discussed in the book on pp. 97–8. The problem is that the 1s orbitals overlap to form a bonding (σ) and an antibonding (σ^*) orbital but both would be filled in the He_2 molecule and the bond order is zero. If there is one fewer electron, only one electron need go into the antibonding orbital (σ^*) and there is a bond order of one-half. The ion He_2^+ does exist.

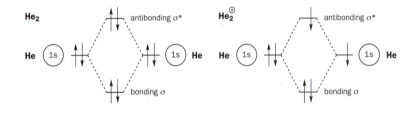

Problem 5

You may be surprised to know that the molecule CH_2, with divalent carbon, can exist. It is, of course, very unstable but it is known and it can have two different structures. One has an H–C–H bond angle of 180° and the other an angle of 120°. Suggest structures for these species and say which orbitals will be occupied by all bonding and nonbonding electrons. Which structure is likely to be more stable?

Purpose of the problem

To demonstrate that a simple MO treatment can be applied to strange and unknown molecules.

Suggested solution

The basic arrangements of the orbitals to get the 180° and 120° bond angles must be an sp hybridized carbon for 180° and an sp^2 carbon for 120°. This leaves over two p orbitals in the first case but just one p orbital and an sp^2 orbital in the second.

The orbitals for the σ structure are straightforward.

There will be four electrons in each of these σ bonds leaving two electrons to make up the six required (divalent carbon has only six valency electrons and cannot achieve a noble gas structure). In the 180° case, the two remaining p orbitals are degenerate so we can put one electron in each. In the 120° case, the remaining sp^2 orbital is of lower energy than the p orbital and will have both electrons.

Structures with unpaired degenerate electrons are usually more stable than those with a full and an empty orbital. However, we have told you only part of the story and we shall return to these 'carbenes' in Chapter 40.

Problem 6

Construct an MO diagram for the molecule LiH and suggest what type of bond it might have.

Purpose of the problem

To demonstrate that a simple MO treatment can be applied to ionic as well as covalent structures.

Suggested solution

H has, of course, only one electron in a 1s orbital. Li has three – a full 1s shell and one electron in the 2s orbital. Li is also very electropositive so its 2s orbital is high in energy (much higher than that of F – see the answer to Problem 3). The result is that lithium gives its 2s electron to the 1s orbital of H and an ionic compound results with both ions having the same electronic configuration: $1s^2$. The hydrides of the alkali metals are useful sources of hydride ion (H^-).

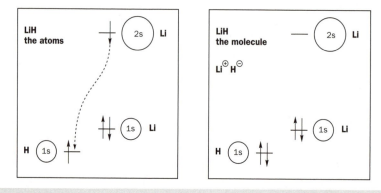

Problem 7

Deduce the MOs for the oxygen molecule. What is the bond order in oxygen and where are the 2p electrons?

Purpose of the problem

To let you try out your skill in a simple diatomic molecule that has a curious structure.

Suggested solution

■ You should note that oxygen has four 2p electrons and it will have two in one p orbital and one each in the other two to avoid repulsion.

Simply dock two oxygen atoms side by side and overlap the orbitals. The 1s and 2s (only the 2s is shown below) interact as usual but both bonding and antibonding MOs are occupied so there is no bonding. When we overlap the partly filled p AOs we find we can make three bonds – one 2pσ and two 2pπ bonds. Now we have two electrons left over and they have to go into antibonding 2pπ* MOs. The first two up are degenerate so it is better to put one electron in each and avoid the repulsion from two electrons in the same orbital.

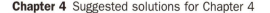

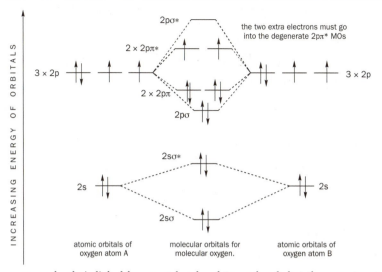

The oxygen molecule is linked by one σ bond and two π bonds but there are two unpaired electrons in a π* antibonding MO. These reduce the π bond order to 1 and the unpaired electrons are more or less located one on each O atom. These unpaired electrons make oxygen very reactive.

Problem 8

Construct MOs for acetylene (ethyne) without hybridization.

Purpose of the problem

It is not easy to find molecules where you can construct MOs from AOs without hybridization but ethyne is one such. You should not be ashamed if you failed to do this problem and you should be equally proud if you succeeded. We shall not in general construct MOs in this way as it is too difficult but if you simply set up an energy diagram of the AOs, in ascending order as usual, you can derive satisfactory MOs.

Suggested solution

Ethyne (acetylene) has a C–C triple bond. Each carbon bonds to only two other atoms – the other C and one of the Hs. Using MO theory, we can see that only the carbon 2s and $2p_x$ have the right symmetry to bind to two atoms at once which leaves the $2p_y$ and $2p_z$ to form π MOs with the 2p orbitals on the other carbon atom.

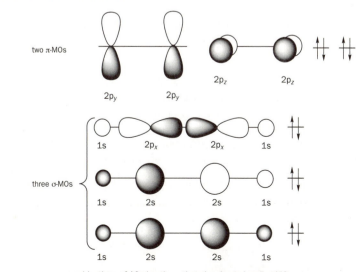

combinations of AOs in ethyne that give rise to bonding MOs

The three lowest orbitals are σ MOs because they are symmetrical about the line H–C–C–H. They are all bonding and all filled but they do not correspond to particular bonds in the molecule. The lowest is bonding right through, the second is bonding for both C–Hs but antibonding for C–C, and the third is again bonding for all three bonds. The two top orbitals are π MOs because they have a node (zero electron density) passing through the atoms. They are degenerate and correspond roughly to the two C–C π bonds. The total number of electrons (10) is right for 5 bonds (2 CHs, 1 σ CC, and 2 π CCs).

Problem 9

What is the shape and hybridization of each carbon atom in these molecules?

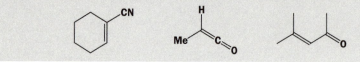

Purpose of the problem

To give you practice in selecting the right hybridization state for carbon atoms in molecules.

Suggested solution

Simply count the number of σ bonds and hybridize that many AOs: if two, then the C atom is sp hybridized (linear); if three, sp² (trigonal); and, if four, sp³ (tetrahedral). A simple statement of the answer should be enough. The atoms marked with an arrow are most likely to give you trouble: make sure you understand why they are as shown.

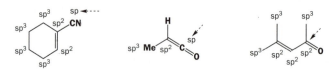

Problem 10

Suggest detailed structures for these molecules and predict their shapes. We have deliberately made noncommittal drawings to avoid giving away the answer to the question. Don't use these sorts of drawings in your answer.

$$CO_2, \ CH_2=NCH_3, \ CHF_3, \ CH_2=C=CH_2, \ (CH_2)_2O$$

Purpose of the problem

To give you practice in selecting the right hybridization state for carbon atoms and translating that information into three-dimensional structures for molecules.

Suggested solution

Carbon dioxide is linear as it has only two C–O σ bonds and no lone pairs on C. The C atom must be sp hybridized and the only trick is to get the two π bonds orthogonal to each other. They must be like that because the p orbitals on C involved in the two π bonds are themselves orthogonal ($2p_y$ and $2p_z$). Most people would draw the O atoms as sp² hybridized, rather than sp or even unhybridized, but this is unimportant as you can't really tell.

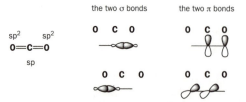

The imine has a C=N double bond so it must have sp^2 hybridized C and N. This means that the lone pair on N is in an sp^2 orbital and not in a p orbital. The molecule is planar (except for the methyl group which is, of course, tetrahedral) and bent at the nitrogen atom.

Trifluoromethane is, of course, tetrahedral with an sp^3 hybridized C atom. The arrangement of the lone pairs around the fluorine atoms (not shown) is also probably tetrahedral.

The next molecule, $CH_2=C=CH_2$ is allene and it has the same shape as CO_2 and for the same reasons, except that we can now be sure that the end carbon atoms are sp^2 hybridized as they are planar. However, the two planes are orthogonal so the molecule as a whole is not planar.

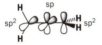

Finally, $(CH_2)_2O$ must be a three-membered ring and therefore the C–C–O skeleton of the molecule must be planar (three points are always in a plane!). However, the two carbon atoms are sp^3 hybridized (four σ bonds) and are approximately tetrahedral with the H atoms above and below the plane. The oxygen atom might also be sp^3 hybridized, but who knows?

Suggested solutions for Chapter 5

5

Problem 1

Each of these molecules is electrophilic. Identify the electrophilic atom and draw a mechanism for reaction with a generalized nucleophile Nu⁻, giving the product in each case.

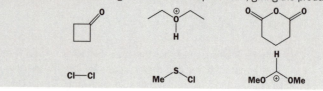

Purpose of the problem

The recognition of electrophilic sites is half the battle in starting to understand mechanisms. Here is some practice: later you will do this automatically.

Suggested solution

Here we have two cations, two compounds with π bonds, and two with nothing but σ bonds. One of the cations has only three bonds to positively charged carbon so it will react by addition of the nucleophile. The other has three-valent oxygen and so cannot add a nucleophile (four-valent oxygen would have 10 valency electrons and is unknown). The nucleophile must attack the proton instead. Some nucleophiles might attack the carbon atom next to the oxygen.

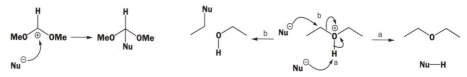

The two carbonyl compounds will be attacked at the carbon atom with cleavage of the π bond. In general, π bonds are more easily broken than σ bonds and negative charges end up on electronegative atoms. If you proposed that each compound reacted further, you were right, but the expected answers do not include these extra steps shown in square brackets. A full discussion is found in Chapter 12 of the main text.

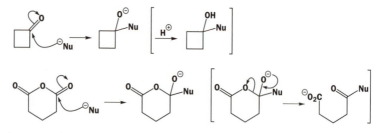

The last two molecules are forced to break σ bonds. In the case of chlorine, the two ends are the same so you can attack either Cl atom. In MeSCl, the S–Cl bond is weaker than the C–S bond and Cl is the most electrophilic atom in the molecule. The S–Cl bond is broken and the negative charge ends up as chloride ion.

Problem 2

Each of these molecules is nucleophilic. Identify the nucleophilic atom and draw a mechanism for reaction with a generalized electrophile E⁺, giving the product in each case.

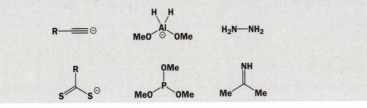

Purpose of the problem

The recognition of nucleophilic sites is the other half of the battle in starting to understand mechanisms. Here is some practice: later you will do this too automatically. Reactions occur when the two meet.

Suggested solution

This time there are three anions but only two of them (the alkyne and the sulfur anions) have lone pair electrons. We can start our arrows from these and they are the points where the electrophile will attach itself.

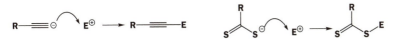

The last anion is like the BH_4^- anion we discussed on p. 125 of the main text. The negative charge does not show a pair of electrons on Al but just an imbalance of protons and electrons. All the valency electrons are in the bonds and we must use these σ electrons in the reaction. The arrow should start halfway along the σ bond and emerge through the H atom.

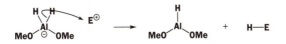

One nucleophile has a π bond from C to N. The nitrogen atom also has a lone pair of electrons and you could start your arrow either there or halfway down the π bond – it doesn't matter which.

■ This point is explored in Chapter 7.

The remaining two nucleophiles have lone pairs. One is symmetrical ($NH_2–NH_2$, hydrazine) and will attack through one nitrogen atom. You may have drawn the product as a cation or you may have removed a proton from it. Either answer is correct.

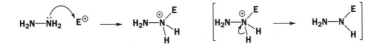

Finally, the phosphorus compound has four atoms with lone pairs! There are three OMe groups and the phosphorus atom itself. However, the lone pairs on the oxygen atoms are probably in $2sp^3$ orbitals (and are certainly in some kind of orbital with the principal quantum number 2) while that on the phosphorus atom is in a $3sp^3$ orbital and is of higher energy. It reacts.

Problem 3

Complete these mechanisms by drawing the structure of the products in each case.

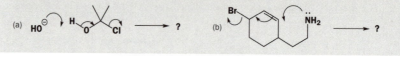

Purpose of the problem

First practice in interpreting curly arrows and drawing the products. Once the arrows are drawn, there is no more scope for decision-making. You must draw the products.

Suggested solution

Just break the bonds that are being broken and make the bonds that are being formed. It's as simple as that, though you might straighten out the products a bit so that there aren't any funny angles.

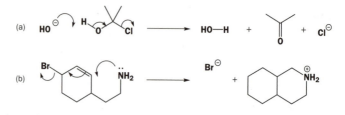

Problem 4

Each of these electrophiles could react with a nucleophile at (at least) two different atoms. Identify these atoms and draw a mechanism for each reaction together with the products from each.

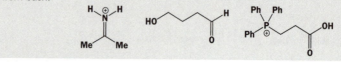

Purpose of the problem

First practice in considering different possible reactions. One of the reactions might seem trivial but it isn't.

Suggested solution

In each case one of the electrophilic sites is an acidic proton. There is also an electrophilic π bond in each case (C=O or C=N$^+$). We might as well use the same abbreviation (Nu$^-$) for the nucleophile that we used in Problem 1. For the first case, we draw the two reactions separately.

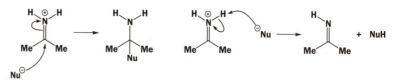

For the second case, it should be clear enough if we draw the two alternatives on the same diagram.

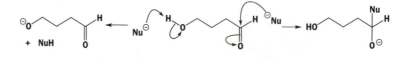

The last compound also has an electrophilic P atom so there are three possibilities. Don't worry if you missed this last one but phosphorus comes below nitrogen in the periodic table and, unlike nitrogen, can have five σ bonds as in PCl$_5$.

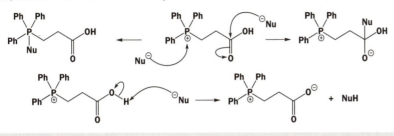

Problem 5

Put in the arrows on these structures (which have been drawn with all the atoms in the right places!) to give the products shown.

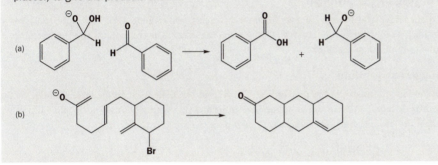

Purpose of the problem

To encourage you to draw arrows for unknown reactions and to show you how easy it is.

Suggested solution

All you have to do is to see which new bonds are formed and which old bonds are broken and draw arrows out of the one into the other. Which way should the arrows go? Take them from an electron-donating atom (an anion in both these examples) towards an electron-accepting atom (O and Br here). In the first example, a hydrogen atom has moved from the left- to the right-hand molecule and this is best shown by an atom-specific arrow.

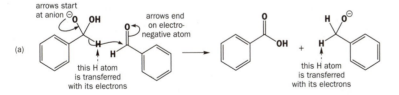

Don't worry if your arrows are not exactly the same as ours – as long as they start and finish in the right place and move the right H atom, they're all right. The notes on the mechanism are for your guidance – you should not usually include them. The second reaction looks more complicated but the problem is easier – just move electrons through the molecule.

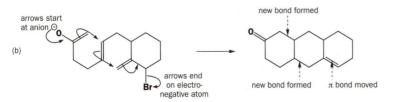

Problem 6

Draw mechanisms for these reactions. The starting materials have not necessarily been drawn in a helpful way.

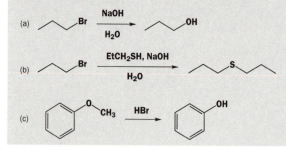

Purpose of the problem

To encourage you to draw arrows for unknown reactions without help. This time you have to decide how to draw the molecules so that reaction can occur. The compounds and the reactions are much simpler than the last set.

Suggested solution

In the first example, OH has replaced Br. The reagent NaOH is a salt so the reactive species is the hydroxide anion. We can draw the mechanism in one step with HO⁻ as a nucleophile displacing stable Br⁻ from the organic molecule.

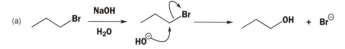

The same reagents are used in the second example with the addition of 'EtCH$_2$SH'. This change is obviously important because the product contains this unit rather than OH. We should first draw out this compound and use NaOH as a base to remove a proton from the SH group. The second step is example (a) with a different nucleophile.

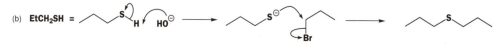

Example (c) uses HBr as the reagent. This is a strong acid with an electrophilic proton. The best nucleophile in the organic compound is the oxygen atom so the first step is a proton transfer and the second step uses the bromide ion, as a nucleophile for the methyl group. Direct attack at O⁺ would give impossible four-valent oxygen.

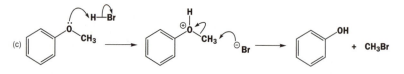

Problem 7

Draw a mechanism for this reaction.

PhCHBr.CHBr.CO$_2$H + NaHCO$_3$ ⟶ PhCH=CHBr

Hints. First draw good diagrams of the reagents. NaHCO$_3$ is a salt and a weak base – strong enough only to remove which proton? Then work out which bonds are formed and which broken, decide whether to push or pull, and draw the arrows. What are the other products?

Purpose of the problem

To develop your arrow-drawing skill in a more difficult example.

Suggested solution

Best to follow the advice given in the hint. First, draw the molecules better.

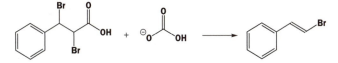

If $NaHCO_3$ is a salt we need draw only the anion, as the very stable Na^+ won't do anything. If the anion is a weak base it can remove only the most acidic proton from the organic molecule and that must be the CO_2H proton.

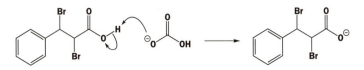

Now we need to lose one Br atom – the one nearer to the benzene ring – and the whole CO_2^- group which must fall off as CO_2. We start our arrows on the negative charge, form the new π bond, and lose Br as the stable anion.

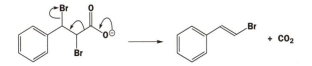

Suggested solutions for Chapter 6

6

Problem 1

Draw mechanisms for these reactions.

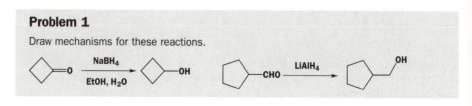

Purpose of the problem

Rehearsal of a simple but important mechanism that works for all aldehydes and ketones.

Suggested solution

Draw out the BH_4 or AlH_4 anion, and the compound if necessary, and transfer the hydride ion. A second protonation step is also needed – during the work-up in the second case. It is not necessary to draw out the whole metal hydride anion but you must draw out one metal-hydrogen bond as you need to take electrons from that bond.

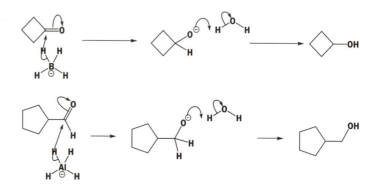

Problem 2

Cyclopropanone exists as the hydrate in water but 2-hydroxyethanal does not exist as its hemiacetal. Explain.

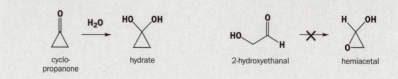

Purpose of the problem

To get you thinking about equilibria and hence stability of compounds rather than reaction mechanisms.

Suggested solution

Hydration is an equilibrium reaction so the mechanism is not strictly relevant to the question. If you have drawn the mechanism, you should be proud rather than ashamed because it is an important mechanism. To answer the question we must consider the effect of the three-membered ring. All three-membered rings are unstable because the ring angles are about 60° instead of the usual angle. Cyclopropanone is very strained because the trigonal (sp²) carbon would like an angle of 120° and there is '60° of strain'. In the hydrate the trigonal carbon is tetrahedral (sp³) and there is only '49° of strain'. The hydrate is more stable than the ketone.

The second case is totally different. The hydroxy-aldehyde is not strained but the hemiacetal has 49° of strain at each corner. Even without strain, hydrates and hemiacetals are normally less stable than aldehydes or ketones because one C=O bond is worth more than two C–O single bonds. In this case the hemiacetal is even less stable because of strain.

Problem 3

One way to make cyanohydrins is illustrated here. Suggest a detailed mechanism for the process.

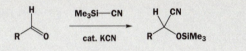

Purpose of the problem

To help you get used to mechanisms involving silicon and to revise an important way to promote additions to C=O groups.

Suggested solution

The silyl cyanide is an electrophile while the cyanide ion in the catalyst is a nucleophile. Cyanide adds to the carbonyl group and the oxyanion product is captured by silicon so liberating another cyanide ion for the next cycle.

Problem 4

There are three possible products from the reduction of this compound with sodium borohydride. What are their structures? How would you distinguish them spectroscopically, assuming you can isolate pure compounds?

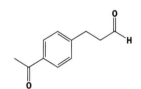

Purpose of the problem

To let you think practically about reactions that may give more than one product.

Suggested solution

The three compounds are easily drawn: one or other C=O or both may be reduced.

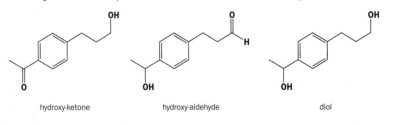

hydroxy-ketone   hydroxy-aldehyde   diol

The diol has no C=O group in the ^{13}C NMR or infrared and has a molecular ion in the mass spectrum at two mass units higher than the other two products. Distinguishing them is more tricky. The hydroxy-ketone has a conjugated carbonyl group (about 1680 cm^{-1} in the infrared) but the hydroxy-aldehyde is not conjugated (about 1730 cm^{-1} in the infrared). The chemical shift of the C–OH carbon atoms in the 100–150 p.p.m. region will also be different because the benzene ring is next to this atom in the hydroxy-ketone. Calculations from tables in Williams and Fleming suggest about 80 p.p.m. for the hydroxy-ketone and about 60 p.p.m. for the hydroxy-aldehyde. The mass spectra will also be different – simple α-cleavage gives quite different fragments.

■ D. H. Williams and I. Fleming (1995). *Spectroscopic methods in organic chemistry* (5th edn). McGraw-Hill, London.

Problem 5

The triketone shown here is called 'ninhydrin' and is used for the detection of amino acids. It exists in aqueous solution as a monohydrate. Which of the three ketones is hydrated and why?

Purpose of the problem

To start you thinking about why some carbonyl groups are more stable than others.

Suggested solution

ninhydrin hydrate

The two ketones next to the benzene ring are conjugated with it and thereby stabilized though they are also destabilized by the middle carbonyl group – two electron-withdrawing groups next to each other is a bad thing. The central carbonyl group has no stabilization from the benzene ring and a double dose of destabilization from its neighbours.

Problem 6

This hydroxy-ketone shows no peaks in its infrared spectrum between 1600 and 1800 cm^{-1} but it does show a broad absorption at 3000 to 3400 cm^{-1}. In the ^{13}C NMR spectrum, there are no peaks above 150 p.p.m. but there is a peak at 110 p.p.m. Suggest an explanation.

Purpose of the problem

Revision of Chapter 3 with a reaction from this chapter.

Suggested solution

The evidence shows that there is no carbonyl group in this molecule but that there is an OH group. The peak at 110 p.p.m. looks at first sight like an alkene but that is not possible (try to draw any alkene structures and you will see why) so it must be an unusual saturated carbon atom (perhaps one with two oxygen atoms). You might also argue that an alcohol and a ketone could react to give a hemiacetal, and that, of course, is what it is. The compound exists as the stable cyclic hemiacetal – stable because of the ring size.

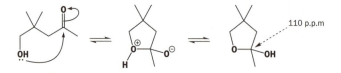

Problem 7

Each of these compounds is a hemiacetal and therefore formed from an alcohol and a carbonyl compound. In each case give the structure of these original materials.

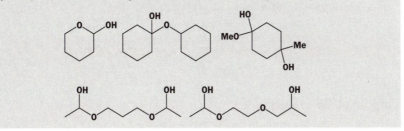

Purpose of the problem

To give you practice in seeing the underlying structure of a hemiacetal.

Suggested solution

Each OH group represents a carbonyl group in disguise (marked with a black blob). Just remove the other oxygen atom with whatever is attached to it and you have the two components: an alcohol and an aldehyde or a ketone. The first example shows how it is done.

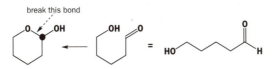

The next is similar but the alcohol is from a different molecule.

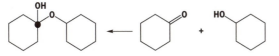

Do not be deceived by the next: it is not symmetrical. There is one hemiacetal (two oxygens on the same carbon atom) but the other end of the molecule is a simple tertiary alcohol.

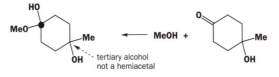

Similarly, the last two examples are not quite the same. The first is indeed symmetrical but the second has one oxygen atom in a different position. There is only one hemiacetal.

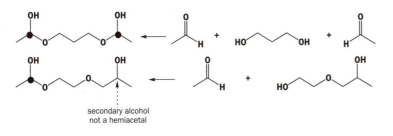

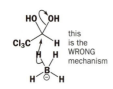

this is the WRONG mechanism

Problem 8

Trichloroethanol may be prepared by the direct reduction of chloral hydrate in water with sodium borohydride. Suggest a mechanism for this reaction. (Warning! Sodium borohydride does *not* displace hydroxide from carbon atoms!)

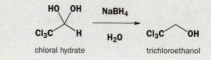

chloral hydrate   trichloroethanol

Purpose of the problem

To help you detect bad mechanisms and find concealed good ones.

Suggested solution

The wrong mechanism, the one the question warns you to avoid, is shown in the margin just to clear the decks. If $NaBH_4$ doesn't displace like this, then what does it do? We know it attacks carbonyl groups to give alcohols and to get trichloroethanol we should have to reduce chloral and we have chloral hydrate. Hydrates are in equilibrium with their carbonyl compounds, so this is the answer!

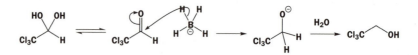

Problem 9

It has not been possible to prepare the adducts from simple aldehydes and HCl. What would be the structure of such compounds, if they could be made, and what would be the mechanism of their formation? Why cannot these compounds in fact be made?

Purpose of the problem

More revision of equilibria to help you develop a judgement about stability.

Suggested solution

This time we need a mechanism so that we can work out what would be formed. Protonation of the carbonyl group and then nucleophilic addition of chloride ion gives the supposed product.

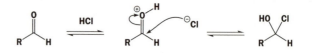

There's nothing wrong with the mechanism; it's just that the reaction is an equilibrium that will run backwards. Hemiacetals are unstable because they decompose back to carbonyl compound and alcohol. Chloride ion is very stable and this reaction will run backwards even more readily.

Problem 10

What would be the products of these reactions? In each case give a mechanism to justify your predictions.

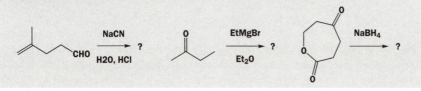

Purpose of the problem

To give you practice in the art of predicting products, more difficult than simply justifying a known answer.

Suggested solution

Each of these reactions is straight out of the textbook and each is a simple addition to the carbonyl group. The first is cyanohydrin formation and you need to draw out the aldehyde group to make a good job of the mechanism.

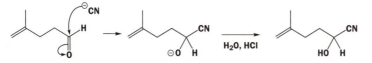

The second is a standard Grignard reaction and you just need to remember that the aqueous work-up step is not usually written down but is still needed.

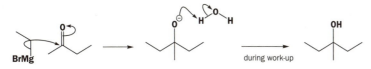

The only trap is in the reduction of the cyclic keto-ester where you need to recall that NaBH$_4$ reduces ketones but doesn't reduce esters. Correct identification of functional groups matters.

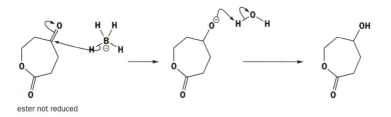

ester not reduced

Problem 11

The equilibrium constant K_{eq} for formation of the cyanohydrin of cyclopentanone and HCN is 67, while for butan-2-one and HCN it is 28. Explain.

Purpose of the problem

More revision of equilibria, this time with some numbers.

Suggested solution

We need first to state the problem in chemistry rather than in writing.

The two ketones are very similar, the ring being the main difference. The two equilibrium constants are also very similar as K_{eq} is a simple ratio of [cyanohydrin] to [ketone] and is not on a log scale. There is a factor of about 2.5 between the two examples. There is, of course, some strain in the five-membered ring (angle 108°) but only at the carbonyl group (ideal angle 120°). Replacing the trigonal centre by a tetrahedral centre so that the ideal angle (109°) is almost identical to the actual angle is enough to explain this small factor.

Suggested solutions for Chapter 7

7

Problem 1

Are these molecules conjugated? Explain your answer in any reasonable way.

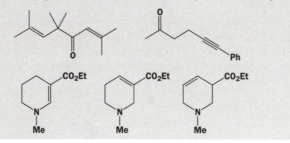

Purpose of the problem

Revision of the basic kinds of conjugation and how to show conjugation with curly arrows.

Suggested solution

The first two are straightforward with one conjugated system (an enone) in the first example and one (a phenyl alkyne) in the second. Of course, the benzene ring is itself conjugated in the second example. You can show either or both with curly arrows: we have done so for the first example only.

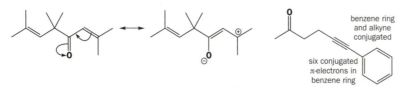

The last three compounds obviously form a group with the same skeleton and only the alkene moved around. There is, of course, ester conjugation in all three but this is the only kind in the last compound. The first is most conjugated with the lone pair on nitrogen delocalized into the carbonyl group. The middle compound just has the alkene and the ester conjugated. This time we have used curly arrow representations for all three compounds and a dotted-line electron distribution summary for the first.

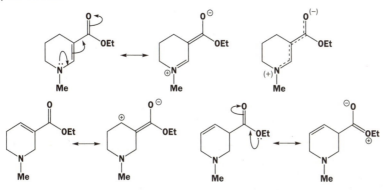

Problem 2

Draw a full orbital diagram for all the bonding and antibonding π orbitals in the three-membered cyclic cation shown here. The molecule is obviously very strained. Might it survive by also being aromatic?

Purpose of the problem

Revision of MO diagrams for conjugated systems with aromaticity.

Suggested solution

There are only two electrons in this simple cation but we need to mix the π bond (π and π^* orbitals) with the empty p orbital to give the MOs. One MO will be bonding all round the ring and this is the only one that matters to the structure. The others may have given you problems. We can mix the p-AO with the π-MO as they have the same symmetry in a three-membered ring but we cannot mix p with π^*. So our three MOs are $\pi + p$, $\pi - p$, and π^*.

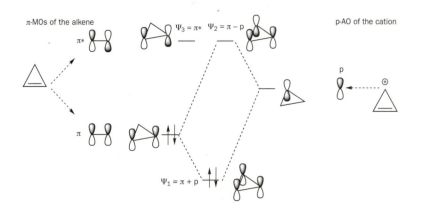

The cyclic cation is stable and can be made because of the gain in energy in populating only the lowest all-bonding orbital. As it happens Ψ_2 and Ψ_3 are degenerate. If you count the number of bonding and antibonding interactions in each you will see that the net result is one antibonding interaction in both orbitals. It is also aromatic, having $4n + 2$ π-electrons where $n = 0$.

You get the same result if you mix three p-AOs, one on each carbon atom. Then it is easier if we look down on the ring showing the top lobe of the p orbital at each atom. The lowest, all-bonding orbital has no nodes (except the plane of the ring) and the two degenerate, antibonding, and unoccupied orbitals have one node each.

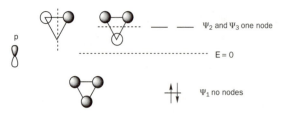

Problem 3

How extensive are the conjugated systems in these molecules?

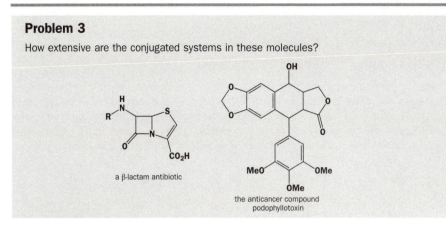

a β-lactam antibiotic

the anticancer compound
podophyllotoxin

Purpose of the problem

A chance to delve more deeply into what is meant by conjugation.

Suggested solution

The β-lactam has two clearly defined conjugating systems: the amide and the more extended unsaturated acid going right through to the sulfur atom. These are shown by curly arrows on the first diagram. These systems are joined by a single bonds so they really are one system: all the p orbitals on the ringed atoms in the second diagram are more or less parallel and all are conjugated.

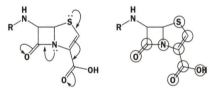

Podophyllotoxin has the obvious two benzene rings and cyclic ester conjugation, shown by curly arrows on the first diagram. Each benzene ring has substituents with lone pairs shown on the second diagram so this molecule has no less than six lone pairs of electrons involved in extended conjugation. There are three separate conjugated systems shown in boxes on the second diagram.

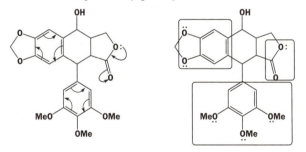

Problem 4

Draw diagrams to illustrate the conjugation present in these molecules. You should draw three types of diagram: (a) conjugation arrows to give at least two different ways of representing the molecule joined by the correct 'reaction' arrow; (b) a diagram with dotted lines and partial charges (if any) to show the double bond and charge distribution (if any); and (c) a diagram of the atomic orbitals that make up the lowest-energy bonding molecular orbital.

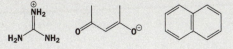

Purpose of the problem

A more exacting exploration of the precise details of conjugation.

Suggested solution

Treating each molecule separately in the styles demanded by the question, the first compound is the guanidinium ion, a very stable cation because of conjugation. The curly arrows in the first set of diagrams show that each nitrogen has an equal amount of positive charge and this is reflected in the second diagram showing exactly one-third + on each N. The third diagram, of the lowest-energy MO, looks down on the molecule and shows the top lobe of the p orbital on each atom.

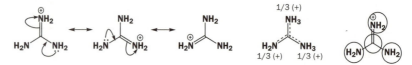

The second molecule is what we shall learn to call an enolate ion. The negative charge is delocalized throughout the conjugated system but is particularly on the two oxygen atoms and, to a lesser extent, on the central carbon atom. It is difficult to represent this accurately with partial charges but the charge on each oxygen atom is nearly a half. The lowest-energy MO has all the p orbitals in phase.

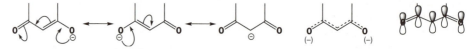

The third compound is naphthalene. The structure drawn is the best; curly arrow diagrams do not do this molecule justice and other versions are less convincing. In particular, they do not show that the middle bond common to both rings is the shortest bond. There are no charges anywhere and all the p orbitals are conjugated. This is the basic ten-electron aromatic system.

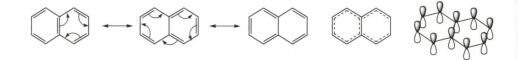

Problem 5

Which of these compounds are aromatic? Justify your answer with some electron counting. You are welcome to treat each ring separately or two or more rings together, whichever you prefer.

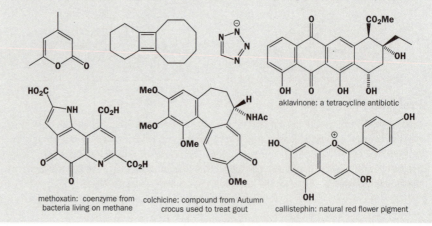

methoxatin: coenzyme from bacteria living on methane

colchicine: compound from Autumn crocus used to treat gout

aklavinone: a tetracycline antibiotic

callistephin: natural red flower pigment

Purpose of the problem

A simple exploration of the idea of aromaticity: can you count to six?

Suggested solution

The first three compounds are straightforward providing you count lone pair electrons on atoms in the ring and do not count electrons outside the ring such as those in the carbonyl π bond in the first compound. Nor should you count the lone pair represented by the negative charge in the third compound. They are in an sp² orbital in the plane of the ring.

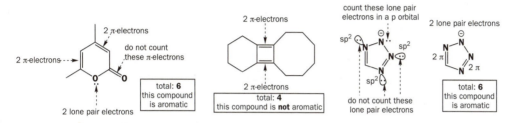

The rest offer variations on the benzene ring and each ring must be considered separately. Methoxatin has five- and six-membered rings with nitrogen in them. Count the lone pair in a p orbital on the nitrogen atom in the five-membered ring but not those in an sp² orbital in the six-membered ring (pyridine). Both are aromatic. Colchicine has an aromatic seven-membered ring with six electrons (don't count the C=O group) while callistephin has an interesting positively charged aromatic ring with three double bonds. We summarize these answers briefly by giving the number of electrons in each conjugated ring.

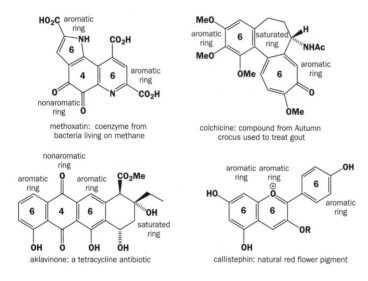

methoxatin: coenzyme from bacteria living on methane

colchicine: compound from Autumn crocus used to treat gout

aklavinone: a tetracycline antibiotic

callistephin: natural red flower pigment

Problem 6

A number of water-soluble pigments in the green/blue/violet ranges used as food dyes are based on cations of the type shown here. Explain why the general structure shows such long wavelength absorption and suggest why the extra functionality (OH group and sulfonate anions) is put into 'CI food green 4', a compound approved by the EU for use in food under E142.

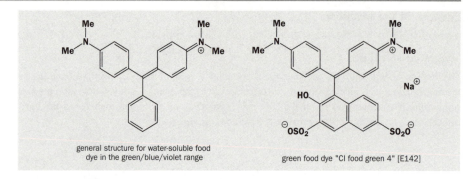

general structure for water-soluble food
dye in the green/blue/violet range

green food dye "CI food green 4" [E142]

Purpose of the problem

To get you thinking about the relationship between conjugation and light absorption.

Suggested solution

Each dyestuff has essentially the same conjugated system shown here isolated from extra substituents. The positive charge is shared between the two nitrogen atoms and the whole system is symmetrical. The colour is due to excitation of the lone pair electrons, which are delocalized over the whole system, into an antibonding π^* orbital of the aromatic system. There will be many bonding π orbitals and the same number of antibonding π^* orbitals. The lowest (the LUMO) will be unusually low in energy and the n, π^* is in the visible region.

■ Did you notice that only one sodium ion is necessary as there is already one positive charge on the molecule?

The substituent on the middle atom is different in the two dyes and is also conjugating. This extra conjugation decides the exact shade of colour with the highly conjugating naphthalene ring in E142 taking the absorption all the way through to red light so the dye appears green to the eye. The OH group also helps as a lone pair is also conjugated into the aromatic rings. The sulfonate groups make the compound water-soluble.

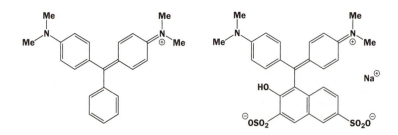

Problem 7

Turn to Chapter 1 and look at the structures of the dyes in the shaving foam described on p. 7. Comment on the structures in comparison with those in Problem 6 and suggest where they get their colour from and why they too have extra functional groups. Then turn to the beginning of Chapter 1 (p. 3) and look at the structures of the compounds in the 'spectrum of molecules'. Can you see what kind of absorption leads to each colour? You will want to think about the conjugation in each molecule but you should not expect to correlate structures with wavelengths in any even roughly quantitative way.

Purpose of the problem

Further thoughts on the relationship between conjugation and light absorption plus revision of Chapter 1.

Suggested solution

One of the dyestuffs in the shaving foam, Fast Green FCF, has a conjugated system very similar to those we have just been discussing in Problem 6, though it has extra solubilizing substituents on the nitrogen atom.

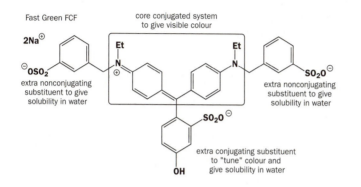

Quinoline yellow is quite different. It has two conjugated ring systems linked by a single bond. However, they are also linked by conjugation from the OH group in one ring to the nitrogen atom in the other. In fact, the whole molecule is conjugated. Because the conjugation is more limited, this molecule absorbs violet light and so appears yellow. It has the usual sulfonate solubilizing groups. The 'quinoline' part of the name refers to the left-hand ring system.

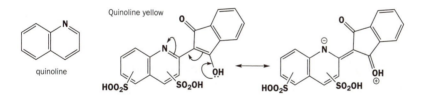

The compounds in the 'spectrum of molecules' are much simpler and you may well have been surprised after reading Chapter 7 that they are coloured at all.

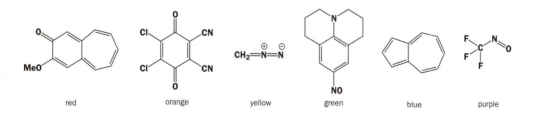

You probably found 'red' easiest to deal with as there is an obvious extended conjugation throughout the whole molecule. The conjugation in 'blue' is less obvious but clearly more effective (a blue compound absorbs longer wavelength light than a red compound) and relies on a redistribution of electrons that makes both rings aromatic. This compound is called azulene (an azure alkene) as a blue hydrocarbon is very remarkable.

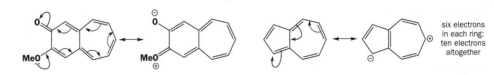

six electrons
in each ring:
ten electrons
altogether

The orange and yellow compounds owe their colour to functionality. 'Orange' is a quinone – a conjugated six-membered ring compound with four electrons – and has a low-energy unoccupied orbital (a π* orbital, the LUMO) from the two conjugated carbonyl groups, which is easily occupied by excitation of a π electron. 'Yellow' is diazomethane, an unstable toxic and even explosive gas, which has some conjugation in its very small framework but owes its colour mainly to the N=N functionality which again has a low-energy π* orbital.

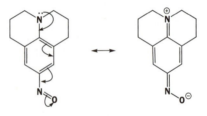

The green compound, delightfully named 9-nitrosojulolidine, has conventional conjugation of an efficient sort between electron-donating N and the electron-accepting nitroso group (N=O). Nitroso compounds can be brightly coloured without any conjugation as is the case with 'purple'. It is very unusual to have blue or purple gases and this compound has a CF₃ group to stabilize the nitroso group which, when not conjugated, is an odd-electron compound. The colour comes from excitation of a lone pair electron into the N=O antibonding orbital.

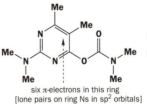

Problem 8

Go through the list of aromatic compounds at the end of the chapter and see how many electrons there are in the rings taken separately or taken together (if they are fused). Are all the numbers of the $(4n + 2)$ kind?

Purpose of the problem

Revision of material in the chapter explored a little more deeply to reinforce the concept of aromaticity.

Suggested solution

The first two compounds are straightforward providing you put lone pairs in the right orbitals and don't count C=O groups.

coumarin – the smell of "new mown hay" also found in lavender

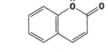

pirimicarb – a selective insecticide that kills aphids but not ladybirds

six π-electrons   four π-electrons and   ten electrons   six π-electrons in this ring
in this ring   two from O lone pair   in both rings together   [lone pairs on ring Ns in sp² orbitals]
        in this ring

LSD and Omeprazole have in common a fused aromatic ring system with one benzene ring and one heterocyclic ring. You can think of these as two six-electron rings sharing two electrons or as one ten-electron system. Omeprazole has a pyridine ring as well.

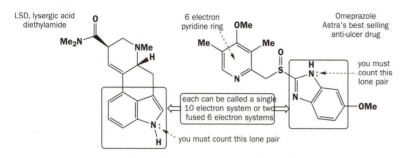

Viagra has an ordinary benzene ring and an interesting double heterocyclic ring. You can simply say that there are two fused rings with six electrons each but a couple of delocalizations show clearly which lone pairs can be delocalized and that the resulting system has 10 electrons delocalized round its rim. The final example, haem, was fully analysed in the chapter as an 18-electron aromatic system.

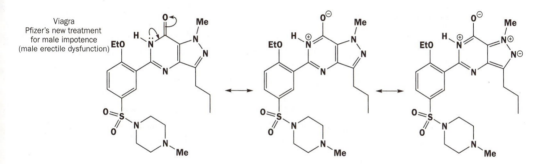

Suggested solutions for Chapter 8

Problem 1

If you wanted to separate a mixture of naphthalene, pyridine, and p-toluic acid, how would you go about it?

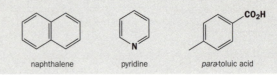

naphthalene   pyridine   para-toluic acid

Purpose of the problem

Revision of simple acidity and basicity in a practical situation.

Suggested solution

Pyridine is a weak base (pK_a 5.5) and can be dissolved in aqueous acid. p-Toluic acid is a weak acid (pK_a about 4.5) and can be dissolved in aqueous base. Naphthalene is neither an acid nor a base and remains insoluble in water at any pH. So, dissolve the mixture in an organic solvent immiscible with water (say ether, Et_2O, or CH_2Cl_2) and extract with aqueous acid (pH < 4); this will remove the pyridine as an aqueous solution of its protonated form. Then extract the remaining organic layer with aqueous base (pH > 6) which will extract the p-toluic acid as its anion. You now have three solutions. Evaporate the organic solution of naphthalene and recrystallize it to get pure naphthalene. Acidify the basic solution of p-toluic acid to precipitate the free acid and recrystallize it. Finally, add base to the pyridine solution, extract the pyridine with an organic solvent, (say ether, Et_2O, or CH_2Cl_2), evaporate the solvent, and distil the pyridine. It is just as good to extract with base first and acid second.

Problem 2

In the separation of benzoic acid from toluene we suggested using NaOH solution. How concentrated a solution would be necessary to ensure that the pH was above the pK_a of benzoic acid (pK_a 4.2)? How would you estimate how much solution to use?

Purpose of the problem

You can check your understanding of the relationship between pH and concentration.

Suggested solution

Even a very weak solution of NaOH will have a pH > 4.2. By the calculation on p. 184 of the chapter, for a pH of 5 we should need $[H_3O^+] = 10^{-5}$ mol dm^{-1}. We know that the ionic product of water is $[H_3O^+] [OH^-] = 10^{-14}$ and so for pH 5 we need 10^{-5} mol dm^{-1} of NaOH. This is very dilute! The trouble is that you need one hydroxide ion for every molecule of benzoic acid so if you had, say, 1.22 g $PhCO_2H$ (= 0.01 equiv.) you would need 1000 litres (dm^{-1}) of your NaOH solution. It makes more sense to have a much more concentrated solution, say, 0.1M. This would give an unnecessarily high pH of 13 but you would need only 100 ml (cm^3) to extract your benzoic acid.

Problem 3

What species would be present if you were to dissolve this hydroxy-acid in: (a) water at pH 7; (b) aqueous alkali at pH 12; or (c) a concentrated solution of a mineral acid?

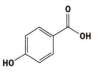

Purpose of the problem

Revision of simple acidity and basicity requiring pK_a values in a practical situation.

Suggested solution

The CO_2H group will have a pK_a of about 4–5 and the OH group a pK_a of about 10. This gives a comfortable margin between the two ionizations. At pH 7 the acid will be ionized and at pH 12 both groups will be ionized. In concentrated mineral acid both will have their protons and the carboxylic acid will start to be protonated. It is more readily protonated than the isolated OH group as it gives a delocalized cation.

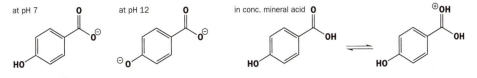

Problem 4

What would you expect to be the site of (a) protonation and (b) deprotonation if the compounds below were treated with an appropriate acid or base? In each case suggest a suitable acid or base for both purposes.

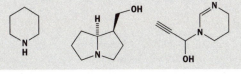

Purpose of the problem

Progressing towards more taxing judgements on more interesting molecules.

Suggested solution

The simple amine, piperidine, will have a normal amine pK_{aH} of about 11 so it will be easy to protonate with even very weak acids. Any mineral acid like HCl will do as would weaker acids like RCO_2H. Deprotonation will remove the NH proton as nitrogen is more electronegative than carbon but a very strong base such as BuLi will be needed as the pK_a will be about 30–35.

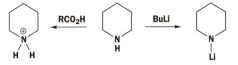

The next example is more complicated but actually contains a normal tertiary amine (pK_{aH} about 11) and a normal primary alcohol (pK_a about 16). Protonation at nitrogen will occur with any acid – let's choose TsOH this time – and deprotonation will occur at the OH group and will need a strong base – let's say NaH.

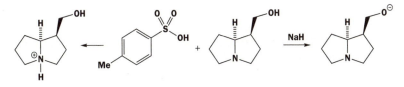

The final example is the most complicated one. There is a normal OH group (pK_a about 16) and a slightly acidic alkyne (pK_a about 32) so the OH will be deprotonated first. The basic group is not a normal amine but an amidine in which the two nitrogen atoms can combine (see arrows) to capture a proton. It will have a pK_a of about 12 and so would even be protonated by a phenol though most chemists would use a carboxylic acid.

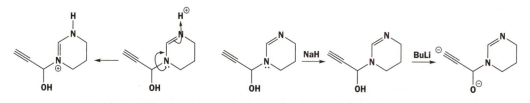

Problem 5

Suggest what species would be formed by each of these combinations of reagents. You are advised to use pK_a values to help you and to beware of some cases where 'no change' might be the answer.

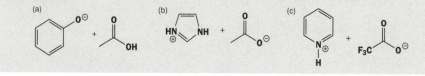

Purpose of the problem

Learning to compare species of similar acidity or basicity.

Suggested solution

In each case one of the reagents might take a proton from the other. In example (a), would the anion of the phenol remove a proton from the carboxylic acid? We can answer that by considering pK_a values. The answer is yes. There are five pH units between the two acids so the equilibrium will be well to the right. The equilibrium constant is about 10^5.

Example (b) has a similar possible reaction. This time the difference is much smaller and the other way. The carboxylic acid is slightly stronger than protonated imidazole and the equilibrium constant is about 10^{-2}.

Example (c) also involves a carboxylic acid but this one is rather different. The three fluorine atoms make CF_3CO_2H a very strong acid, comparable to mineral acids. This equilibrium is well over to the left.

Problem 6

What is the relationship between these two molecules? Discuss the structure of the anion that would be formed by the deprotonation of each compound.

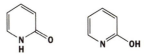

Purpose of the problem

To help you recognize that conjugation may be at the back of some related structures.

Suggested solution

They are tautomers. They differ only in the position of a proton. Each is an acid – the first has an NH group as part of an amide and the second has an OH group. Deprotonation may at first appear to produce two different anions but they are actually the same because of delocalization of the anion.

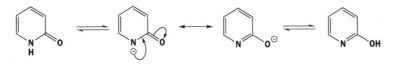

Problem 7

What species would be formed by treating this compound with: (a) one equivalent; (b) two equivalents of NaNH$_2$ in liquid ammonia?

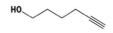

Purpose of the problem

A simple problem to help you think about the possibility of selective deprotonation.

Suggested solution

The more acidic group is the OH (pK_a about 16) and the less acidic is the alkyne (pK_a about 35). The alkoxide will be formed first, and then a second deprotonation is possible because the two negative charges are far apart in the dianion and will not significantly destabilize each other.

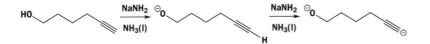

Problem 8

The carbon NMR spectra of these compounds could be run in D$_2$O under the conditions shown. Why were these conditions necessary and what spectrum would you expect to observe?

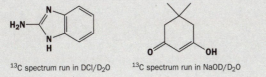

^{13}C spectrum run in DCl/D$_2$O ^{13}C spectrum run in NaOD/D$_2$O

Purpose of the problem

NMR revision and practice at judging the states of compounds at different pHs. Hidden symmetry from conjugation.

Suggested solution

Both compounds are quite polar and not very soluble in the usual organic NMR solvents. In addition, they have NH or OH protons that exchange and broaden the signals. If the exchange is made rapid by acid or base catalysis and the NHs and OHs exchanged for deuterium, all these problems disappear and clean sharp spectra result. The first compound is a guanidine and forms a cation in acid. The cation is symmetrical and a very simple spectrum results – just three types of carbon in the aromatic region and one very low field carbon (at large δ) for the carbon in the middle of the cation.

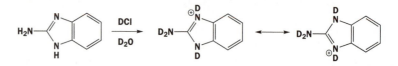

The second compound loses a proton from the OH group in base to give a delocalized symmetrical anion. There will be five signals in the NMR spectrum: the two methyl groups are the same and the two CH₂ groups in the ring are equivalent. There is one unique carbon atom joined to the two methyl groups and another in the middle of the anion and, finally, the two carbonyl groups are equivalent.

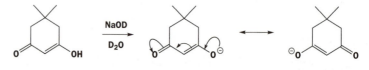

Problem 9

The phenols shown here have approximate pK_a values of 4, 7, 9, 10, and 11. Suggest with explanations which pK_a value belongs to which phenol.

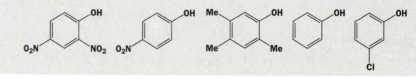

Purpose of the problem

A problem requiring detailed interpretation of electronic effects.

Suggested solution

Electron-withdrawing groups (Cl and even more so NO_2) make phenols more acidic but electron-donating groups (Me) make them less acidic. The nitro group is very effective if it can stabilize the anion by conjugation. A pK_a value of about 10 is normal for a phenol, so it is phenol itself that has that value. The compound less acidic than phenol (pK_a 11) must be the trimethyl compound. One chlorine atom will make phenol slightly more acidic so pK_a 9 must be 3-chlorophenol and the compounds with one nitro group (pK_a 7) or two (pK_a 4) must be the very acidic phenols.

Problem 10

Discuss the stabilization of the anions formed by the deprotonation of (a) and (b) and the cation formed by the protonation of (c). Consider delocalization in general and the possibility of aromaticity in particular.

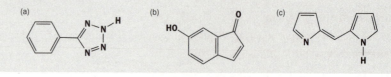

Purpose of the problem

To get you thinking about delocalized cations and anions.

Suggested solution

There is only one remotely acidic proton in (a), the NH proton. Removal of this gives an anion that can be delocalized on to all the four nitrogen atoms in the five-membered ring. The combination of four electronegative atoms, a delocalized anion, and an aromatic anion (see diagram in margin) makes this a very stable anion and the original amine is a real acid with a pK_a (5) about the same as that of a carboxylic acid (RCO_2H).

(a)

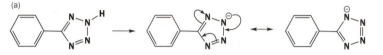

The most acidic proton in (b) is that of the phenol and the anion is delocalized round the benzene ring, into the fused five-membered ring, and even out on to the ketone oxygen atom so it is a stable anion. However, there are only eight delocalized electrons altogether so, although the benzene ring remains aromatic, there is no extra stabilization from any aromaticity in the delocalized five-membered ring or in the system as a whole. It has a pK_a value of about 8.5.

(b)

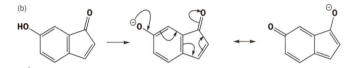

The diamine (c) will be protonated on the trigonal (sp^2) nitrogen using the delocalized lone pair in a p orbital on the other nitrogen. The resulting cation is delocalized in exactly the same way so that the two nitrogen atoms are the same. Both rings are aromatic (though neither valence bond diagram shows both rings as aromatic) as there are 12 electrons delocalized – six for each ring. The system as a whole isn't aromatic because it is not *one* conjugated ring system.

(c)

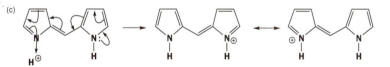

Problem 11

The pK_a values for the amino acid cysteine are 1.8, 8.3, and 10.8. Assign these pK_a values to the functional groups in cysteine and draw the structure of the molecule in aqueous solution at the following pHs: 1, 5, 9, and 12.

cysteine

Purpose of the problem

Further revision in thinking about the acidity and basicity of functional groups – amino acids are particularly important.

Suggested solution

You would expect the NH_2 group to have a pK_{aH} of about 11 and the SH group a pK_a of about 8 so they look all right. But you would also expect the CO_2H group to have a pK_a of about 5 and that looks all wrong! We need to start at high pH and work downwards to see what exactly we have by the time we come to pH 5. This is what happens.

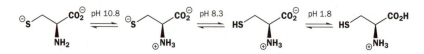

By the time we want to protonate the CO_2^- ion, there is an NH_3^+ group next door so it will be much more difficult to make the cation and the pK_a value is lower because we must go to more acidic solutions to get protonation. The structure of the molecule at pHs 1 (cation), 5 (zwitterion), 9 (monoanion, that is, dianion and monocation in the same molecule), and 12 (dianion) is as shown.

Problem 12

Explain the variations in the pK_a values for these carbon acids.

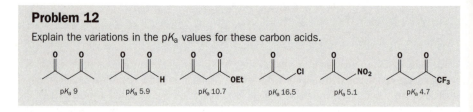

Purpose of the problem

To get you thinking about carbon acids, how their anions are delocalized, and what makes the anions stable.

Suggested solution

All these carbon acids give anions whose negative charge is mainly on oxygen. Similar ones are described in the chapter (Table 8.2, p. 193) and we expect simple carbonyl compounds to have pK_as of about 15–20 and 1,3-dicarbonyl compounds to have pK_as of about 5–15. The first three compounds are all ketones on the left and have ketone, aldehyde, and ester groups on the right, respectively. The charge is delocalized over both carbonyl groups. The aldehyde is most effective as it is a simple carbonyl group. The ketone has σ-conjugation, which reduces the effectiveness of the carbonyl delocalization, and the ester is worst as it has competing conjugation from the ester oxygen atom.

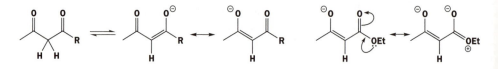

The next compound has only one carbonyl group but gains a little acidity from the electronegative chlorine atom – it is a proton from that side of the ketone that is lost. Delocalizing the charge on to the carbon atoms shows this.

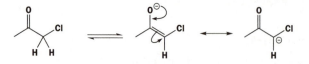

In the last two compounds the charge is supported by very electron-withdrawing groups – nitro and trifluoroacetyl. Delocalization is very effective and most of the negative charge will be on one of the two oxygen atoms.

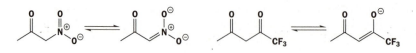

Problem 13

Explain the various pK_a values for these derivatives of the naturally occurring amino acid glutamic acid. Say which pK_a belongs to which functional group and explain why pK_as vary among the different derivatives.

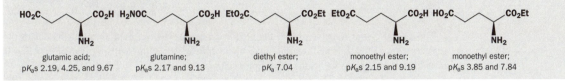

| glutamic acid; | glutamine; | diethyl ester; | monoethyl ester; | monoethyl ester; |
| pK_as 2.19, 4.25, and 9.67 | pK_as 2.17 and 9.13 | pK_a 7.04 | pK_as 2.15 and 9.19 | pK_as 3.85 and 7.84 |

Purpose of the problem

To get you thinking about carbon acids, how their anions are delocalized, and what makes the anions stable.

Suggested solution

The three pK_a values for glutamic acid itself must be like this (the remote CO_2H group is normal but the one near the amino group is unusual as we discussed in the answer to Problem 11).

■ Notice that at no pH does glutamic acid exist as the form we normally draw, though at pHs between 2.2 and 4.2 this form is in equilibrium with the zwitterion.

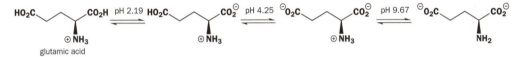

In glutamine, the remote CO_2H group is missing and the other two pK_a values are hardly changed. The dimethyl ester has only the amino group left and its pK_{aH} value is less because we have replaced the two CO_2^- anions with electron-withdrawing ester groups. The two monoesters are most interesting. The first is essentially the same as glutamine – the remote ester has little effect on the other two groups. The second has, as we should expect, a remote CO_2H group like that of glutamic acid itself but an amino group like that of the diester showing that the nearer of the two ester groups in the diester has the bigger effect.

Problem 14

Neither of these methods of making pentan-1,4-diol will work. Explain why not – what will happen instead?

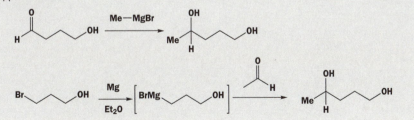

Purpose of the problem

To help you appreciate the disastrous effects that innocent-looking groups may have because of their weak acidity.

Suggested solution

The OH group is the Wicked Witch of the West in this problem. Whoever planned these syntheses expected it to lie quietly and do nothing. But, although an OH group is only a weak acid (pK_a about

16), it will give up its proton to the very basic Grignard reagent. In the first case, one equivalent of Grignard reagent is destroyed but addition of a second would save the day.

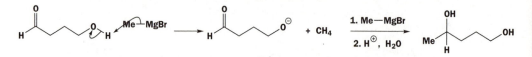

The second case is more serious as the Grignard reagent destroys itself as it is formed with the replacement of Br with H and then nothing can be done to rescue the synthesis. The only answer is to block the OH group with something non-acidic (see Chapter 24).

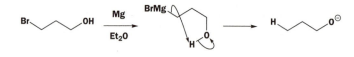

Suggested solutions for Chapter 9

9

Problem 1

Propose mechanisms for the first four reactions in the chapter.

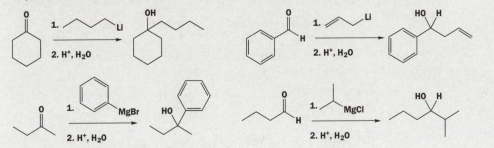

Purpose of the problem

Rehearsal of the basic reactions of Chapter 9.

Suggested solution

Each reaction simply involves the nucleophilic attack of the organometallic reagent on the aldehyde or ketone followed by protonation. You may draw the intermediate as an anion or with an O–metal bond as you please. Note the atom-specific arrows to show which atom is nucleophilic. In the second example the allyl-lithium might attack through its other end.

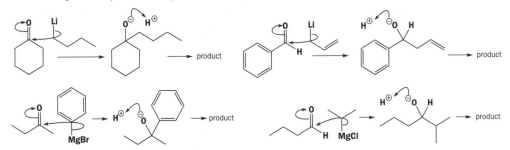

Problem 2

When this reaction is carried out with allyl bromide labelled as shown with ^{13}C, the label is found equally distributed between the ends of the allyl system in the product. Explain how this is possible. How would you detect the ^{13}C distribution in the product?

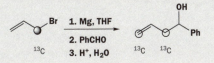

Purpose of the problem

Reminder of the way allyl-metals react and revision of NMR.

Suggested solution

One explanation is that allyl Grignard reagents might react as nucleophiles at either end of the allyl system.

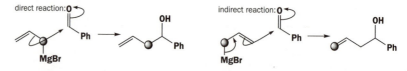

The mechanism given in the chapter (p. 224) suggests that allyl Grignards always react at the remote end by a cyclic mechanism. If that is the case, the formation of the Grignard must occur at both ends.

This could occur during the formation by attack of the metal at the other end of the allylic system, or after formation by equilibration with the allyl anion.

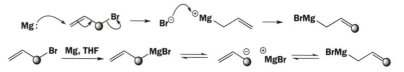

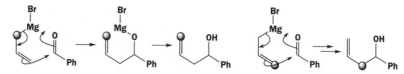

In either case the two allyl Grignard reagents would have equal energies and would be formed in equal amounts. The reaction with benzaldehyde would then follow a cyclic mechanism.

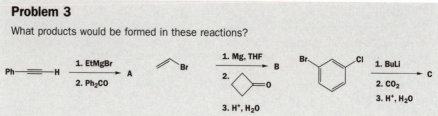

Your explanation may not be exactly the same as either of these – the only important thing is that you should have some way of getting the reaction to occur at the other end of the allylic system. The ^{13}C distribution in the product would be determined by NMR spectra. Each labelled atom would appear as a much stronger signal (the natural abundance of ^{13}C is 1.1%) and each signal will be in a different part of the spectrum. The chemical shifts are more accurately estimated as 115–120 and 45–50 p.p.m.

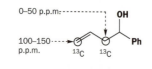

Problem 3

What products would be formed in these reactions?

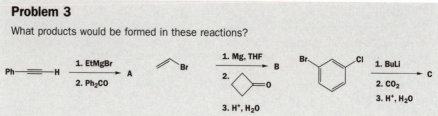

Purpose of the problem

The toughest test – predicting the product. The sooner you get practice the better.

Suggested solution

Though prediction is harder than explanation, you should get these right first time as only the last one has a hint of difficulty. In the first example, ethyl Grignard reagent acts as a base to remove a proton from the alkyne. Whether you draw the intermediate as an alkyne anion or Grignard reagent is up to you and you need not necessarily include the mechanism for the protonation step.

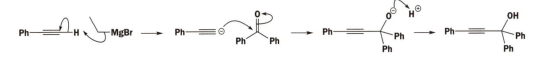

For the second example just make the organometallic compound and add it to the carbonyl group. Cyclobutanone is an electrophilic ketone because of the strain of a carbonyl group in a four-membered ring. Note that the Grignard reagent is vinyl and not allyl so the ambiguities in Problem 2 do not arise.

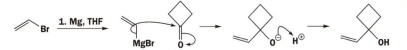

The third example raises the question of halogen replacement or *ortho*-lithiation and, if the former, which halogen? When bromine is one of the halogens, halogen replacement is usually the reaction which occurs but if you did ortho-lithiation between Cl and Br, you should not be ashamed.

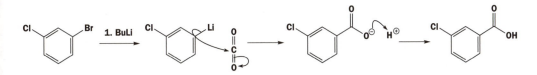

Problem 4

Suggest alternative routes to fenarimol – that is, routes different from the one shown in the chapter (p. 216).

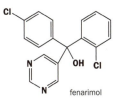

fenarimol

Purpose of the problem

Practice in choosing alternative routes.

Suggested solution

The route shown in the chapter (p. 216) is addition of the lithium derivative of the heterocycle to a diaryl ketone.

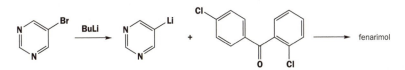

The alternatives are the addition of either of the aryl groups as a Grignard or organolithium reagent to a ketone made up from the rest of the molecule. In both cases the organometallic reagent will have a chlorine atom on the benzene ring so you should choose a more reactive halogen (Br or I) for replacement by the metal. In the structures M represents either Li or MgBr, depending on the method chosen.

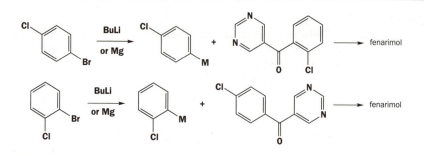

Problem 5

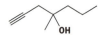

The synthesis of the gastric antisecretory drug rioprostil requires this alcohol. (a) Suggest possible syntheses starting from ketones and organometallics and (b) suggest possible syntheses of the ketones in part (a) from aldehydes and organometallics (don't forget about CrO_3 oxidation!).

Purpose of the problem

Your first introduction to sequences of reactions where more complex molecules of the same type are created so that the sequence can be repeated.

Suggested solution

There are three one-step syntheses from ketones and organometallic compounds.

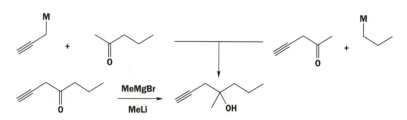

Each of these three ketones can be made by oxidation of an alcohol, which in turn can be made from an organometallic compound and an aldehyde(or in other cases from another ketone).

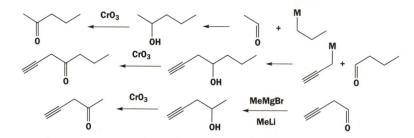

Problem 6

Suggest two syntheses of the bee pheromone heptan-2-one.

Purpose of the problem

Your chance to propose not one but two syntheses for a natural compound with biological activity.

Suggested solution

There are two obvious routes, each from an aldehyde and an organometallic compound followed by oxidation.

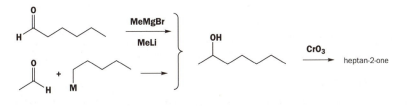

Problem 7

How could you prepare these compounds using *ortho*-lithiation procedures?

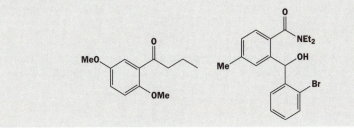

Purpose of the problem

An initial exploration of the principles behind *ortho*-lithiation.

Suggested solution

The ketone side chain is obviously added to the aromatic ring, which then has four identical sites for lithiation activated by the MeO groups. Oxidation is again needed to make the ketone.

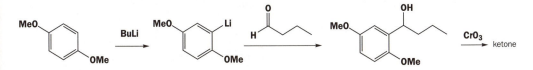

The second synthesis offers a choice between the lithiation of two aromatic rings. One has an excellent activating group, an amide, while the other has a bromine atom that will exchange rather than activate towards *ortho*-lithiation.

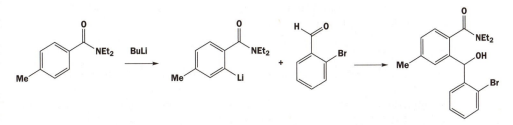

Problem 8

Why is it possible to make the lithium derivative A by Br/Li exchange, but not the lithium derivative B?

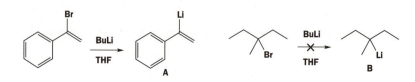

Purpose of the problem

Revision of the stability of carbanions and its relevance to lithium/bromine exchange.

Suggested solution

The first example is a vinyl bromide and vinyl (sp^2) carbanions are more stable than saturated (sp^3) carbanions because of the greater s character in the C–Li σ bond. The second example is saturated like BuLi but, unlike BuLi, it is a *tertiary* alkyl bromide. The *t*-alkyl carbanion would be less stable than the primary BuLi and is not formed.

Problem 9

Comment on the selectivity (that is, say what else might have happened and why it didn't) shown in this Grignard addition reaction used in the manufacture of an antihistamine drug.

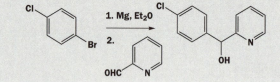

Purpose of the problem

First steps in thinking about selectivity – a most important concept in organic chemistry.

Suggested solution

What happens must be this: a Grignard reagent is formed with the aryl bromide and adds to the aldehyde.

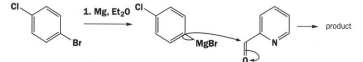

What does *not* happen is Grignard formation at the aryl chloride – bromides are more reactive than chlorides towards Grignard formation. Furthermore, nucleophilic attack occurs at the carbonyl group and not at the C=N group of the electrophile. This is because C=N is less electrophilic than C=O (N is less electronegative than O) and also because attack at C=N would destroy the aromaticity of the pyridine ring.

Problem 10

The antispasmodic drug biperidin is made by the Grignard addition reaction shown here. What is the structure of the drug? Do not be put off by the apparent complexity of the compounds – the chemistry is the same as you have seen in this chapter. How would you suggest that the drug procyclidine should be made?

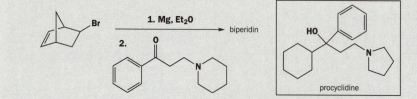

Purpose of the problem

Product prediction in a more complicated case and logical extension to a new situation.

Suggested solution

A Grignard reagent must be formed from the alkyl bromide and this must add to the ketone. Aqueous acidic work-up (not mentioned in the question as is often the case) now gives a tertiary alcohol that must be biperidine.

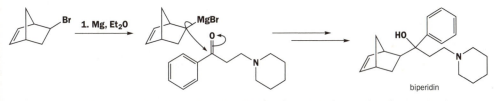

biperidin

To get procyclidine we need to change both the alkyl halide and the ketone, but the reaction is very similar.

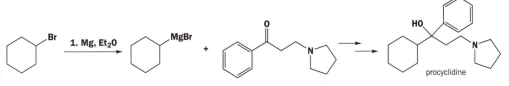

procyclidine

Problem 11

Though heterocyclic compounds, such as the nitrogen ring system in this question, are introduced rather later in this book, use your knowledge of Grignard chemistry to draw a mechanism for what happens here. It is important that you prove to yourself that you can draw mechanisms for reactions on compounds that you have never met before.

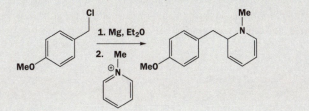

Purpose of the problem

Just to prove that the reactions that didn't happen in Problem 9 can actually take place if there's no competition.

Suggested solution

Though this is chemistry you've not seen, the mere structure of the product tells you what happens – the alkyl (benzylic actually) chloride forms a Grignard reagent and this adds to C=N$^+$. Note that + charge – it makes all the difference as to how electrophilic a C=N group might be.

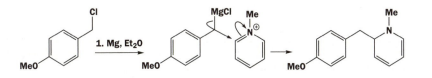

Suggested solutions for Chapter 10

10

Problem 1

Draw mechanisms for this reaction and explain why this particular product is formed.

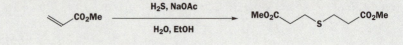

Purpose of the problem

Rehearsal of the basic reaction of Chapter 10 and revision of pK_a values from Chapter 8.

Suggested solution

The first step must be a conjugate addition but will it be H_2S or the HS^- anion that does the job? The base present is NaOAc (pK_a about 5) while H_2S has a pK_a of 7.0 so that very little of it is ionized. We should use neutral H_2S.

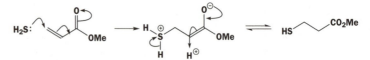

■ Did you have enough carbon atoms in your intermediates? It is a very common mistake to lose a carbon atom when drawing the mechanism for a conjugate addition.

The enolate intermediate captures a proton from the H_2S^+ group of the same or another molecule and the first product of conjugate addition is formed. This thiol (–SH) has another removable proton so it can add again to another molecule of the unsaturated ester to give the final product. Sulfur nucleophiles are excellent at conjugate addition.

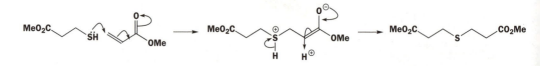

Problem 2

Which of the two routes shown here would actually lead to the product? Why?

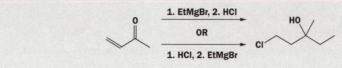

Purpose of the problem

To check that you understand the functional groups needed for a conjugate addition.

Suggested solution

■ We shall not always give proton transfer steps in detail from now on.

The ethyl Grignard reagent must add directly to the carbonyl group but the HCl must do conjugate addition. Direct addition destroys the carbonyl group but conjugate addition does not. The carbonyl group must be present during the conjugate addition so that must be done first. The second route would lead to the required product.

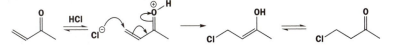

After the conjugate addition has removed the alkene, direct addition is the only option for the second nucleophile. The alternative order of events would probably lead to direct addition by the Grignard reagent followed by protonation of the double bond. You were not asked what the product of the wrong reaction sequence is, but it might be this.

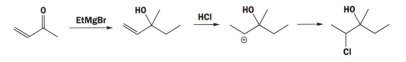

Problem 3

Suggest reasons for the different outcomes of the following reactions (your answer must, of course, include a mechanism for each reaction).

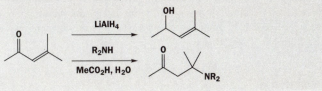

Purpose of the problem

Revision of the relationship between type of nucleophile and type of reaction.

Suggested solution

The Michael acceptor is a conjugated ketone that is able to do either direct or conjugate addition. Amines are softer nucleophiles able to add reversibly and more likely to give conjugate addition by thermodynamic control. LiAlH₄ is very hard and completely irreversible so it gives direct addition by kinetic control. If you drew a full mechanism for each reaction, your answer is better than ours.

Problem 4

Addition of dimethylamine to the unsaturated ester A could give either product B or C. Draw mechanisms for both reactions and show how you would distinguish them spectroscopically.

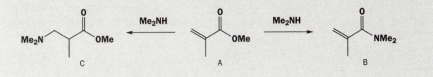

Purpose of the problem

To get you to draw the mechanisms for both reactions and to revise NMR.

Suggested solution

The mechanisms are standard and this is an enone that really might do either. Product B is the result of direct addition followed by loss of a leaving group – a likely reaction as amines attack saturated esters in this way. This reaction is fully treated in Chapter 12.

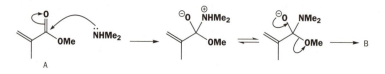

Product C is the result of conjugate addition – again a likely reaction as esters are prone to conjugate addition and amines are good nucleophiles for that reaction.

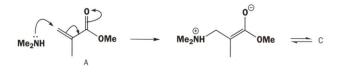

The compounds could be distinguished easily by ^{13}C NMR. Product B has the two carbon atoms of the alkene in the 100–150 p.p.m. region while product C has no signals in this region but does have a signal for the OMe group in the 50–100 p.p.m. region as well as more signals in the 0–50 p.p.m. region. In addition, the molecular ions are different in the mass spectrum as B is $C_6H_{11}NO$ while C is $C_7H_{15}NO_2$.

Problem 5

Suggest mechanisms for the following reactions.

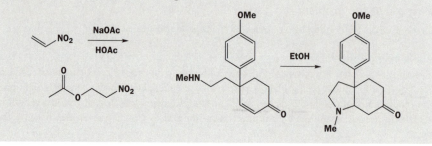

Purpose of the problem

To get you to draw mechanisms for more exotic examples of conjugate addition.

Suggested solution

The nitro group is one of the best activating groups for conjugate addition as it almost never gives direct addition. Even the weakly nucleophilic acetate ion will react as the nitro-stabilized anion is very stable indeed.

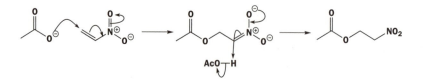

The second example is an intramolecular conjugate addition that makes an interesting natural amine of a kind that occurs in daffodil plants. Direct addition would be difficult in this example as the nitrogen atom can hardly reach the carbonyl group.

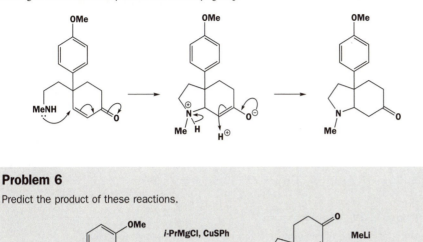

Problem 6

Predict the product of these reactions.

Purpose of the problem

Practice at prediction when the choice is defined – direct or conjugate addition?

Suggested solution

Both reactions involve addition of organometallic compounds to unsaturated carbonyl compounds. The key difference is in the metal. With Cu(I) as catalyst, the Grignard reagent will give conjugate addition in the first example but MeLi will give direct addition in the second.

A = ![structure A]   B = ![structure B]

Problem 7

Two routes are proposed for the preparation of this amino-alcohol. Which do you think is more likely to succeed and why?

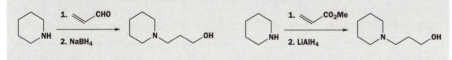

Purpose of the problem

Practical application of the choice of reagent to ensure conjugate addition.

Suggested solution

Addition of the amine to either an unsaturated aldehyde or ester gives a compound that can be reduced to the amino-alcohol. Which is more likely to give conjugate addition? As explained in the chapter (p. 236) the ester is better.

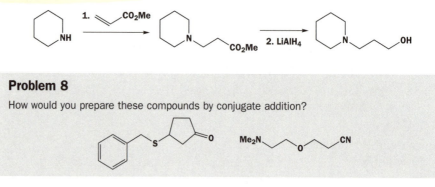

Problem 8

How would you prepare these compounds by conjugate addition?

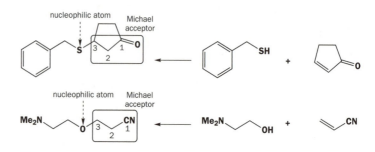

Purpose of the problem

Practical application of conjugate addition in making interesting molecules.

Suggested solution

All you need to do is to identify the 1,3-relationship between the nucleophile and the anion-stabilizing group. Alternatively, you might try to spot the Michael acceptor (shown in a frame on the diagrams) inside the molecule you're trying to make. The nucleophiles are heteroatoms and easy to spot.

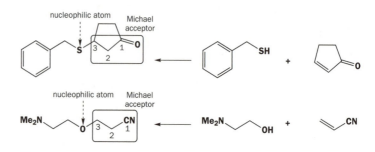

Problem 9

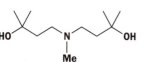

How might this compound be made using a conjugate addition as one of the steps? You might find it helpful to consider the preparation of tertiary alcohols as described in Chapter 9 and also to refer back to Problem 1 in this chapter.

Purpose of the problem

A more advanced example of the use of conjugate addition with revision of Chapter 9.

Suggested solution

We need carbonyl groups to do conjugate addition so the *tertiary* alcohols will have to be made from them. In Chapter 9 (p. 222) we saw how to make such *tertiary* alcohols from esters and organometallic compounds and in Problem 1 of this set we saw how to do double conjugate additions on unsaturated esters – all the pieces fit together.

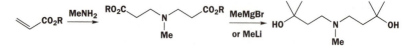

Problem 10

When we discussed reduction of cyclopentenone to cyclopentanol, we suggested that conjugate addition of borohydride must occur before direct addition of borohydride; in other words, this scheme must be followed:

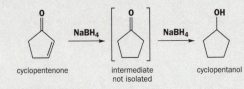

cyclopentenone   intermediate   cyclopentanol
                          not isolated

What is the alternative scheme? Why is the scheme shown above definitely correct?

Purpose of the problem

Serious thinking on mechanisms is an advantage when reactions get more complex.

Suggested solution

The alternative scheme would be to reduce the ketone first and the alkene second.

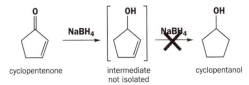

cyclopentenone   intermediate   cyclopentanol
                          not isolated

This scheme is wrong because simple alkenes are not reduced with NaBH$_4$, which is a nucleophilic reducing agent and attacks alkenes only if they are conjugated with an electron-withdrawing group. The conjugate addition must always be done first while the carbonyl group is intact. See Problem 2 for a simpler example.

Problem 11

Suggest a mechanism for this reaction. Why does conjugate addition occur rather than direct addition? Why is the product shown as a cation? If it is indeed a salt, what is the anion?

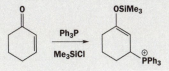

Purpose of the problem

Extension of simple ideas to more exotic reagents with trapping of the intermediate.

Suggested solution

The phosphine is a soft reversible nucleophile and is good at conjugate addition while unsaturated ketones are good at both types of reaction. On balance, conjugate addition wins. The intermediate enolate is trapped by silicon – an excellent electrophile for oxygen. The product is indeed a cation. The nucleophile was neutral and has lost electrons in the addition step. The counterion is chloride as all the atoms in the reagents are incorporated into the product.

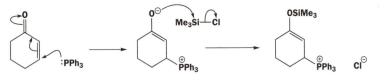

Problem 12

How, by choice of reagent, would you make this reaction give the direct addition product (route A)? How would you make it give the conjugate addition product (route B)?

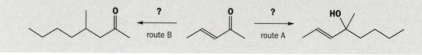

Purpose of the problem

Practising mastery over chemistry by choice of reagents. You need to know how to control this one.

Suggested solution

The added fragment is obviously a butyl group and the nucleophile should be BuLi, BuMgBr, or some other organometallic reagent. The two we have mentioned prefer direct addition and we shall have to add Cu(I) to get conjugate addition.

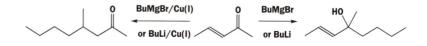

Suggested solutions for Chapter 11

11

In the solutions for this chapter for the first time there are some references to the original reports of these structures. This will become an important feature as the chemistry gets more complicated and is intended to help those who would like to read further into the science.

Problem 1

How many signals will there be in the ^1H NMR spectrum of each of these compounds? Estimate the chemical shifts of the signals.

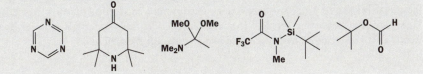

Purpose of the problem

Simple revision of symmetry with adjustment to proton rather than carbon NMR and chemical shift.

Suggested solution

You should find the symmetry easy now and this is the answer with different types of proton marked with different letters.

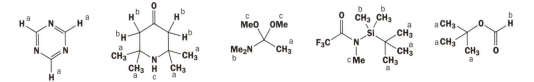

Estimating the chemical shift requires an adjustment to the narrower range of proton shifts plus a new region – aromatic as distinct from alkene. In each case we give a reasonable estimate and then the actual values. If your values agree with our estimates, you have done well. If you get something near the actual values, be very proud of yourself. The first compound has hydrogens on sp^2 carbon atoms bonded to two nitrogen atoms – hence the very large shift. The fourth molecule has two methyl groups directly bonded to electropositive silicon – hence the very small shift. The rest are more easily explained.

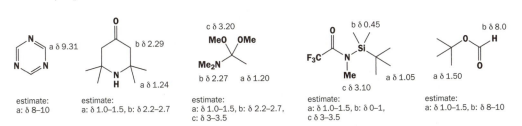

estimate:
a: δ 8–10

estimate:
a: δ 1.0–1.5, b: δ 2.2–2.7

estimate:
a: δ 1.0–1.5, b: δ 2.2–2.7,
c: δ 3–3.5

estimate:
a: δ 1.0–1.5, b: δ 0–1,
c δ 3–3.5

estimate:
a: δ 1.0–1.5, b: δ 8–10

Problem 2

Comment on the chemical shifts of these three compounds and suggest whether there is a worthwhile correlation with pK_a.

Compound	δ_H, p.p.m.	pK_a
CH_3NO_2	4.33	10
$CH_2(NO_2)_2$	6.10	4
$CH(NO_2)_3$	7.52	0

Purpose of the problem

Simple correlation of structure and chemical shift.

Suggested solution

The nitro group is a strongly electron-withdrawing group and would be expected to remove electron shielding from neighbouring hydrogen atoms. A CH_3 group would normally come at about 1.2 p.p.m. so one nitro group has shifted this downfield by about 3 p.p.m. The second adds less than 2 p.p.m. and the third only just 1.4 p.p.m. These diminishing effects are much smaller than the effects on pK_a, which is a log scale, but in the same direction. There is a kind of correlation showing that removing electrons from a C–H bond makes it more acidic and also increases the chemical shift.

Problem 3

One isomer of dimethoxybenzoic acid has the ^{1}H NMR spectrum δ_H (p.p.m.) 3.85 (6H, s), 6.63 (1H, t, J 2 Hz), 7.17 (2H, d, J 2 Hz) and one isomer of coumalic acid has the ^{1}H NMR spectrum δ_H (p.p.m.) 6.41 (1H, d, J 10 Hz), 7.82 (1H, dd, J 2, 10 Hz), and 8.51 (1H, d, J 2 Hz). In each case, which isomer is it? The substituents on bonds sticking into the centre of the rings can be on any carbon atom.

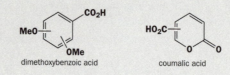

dimethoxybenzoic acid   coumalic acid

Purpose of the problem

First steps in using coupling to decide structure.

Suggested solution

The coupling constants in the first spectrum are all too small to be between Hs on neighbouring Cs and there is symmetry in the molecule. There is only one structure that answers these criteria. The compound is 3,5-dimethoxybenzoic acid.

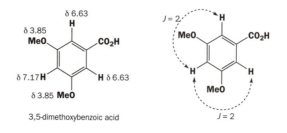

3,5-dimethoxybenzoic acid

The second compound has one coupling of 10 Hz and this must be between neighbours. The other coupling is 2 Hz and this must be *meta* coupling. There are two structures that might be right. In fact, the first is correct and you might have worked that out by the very large chemical shift – almost in the aldehyde region – of the isolated proton with only a 2 Hz coupling. This is on an alkene carbon bonded to oxygen in the first structure but on a simple alkene carbon in the second.

Problem 4

Assign the NMR spectra of this compound (assign means say which signal belongs to which atom) and justify your assignments.

■ In this problem and in other problems where NMR spectra are given in the textbook, we shall not repeat them here. Refer to the text if you wish to see them.

Purpose of the problem

Practice in the interpretation of real NMR spectra – this is harder than if the spectra have already been measured.

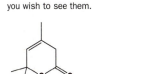

Suggested solution

There is no coupling in the proton NMR spectrum, which makes it much easier, but you should measure the chemical shifts and estimate the number of protons in each signal from the integration.

δ_H (p.p.m.) 1.4 (6H), 1.8 (3H), 2.9 (2H), and 5.5 (1H). The peak at 7.25 p.p.m. is CHCl$_3$ impurity in the CDCl$_3$ solvent.

This is enough to assign the spectrum but we should check that the chemical shifts are right too, and they are.

The carbon spectrum is more familiar to you from Chapter 3 and you will remember that integration means little here. There are three peaks in the 0–50 p.p.m. region corresponding to the methyl group on the alkene, the CH$_2$ group, and the pair of methyls on the same carbon atom. The 1:1:1 triplet at 77 p.p.m. is the solvent CDCl$_3$. The other signal in the 50–100 p.p.m. region must be the carbon next to oxygen (Me$_2$CO). The two signals in the 100–150 p.p.m. region are the two carbons of the alkene and the very small peak above 150 p.p.m. must be the carbonyl group. No further assignment is necessary.

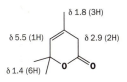

Problem 5

Assign the ^{1}H NMR spectra of these compounds and explain the multiplicity of the signals.

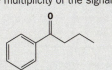

δ_H (p.p.m.) 0.97 (3H, t, *J* 7 Hz), 1.42 (2H, sextuplet, *J* 7 Hz), 2.00 (2H, quintet, *J* 7 Hz), 4.40 (2H, t, *J* 7 Hz)

δ_H (p.p.m.) 1.08 (6H, d, *J* 7 Hz), 2.45 (4H, t, *J* 5 Hz), 2.80 (4H, t, *J* 5 Hz), 2.93 (1H, septuplet, *J* 7 Hz)

δ_H (p.p.m.) 1.00 (3H, t, *J* 7 Hz), 1.75 (2H, sextuplet, *J* 7 Hz), 2.91 (2H, t, *J* 7 Hz), 7.47.9 (5H, m)

Purpose of the problem

First serious practice in correlating splitting patterns and chemical shifts.

Suggested solution

Redrawing the molecules with all hydrogens showing probably helps at this stage, though you will not do this for long. The spectrum of 1-nitrobutane can be assigned by integration and splitting pattern without even looking at the chemical shifts! Just counting the number of neighbours and adding one gives the multiplicity of the signals and leads to the assignment in the frame. It is equally valid to start with the chemical shift, which you would expect to get steadily smaller as the H atoms are more distant from the NO_2 group. Everything fits.

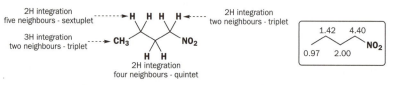

The next compound has an isopropyl group, typically a 6H doublet at about δ_H 1 p.p.m. and a 1H septuplet with a larger chemical shift so they can be quickly found. Assigning the two triplets is not so easy as the chemical shift difference is very small (0.35 p.p.m.) and it is safer not to say which is which. This uncertainty does not affect our identification of the compound.

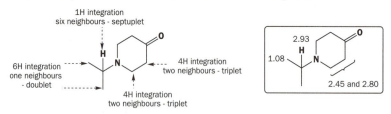

The aromatic ketone happens to have all the aromatic signals overlapping so that they cannot be sorted out. This is not unusual and a signal in the 6.5–8 region described as '5H, m' usually means a monosubstituted benzene ring. The side chain is straightforward with the CH_2 next to the ketone coming at larger chemical shift. All the coupling constants happen to be the same (7 Hz) as we expect for a freely rotating open-chain compound.

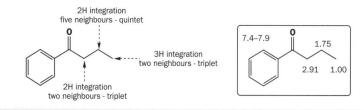

Problem 6

The reaction below was expected to give product 6A and did indeed give a product with the correct molecular formula by mass spectrum. The ^{1}H NMR spectrum of the product was however: δ_H (p.p.m.) 1.27 (6H, s), 1.70 (4H, m), 2.88 (2H, m), 5.4–6.1 (2H, broad s, exchanges with D_2O), 7.0–7.5 (3H, m). Though the detail is missing from this spectrum, how can you already tell that this is not the compound expected?

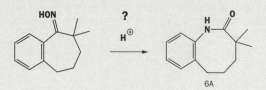

6A

Purpose of the problem

To show that it is helpful to predict the NMR spectrum of an expected product provided that the structure is rejected if the spectrum is 'wrong'.

Suggested solution

The spectrum is all wrong. There are only *three* aromatic Hs (7.0–7.5 p.p.m.), not the four expected. There are *two* exchanging hydrogens, presumably NH_2, and not the one expected. The only thing that is as expected is the chain of three CH_2 groups. You could not be expected to find the right structure, which is actually an amide with a new carbocyclic ring.

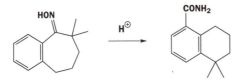

Now that you know what the product is, you might like to assign the spectrum and convince yourself that this does indeed fit the NMR spectrum.

■ This surprising result was reported by B. Amit and A. Hassner, *Synthesis*, 1978, 932. The expected reaction was the Beckmann rearrangement (Chapter 37) but what actually happened was a Beckmann fragmentation (Chapter 38) followed by an intramolecular Friedel-Crafts alkylation (Chapter 22).

Problem 7

Assign the 400 MHz ^{1}H NMR spectrum of this enynone as far as possible, justifying both chemical shifts and coupling patterns.

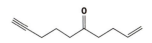

Purpose of the problem

Practice at interpretation of high-field NMR spectra. More complex problems can be solved reasonably easily.

Suggested solution

First measure the spectrum and list the data. The expansions make it much easier to see the coupling but even so we are going to have to call the signal at 5.6 p.p.m. a multiplet. For the rest of the signals, you should have measured the *J* values. Coupling is measured in Hz and at 400 MHz each chemical shift unit of 1 p.p.m. is 400 Hz, so each subunit of 0.1 p.p.m. is 40 Hz. Out with the ruler or the dividers and get measuring!

chemical shift (δ, p.p.m.)	Integration	Multiplicity	Coupling, *J*, Hz	Comments
5.6	1H	m	?	alkene region
5.05	1H	d with fine splitting	16.3	alkene region
4.97	1H	d with fine splitting	10.4	alkene region
2.58	2H	t with fine splitting	6.5	next to C=O or C=C?
2.47	2H	t with fine splitting	6.5	next to C=O or C=C?
2.32	2H	q with fine splitting	6.5	next to C=O or C=C?
2.21	2H	t with fine splitting	6.5	next to C=O or C=C?
1.95	1H	broad s	–	alkyne?
1.77	2H	q	6.5	not next to anything

That gives us three protons in the alkene region, five CH_2 groups, and one solitary proton, which can be only the alkyne proton. If you were surprised by its small chemical shift, check the tables and it is all right. In the alkene region, the multiplet must be H^2 which couples to the CH_2 at C3 and both H^1s. On C1, H^{1a} has a large *trans* coupling (16 Hz) to H^2 while H^{1b} has a smaller *cis* coupling (10 Hz). The coupling between H^{1a} and H^{1b} is very small.

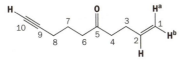

Of the five CH_2 groups, the quintet at small chemical shift must be at C7. Those at C4, C6, and C8 have two neighbours and are basically triplets, but that at C3 couples to three protons and must be the quartet at δ_H 2.32 p.p.m.

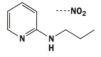

[NO$_2$ somewhere in the molecule]

Problem 8

A nitration product ($C_8H_{11}N_3O_2$) of this pyridine has been isolated which has a nitro (NO_2) group somewhere on the molecule. From the 90 MHz ^{1}H NMR spectrum, deduce whether the nitro group is (a) on the ring, (b) on the NH nitrogen atom, or (c) on the aliphatic side chain and then exactly where it is. Give a full analysis of the spectrum.

Purpose of the problem

Practice at interpretation of high-field NMR spectra. More complex problems can be solved reasonably easily.

Suggested solution

The three types of compound under consideration are these (there are isomers of the first and last).

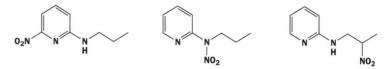

Checking the integral will deal with that problem. The propyl side chain is still there with a CH_3 triplet, a CH_2 quintet, and a CH_2 triplet with a large chemical shift. The broad signal at δ_H 5.9 p.p.m. is typical of an NH group so no reaction has happened there. The small signal – less than a quarter of a proton – at 7.2 p.p.m. cannot be part of the molecule and must be $CHCl_3$. The remaining signals in the aromatic region at 6.3, 8.1, and 9.1 p.p.m. must be three protons on the pyridine ring. The nitro group has replaced one of the original four and is somewhere on the pyridine ring. These are the possibilities.

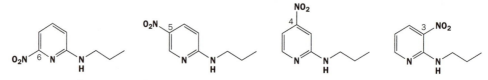

The most significant feature of the aromatic region is a proton at very large chemical shift (9.1 p.p.m.) with no coupling to speak of (there might be some long-range coupling). This proton has no neighbours and that rules out the 3-nitro and 6-nitro compounds. If you check in tables you will see that protons on carbons next to nitrogen (C2 and C6) in pyridine have very large shifts as you would expect from their aldehyde-like nature. The nitro group also increases the shifts of protons on neighbouring carbon atoms. The compound must be 5-nitro and the spectrum can be assigned like this.

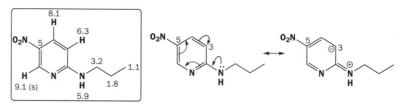

The coupling in the side chain is all about 7 Hz as expected and that between C3 and C4 in the pyridine ring is somewhat larger also as expected for an aromatic ring. The smaller chemical shift of the proton at C3 occurs because it is not affected by the nitro group but it is affected by electron donation from the amine nitrogen atom.

Problem 9

The natural product bullatenone was isolated in the 1950s from a New Zealand myrtle and assigned the structure 9A. Then compound 9A was synthesized and found not to be identical with natural bullatenone. Predict the expected ^{1}H NMR spectrum of 9A. Given the full spectroscopic data available nowadays, but not in the 1950s, say why 9A is definitely wrong and suggest a better structure for bullatenone.

9A
alleged bullatenone

Spectra of bullatenone:

Mass spectrum: m/z 188 (10%) (high resolution confirms $C_{12}H_{12}O_2$), 105 (20%), 102 (100%), and 77 (20%).
Infrared: 1604 and 1705 cm^{-1}.
^{1}H NMR: δ_H (p.p.m.) 1.45 (6H, s), 5.82 (1H, s), 7.35 (3H, m), and 7.68 (2H, m).

Purpose of the problem

Detecting wrong structures is fun and teaches us to be alert to what the spectra are telling us rather than what we expect.

Suggested solution

The mass spectrum and infrared are all right for structure 9A but the NMR shows at once that bullatenone cannot be 9A. There is indeed a monosubstituted benzene ring (the 2H and 3H signals in the aromatic region confirm this) but the aliphatic protons consist of a 6H signal, almost certainly a CMe$_2$ group, and a 1H singlet in the alkene region at 5.82 p.p.m.

The fragments we have are Ph, carbonyl, and CMe$_2$ groups and an alkene with one H on it. That adds up to $C_{12}H_{12}O$ and leaves only one oxygen atom to fit in. There must still be a ring or else there will be too few hydrogen atoms and the ring is five-membered (just try other possibilities yourself). There are three basic rings we can choose and each can have the phenyl group on either end of the double bond making six possibilities in all.

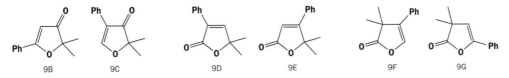

9B　　9C　　9D　　9E　　9F　　9G

The last four are all esters (cyclic esters or lactones) and they would have IR carbonyl stretches at higher frequency in the 1745–1780 cm^{-1} range. The hydrogen on the alkene cannot be on the same carbon atom as the oxygen or it would be at a very large shift indeed whereas it is close to the 'normal' alkene shift of 5.25 p.p.m. For 9C we could estimate 5.25 + 0.64 p.p.m. for O, + 0.36 p.p.m. for Ph, and + 0.87 p.p.m. for C=O making 7.12 p.p.m., a long way from the observed 5.82 p.p.m. Structure 9B is correct and the spectra can be assigned.

Compound 9B was synthesized and proved identical to natural bullatenone.

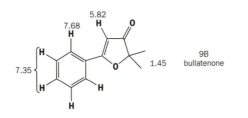

9B
bullatenone

■ You can read the full story in W. Parker *et al., J. Chem. Soc.,* 1958, 3871; see also T. Reffstrup and P. M. Boll, *Acta Chem. Scand.,* 1977, **31B**, 727, *Tetrahedron Lett.,* 1971, 4891, and R. F. W. Jackson and R. A. Raphael, *J. Chem. Soc., Perkin Trans. 1,* 1984, 535.

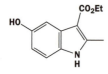

Problem 10

Interpret this ^{1}H NMR spectrum (see text, p. 277, for the 90 MHz NMR of the compound in the margin).

Purpose of the problem

Further correlation of chemical shift and coupling with interpretation of longer-range coupling.

Suggested solution

The ethyl group is easy to find – a typical 3H triplet at 1.2 p.p.m. and a 2H quartet at 4.3 p.p.m. – not just an ethyl group but an OEt group to get the large shift for the CH_2 group. The methyl group is also easy – a 3H singlet at 2.6 p.p.m., typical of a methyl on an alkene. At the other end of the spectrum, the broad signal at 11.0 p.p.m. can be only the NH or the OH – one exchanges and is not seen. That leaves the three signals in the aromatic region: δ_H (p.p.m.) 6.75 (1H, dd, J 9, 2 Hz), 7.15 (1H, d, J 9 Hz), and 7.48 (1H, d, J 2 Hz). The larger coupling is a typical *ortho* and the smaller a typical *meta* coupling so we can assign the whole spectrum.

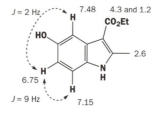

Problem 11

Suggest structures for the products of these reactions, interpreting the spectroscopic data. You are *not* expected to write mechanisms for the reactions and you should resist the temptation to work out what 'should happen' from the reactions. These are all unexpected products.

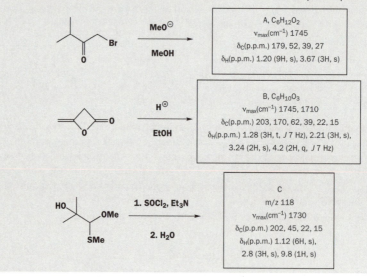

Purpose of the problem

Practice of a favourite exam question as well as a common situation in real life – you carry out a reaction, isolate the product, and find it's something quite different from what you had expected. What is it?

Suggested solution

Compound A has a carbonyl group (IR) which is an acid derivative (179 p.p.m. in ^{13}C NMR). The 9H singlet in the proton NMR must be a *t*-butyl group and the 3H singlet at 3.67 p.p.m. must be an OMe group. Putting those fragments together we get the structure immediately. The IR is typical for an ester ($1715 + 30 = 1745$ cm^{-1}).

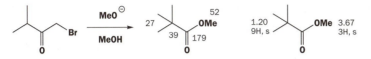

Compound B again has an ester (1745 cm^{-1} and 170 p.p.m.) but it also has a ketone (1710 cm^{-1} and 203 p.p.m.). The proton NMR shows an OEt group (3H triplet and 2H quartet at 4.2 p.p.m.) together with another Me group next to something (which can only be a C=O as there isn't anything else!) and a CH$_2$ at 3.24 p.p.m.. This is 2 p.p.m. away from the 'normal' CH$_2$ but it can't be next to O as we've used up the three O atoms so it must be next to two functional groups. Again, these can only be the carbonyls so this CH$_2$ is between the two carbonyl groups and we have the structure.

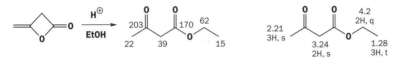

Compound C has no formula given, just a molecular ion in the mass spectrum. You should wonder why. The most obvious formula is $C_5H_{10}O_3$ but sulfur is 32 while O is 16 so two of those oxygen atoms could be one sulfur atom. It might be $C_5H_{10}OS$. We must look at the rest of the spectra for clarification. There is a carbonyl group (1730 cm^{-1}) which is an aldehyde or ketone (202 p.p.m.). The proton NMR shows a CMe$_2$ group (6H, s), a methyl group at 2.8 p.p.m. which doesn't look like an OMe (expected shift about 3–3.5 p.p.m.) but might be an SMe (the carbon spectrum also suggests SMe rather than OMe at δ_Ψ 45 p.p.m.), and one hydrogen atom at 9.8 p.p.m. which looks like an aldehyde. We know we have these fragments.

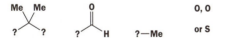

It is not possible to construct a molecule with two extra oxygen atoms but without an OMe group. One possible structure is shown below but it does not fit the data very well. Other possibilities include having a peroxide (O–O) link and are unlikely.

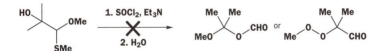

Only one compound is possible if there is a sulfur atom – this fits the data much better and is, in fact, the correct answer. It contains a genuine aldehyde (δ_H expected about 10 p.p.m.) not a formate (methanoate) ester (δ_H expected about 8 p.p.m.; see Problem 1), and SMe is better than OMe for the signal at δ_H 2.8 p.p.m. and δ_Ψ 45 p.p.m.

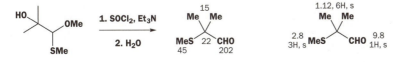

Problem 12

Precocene is a compound that causes insect larvae to pupate and can also be found in some plants (*Ageratum* spp.) where it may act as an insecticide. It was isolated in minute amounts and has the following spectroscopic details. Propose a structure for precocene.
Spectra of precocene:
Mass spectrum: m/z (high resolution gives $C_{13}H_{16}O_3$), M–15 (100%) and M–30 (weak)

Infrared: CH and fingerprint only.

^{1}H NMR: δ_H (p.p.m.) 1.34 (6H, s), 3.80 (3H, s), 3.82 (3H, s), 5.54 (1H, d, *J* 10 Hz), 6.37 (1H, d, *J* 10 Hz), 6.42 (1H, s), and 6.58 (1H, s).

Purpose of the problem

Your first attempt at determining the structure of a natural product without any hints – not even a wrong structure.

Suggested solution

The mass spectrum gives us the formula ($C_{13}H_{16}O_3$), which looks like an aromatic compound as there are so few hydrogens, and the base peak at M – 15 suggests that a methyl group is lost rather easily. The infrared suggests that all three oxygen atoms are present as ethers. The NMR shows us these details:

1.34 (6H, s): two identical methyl groups, probably CMe_2. Rather large shift.

3.80 (3H, s) and 3.82 (3H, s): two different OMe groups.

5.54 (1H, d, *J* 10 Hz) and 6.37 (1H, d, *J* 10 Hz): two *cis* protons on an alkene.

6.42 (1H, s) and 6.58 (1H, s): two isolated protons on an aromatic ring, probably 1,4-related as there is no coupling. Rather small shifts – electron-rich ring.

If all this is true, we have these fragments.

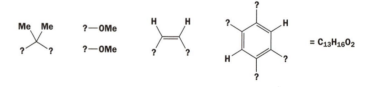

This adds up to $C_{13}H_{16}O_2$, so we have one more oxygen atom to fit in somewhere. It is very important that you now start to join up these fragments and see what you get. You quickly find that no chains of atoms are possible as there are not then enough substituents to go on the benzene ring (see the structure in the margin). We must find four substituents for the ring and there are only four other pieces to be joined up. The only chain-terminating units are the two OMe groups.

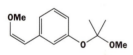

The obvious solution is to put the *cis* double bond in another ring fused on to the benzene ring and hence making two of the required four substituents. We can then combine the other groups in various ways such as these.

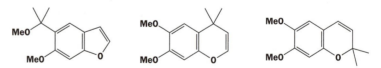

The five-membered ring is no good because the coupling constant between the alkene protons would be much smaller than 10. The two six-membered rings are good because they make the benzene ring electron-rich (three OR substituents). The middle structure is not right because the chemical shifts of the alkene protons would be very different: H^a predicted to be about δ_H 4.5 p.p.m. and H^b about δ_H 6.2 p.p.m.

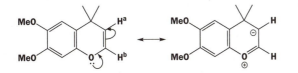

The third structure (which is the correct one) fits all the data. The two alkene protons are at normal shifts for such protons (estimated from tables in Williams and Fleming: 6.35 p.p.m. for H^a and 5.53 p.p.m. for H^b), the two OMe groups on the ring are *slightly* different, and the loss of a methyl group in the mass spectrum is easily explained by removal of an electron from oxygen followed by loss of Me$^\bullet$ to form a stable aromatic cation.

■ D. H. Williams and I. Fleming, *Spectroscopic methods in organic chemistry* (5th edn), McGraw Hill, London (1995).

■ More details of the discovery, biological activity, and structural elucidation of precocene are given by W. S. Bowers *et al. Science*, 1976, **193**, 542.

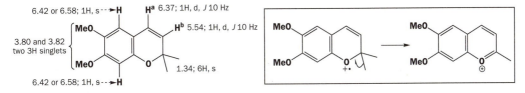

Problem 13

Suggest structures for the products of these reactions, interpreting the spectroscopic data. Though these products, unlike those in Problem 11, are reasonably logical, you will not meet the mechanisms for the reactions until Chapters 22, 29, and 23, respectively, and you are advised to solve the structures through the spectra.

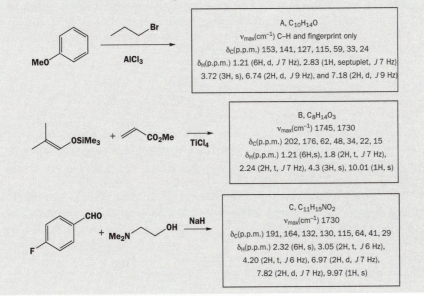

A, $C_{10}H_{14}O$
$\nu_{max}(cm^{-1})$ C–H and fingerprint only
δ_C(p.p.m.) 153, 141, 127, 115, 59, 33, 24
δ_H(p.p.m.) 1.21 (6H, d, J 7 Hz), 2.83 (1H, septuplet, J 7 Hz), 3.72 (3H, s), 6.74 (2H, d, J 9 Hz), and 7.18 (2H, d, J 9 Hz)

B, $C_8H_{14}O_3$
$\nu_{max}(cm^{-1})$ 1745, 1730
δ_C(p.p.m.) 202, 176, 62, 48, 34, 22, 15
δ_H(p.p.m.) 1.21 (6H,s), 1.8 (2H, t, J 7 Hz), 2.24 (2H, t, J 7 Hz), 4.3 (3H, s), 10.01 (1H, s)

C, $C_{11}H_{15}NO_2$
$\nu_{max}(cm^{-1})$ 1730
δ_C(p.p.m.) 191, 164, 132, 130, 115, 64, 41, 29
δ_H(p.p.m.) 2.32 (6H, s), 3.05 (2H, t, J 6 Hz), 4.20 (2H, t, J 6 Hz), 6.97 (2H, d, J 7 Hz), 7.82 (2H, d, J 7 Hz), 9.97 (1H, s)

Purpose of the problem

Your second attempt at determining the structures of reaction products, now of moderate complexity.

Suggested solution

Compound A contains the two reagents combined with the loss of HBr, and the four Hs at 6.74 and
7.18 p.p.m. suggest that the other reagent is attached to the benzene ring. The OMe group is still
there (3H singlet at 3.72 p.p.m.) and the new signals are a coupled 6H doublet and 1H septuplet –
an isopropyl group. The compound is one of these three isomers.

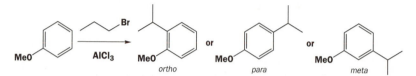

The correct isomer can be determined by the detailed NMR of the aromatic protons. They fall
into two groups of 2H each and each group is coupled with the other with a 9 Hz coupling. This fits
only with the *para* isomer. Both *ortho* and *meta* isomers have four different aromatic protons.
Notice that we did this using the proton NMR only, but we might as well assign the other data.
There is no IR as there are no functional groups except an ether. The four signals in the aromatic
region of the carbon NMR also reflect the symmetry of the molecule.

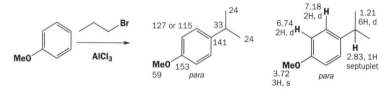

Compound B combines the two reagents with the loss of Me_3Si and the gain of H. Both the
infrared and the ^{13}C NMR show the appearance of a second carbonyl group – the ester (1745 cm^{-1},
176 p.p.m.) has been joined by an aldehyde or ketone (1730 cm^{-1}, 202 p.p.m.). The proton NMR
shows that it is an aldehyde (10.01, 1H, s). There is also a CMe_2 group, but no longer part of an
alkene (proton and carbon NMR also show that the alkene has gone) and the OMe group of the
ester (4.3, 3H, s). Finally, and very helpfully, there are two open-chain CH_2s linked together (the
two triplets with J 7 Hz). One of them (2.24 p.p.m.) is next to something and this has to be one of
the carbonyl groups as there isn't anything else! So we have:

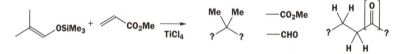

The carbonyl group on the CH_2CH_2 chain is in brackets because we already have it as the aldehyde
or the ester – we just don't know which. There are only two ways to join these fragments up.

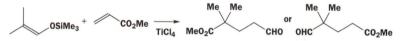

Though we mustn't say what 'ought to happen' we can still see that it makes more sense for the ester
group to stay where it is, on a chain of two carbon atoms, than mysteriously move to the other end of
the molecule next to the two methyl groups. This is not evidence, but we prefer the second structure.
Real evidence comes from the absence of coupling between the aldehyde proton and anything. If the
first structure were correct, the aldehyde proton would be a triplet. The second structure is right.

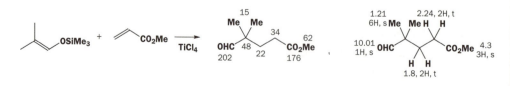

Again, adding up the atoms for compound C, the two reagents have joined together with the loss of HF. The 1,4-disubstituted benzene ring is still there (same NMR pattern as A) as is the aldehyde (1730, 191 p.p.m., and 9.97 p.p.m., s). The NMe_2 group and the CH_2CH_2 chain in the other reagent also survive so the oxygen atom of the alcohol has replaced the fluorine on the benzene ring. Simple!

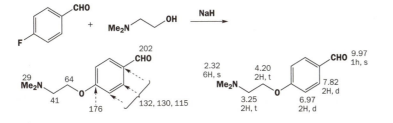

■ In all these questions involving aromatic rings, it may not be possible to assign all the carbon NMR with confidence but you should do the best you can. This does not affect the structure determination.

Problem 14

The following reaction between a phosphonium salt, base, and an aldehyde gives a hydrocarbon C_6H_{12} with the 200 MHz ^1H NMR spectrum shown (p. 278 of the text). Give a structure for the product and comment on its stereochemistry. You are not expected to discuss the chemistry!

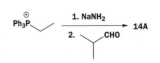

Purpose of the problem

Structure determination from exotic chemistry including a stereochemical extra.

Suggested solution

First analyse the data, measuring chemical shifts, integrals, and J values on the 200 MHz NMR.

δ_H, p.p.m.	Integral	Multiplicity	J values, Hz	Comments
0.97	6H	d	7	CH**Me₂**
1.60	3H	d	5	**Me**CHX
2.70	1H	double septuplet	7, 4	Me₂**CH**–CH
5.15	1H	dd	10, 4	alkene
5.35	1H	1:3:4:4:3:1?	5	alkene

From this alone, we can see an alkene with two Hs, a methyl group, and an isopropyl group. That adds up to C_6H_{12} so we have found everything and we can join them up in two ways (yes, two – the alkene can be *cis* or *trans*!). There is also a puzzle over the J values – there are too many 5 Hz couplings and the second 10 Hz coupling is missing, but we'll unravel that later.

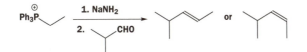

■ You will meet the reaction in Chapter 14.

The isopropyl group contains a 7 Hz coupling between the two methyl groups and the H at 2.70 p.p.m. which is coupled to one of the alkene protons at 5.15 p.p.m. with $J = 4$ Hz. The remaining coupling of the alkene proton at 5.15 p.p.m. (10 Hz) must be to the other alkene proton and that fits with a *cis* double bond. Working from the other end, the methyl group at 1.60 p.p.m. is coupled to the alkene proton at 5.35 p.p.m. with $J = 5$ Hz. Now we can solve the coupling constant mystery. We know that the proton at 5.35 p.p.m. is actually a double quartet and that the doublet coupling happens to be exactly twice that within the quartet. So we have two 1:3:3:1 quartets overlapping so that the inner lines coincide and give six lines in the ratio 1:3:4:4:3:1. The compound 14A is *cis*-4-methylpent-2-ene.

Suggested solutions for Chapter 12

12

Problem 1

Suggest reagents to make the drug 'phenaglycodol' by the route shown.

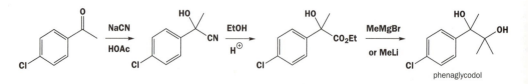

Purpose of the problem

Simple revision of carbonyl addition from Chapter 6 with a bit of substitution added.

Suggested solution

The first step is a simple addition: we used NaCN and HOAc in Chapter 6. The second is a nitrile reaction with ethanol – an acid-catalysed addition introduced on p. 294 of the text. Finally, there is a double addition of a methyl-metal compound such as MeLi or MeMgBr. The first molecule substitutes at the ester and the second adds to the resulting ketone. This reaction appears on p. 297 of the chapter.

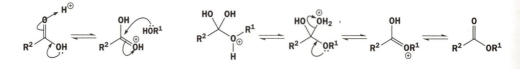

Problem 2

Direct ester formation from alcohols (R^1OH) and carboxylic acids (R^2CO_2H) works in acid solution but does not work at all in basic solution. Why not? By contrast, ester formation from alcohols (R^1OH) and carboxylic acid anhydrides, ($R^2CO)_2O$, or acid chlorides, RCOCl, is commonly carried out in the presence of amines such as pyridine or Et_3N. Why does this work?

Purpose of the problem

This question may sound trivial but students starting organic chemistry often make the error of trying to form esters from carboxylic acids in basic solution. Thinking about the reasons may save you from the error.

Suggested solution

The reaction works in acid solution because the carboxylic acid gets protonated and becomes a good electrophile. Later, the intermediate gets protonated again so that a molecule of water can be lost.

In basic solution, the first thing that happens is the removal of the acidic proton from the carboxylic acid. Nucleophiles will not now attack the carboxylate anion – the strongly nucleophilic alkoxide anion, in particular, will be repelled from another anion.

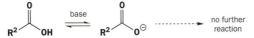

Acid anhydrides and chlorides do not have this acidic hydrogen so the alcohol attacks them readily and the base is helpful to remove a proton from the intermediate and encourage decomposition. The most popular bases are amines such as pyridine or Et_3N, which are not strong enough to remove the proton from the alcohol before the reaction starts. The by-product from the uncatalysed reaction is HCl (or R^2CO_2H from an anhydride) and the base removes that.

Problem 3

Predict the success or failure of these attempted nucleophilic substitutions at the carbonyl group. You should use estimated pK_a or pK_{aH} values in your answer and, of course, draw mechanisms.

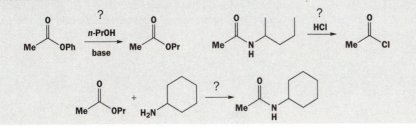

Purpose of the problem

A chance to try out the correlation between leaving group ability and pK_a explained in the chapter.

Suggested solution

You need to draw mechanisms for each reaction and identify the best leaving group by pK_a or pK_{aH} value. Note the difference between the second two reactions. In acid solution the important pK_a of an amine is about 10 but in base about 35. Even protonated nitrogen is not as good a leaving group as chloride.

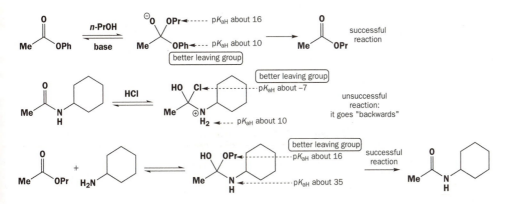

Problem 4

Suggest mechanisms for these reactions.

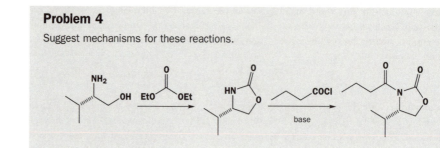

Purpose of the problem

Drawing mechanisms for nucleophilic substitution at the carbonyl group with important compounds.

Suggested solution

In the first reaction there are two nucleophilic substitutions and you must decide which nucleophile attacks first. The amine is a better nucleophile than the alcohol and the cyclization occurs because it is an equilibrium with two equal leaving groups (both alcohols) but one (EtOH) goes away when it leaves while the other is attached and cannot escape. The second reaction is more straightforward. The product is used to control the stereochemistry of new molecules as you will see in Chapter 45. For the first time we are using shorthand mechanisms. Note the double-headed arrow on the carbonyl group and the omission of proton transfer steps. If you drew the full mechanism, you did a better job. If you removed the amide (NH) proton before reaction with the acid chloride in the second step you also did a better job.

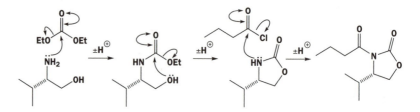

Problem 5

In making esters of the naturally occurring amino acids (general formula below) it is important to keep them as their hydrochloride salts. What would happen to these compounds if they were neutralized?

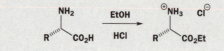

Purpose of the problem

Exploration of a reaction that can go seriously wrong if we do not think about what we are doing.

Suggested solution

You might well draw the mechanisms of the acid-catalysed esterification as part of your answer. That is in the chapter (p. 289) and we shall just answer the main question briefly here. Throughout the esterification process (check that this is so in your mechanism) the amino group remains protonated and plays no part in the reaction. If it is neutralized, it becomes the best nucleophile in the system, and the amino group of one molecule will attack the ester group of another. The result is the formation of a 'diketopiperazine' – a notorious by-product in amino acid chemistry. The second substitution is faster than the first because the first is bimolecular but the second is a unimolecular cyclization to form a stable (six-membered) ring.

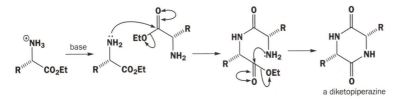

a diketopiperazine

Problem 6

It is possible to make either the diester or the monoester of butanedioic acid (succinic acid) from the cyclic anhydride as shown. Why does the one method give the monoester and the other the diester?

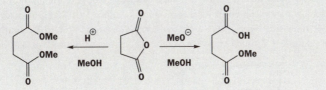

Purpose of the problem

An exploration of selectivity in carbonyl substitutions. Mechanistic thinking allows you to say confidently whether a reaction will happen or not. This problem builds on Problems 2 and 5.

Suggested solution

In basic solution the nucleophile is the excellent methoxide ion, which attacks the anhydride to give a tetrahedral intermediate having the carboxylate anion as its best leaving group. This carboxylate cannot be protonated and is not attacked by methoxide anion.

In acid solution the first reaction is similar except that the intermediate is neutral but the carboxylate is still the better leaving group. The second esterification is now all right because methanol can attack the protonated carboxylic acid and water can be driven out after a second protonation. This second step is an equilibrium with water and methanol about equal in leaving group ability but MeOH is in large excess as solvent. We have omitted proton transfer steps.

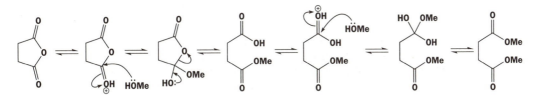

Problem 7

Suggest mechanisms for these reactions, explaining why these particular products are formed.

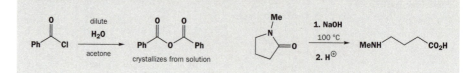

Purpose of the problem

A contrast between a very reactive (acid chloride), a less reactive (anhydride), and an unreactive (amide) carbonyl group.

Suggested solution

The acid chloride reacts rapidly with water and the acid produced rapidly attacks another molecule of acid chloride. The anhydride reacts much more slowly (pK_{aH} of Cl^- is -7, pK_{aH} of RCO_2^- is about $+5$) with water so there is a good chance of stopping the reaction there especially when we add a low concentration of water in acetone solution. This chance is made a certainty because the solid anhydride precipitates from solution and is no longer in equilibrium with the other compounds. It is usually possible to descend the reactivity order of acid derivatives (text, p. 287).

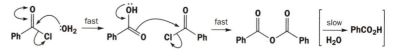

The second reaction is an example of the alkaline hydrolysis of amides (see p. 293 of the text). Though the nitrogen atom is never the best leaving group, it will leave from the dianion and, once gone, it is quickly protonated and does not come back. This example also benefits from the slight release of strain in opening the five-membered ring (C=O angle 120°, angle in five-membered ring 108°).

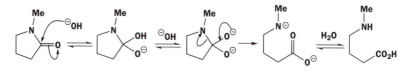

Problem 8

Here is a summary of part of the synthesis of Pfizer's heart drug doxazosin (Cordura®). The mechanism for the first step will be a problem at the end of Chapter 17. Suggest reagent(s) for the conversion of the methyl ester into the acid chloride. In the last step, good yields of the amide are achieved if the amine is added as its hydrochloride salt in excess. Why is this necessary?

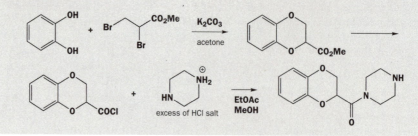

Purpose of the problem

To show that simple carbonyl substitutions are used in the manufacture of medicinal compounds.

Suggested solution

You know how to make acid chlorides from acids so the simplest answer is to hydrolyse the ester first, in acid or alkaline solution, and treat the free acid with PCl_5 or $SOCl_2$.

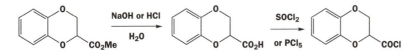

The 'amine' is, in fact, a diamine with two equally reactive amino groups. If one of these is always protonated, it will not react with the acid chloride and clean monosubstiution will be achieved. The product will be formed under the reaction conditions with the other nitrogen protonated.

Problem 9

Esters can be made directly from nitriles by acid-catalysed reaction with the appropriate alcohol. Suggest a mechanism.

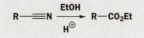

Purpose of the problem

An extension of simple carbonyl substitutions into a related reaction and practice at drawing new mechanisms.

Suggested solution

Initial attack of ethanol on the protonated nitrile is the only reasonable way to start (compare p. 294 of the text). The product is a protonated imine that hydrolyses on work-up to the ester.

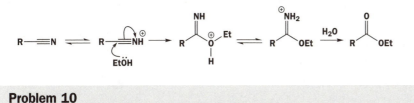

Problem 10

Give mechanisms for these reactions, explaining the selectivity (or lack of it!) in each case.

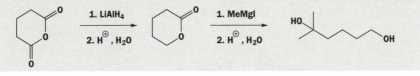

Purpose of the problem

Analysis of a sequence of reactions where the first stops at the halfway stage but the second does not.

Suggested solution

One of the carbonyl groups of the anhydride must be attacked by $LiAlH_4$ and we need to follow that reaction through to see what happens next. Before we do this it seems that the product, an ester, must surely be reduced by the LiAlH$_4$. The first addition of 'H–' produces a tetrahedral intermediate, which decomposes by the loss of the only leaving group, the carboxylate ion, to give an aldehyde. That too is quickly reduced to give the hydroxy-acid as its anion. Even LiAlH$_4$ attacks anions only very slowly so the hydroxy-acid remains until work-up when it cyclizes quickly to form the ester. When water is added, all the LiAlH$_4$ is instantly destroyed so none of it remains to reduce the ester.

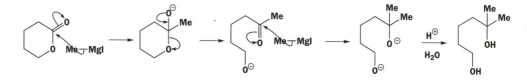

The second reaction starts similarly with the Grignard reagent adding to the ester carbonyl and the tetrahedral intermediate losing the only possible leaving group, the alkoxide anion. Again, a reactive carbonyl compound is produced – a ketone that is more reactive than the original ester so a second Grignard addition occurs faster than the first. Work-up in aqueous acid gives the diol.

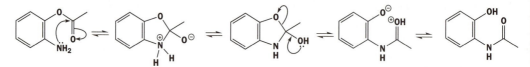

Problem 11

This reaction goes in one direction in acidic solution and in the other direction in basic solution. Draw mechanisms for the reactions and explain why the product depends on the conditions.

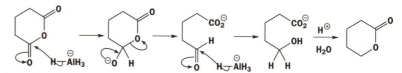

Purpose of the problem

A reminder that carbonyl substitutions are equilibria and that removal of a product from the equilibrium may decide which way a reaction goes. Drawing mechanisms for intramolecular reactions.

Suggested solution

It is best to draw the mechanism for the forward reaction to see what we have to think about.

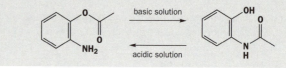

The amine attacks the ester in the usual way and the oxygen leaving group (the 'alcohol' is a phenol so pK_{aH} is now about 10) is much better than the amine (pK_{aH} about 35) so the reaction will go this way. In acid solution, the mechanism is the same except that the nitrogen atom is protonated in the tetrahedral intermediate so the two leaving groups are about the same (pK_{aH} of phenol about 10, pK_{aH} of aromatic amine about 8). What tips the balance is that one possible product is a weakly basic amide and is not protonated under the reaction conditions, but the other is an amine and is protonated under the reaction conditions. This removes it from the equilibrium and drives the reaction 'back'.

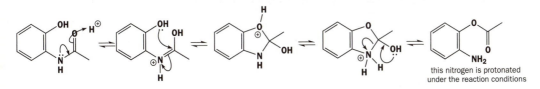

this nitrogen is protonated under the reaction conditions

Problem 12

These reactions do not work. Explain the failures and suggest in each case an alternative method that might be successful.

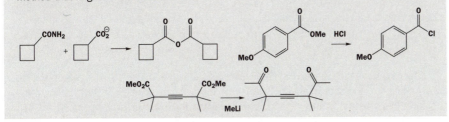

Purpose of the problem

Explaining why reactions *don't* work may be the easist way to see how to *make* them work.

Suggested solution

The first two reactions are cases where the supposed electrophile (amide in the first case and ester in the second) is not reactive enough to combine wth the nucleophile. In the tetrahedral intermediates, the better leaving group is the one that has just been introduced (carboxylate in the first case and chloride in the second) so the reactions could be made to work backwards but not in the direction given.

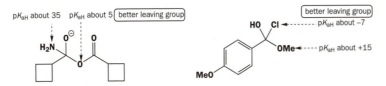

The first reaction could be made to work by using the acid chloride instead of the amide. Chloride is a better leaving group than carboxylate. We cannot use this method in the second case as chloride is the nucleophile. It is best to use the free acid instead of the ester and PCl_5 or $SOCl_2$ to make the acid chloride.

The last example is rather different. The reaction will work – indeed it will work only too well! The product shown, with its two ketone groups, is more electrophilic than the original diester so it will react again and a diol will be formed.

This problem is analysed in detail at the end of the chapter (pp. 299–301) where the alternatives suggested include using the acid chloride with Me_2CuLi, using the free acid or the *N,N*-dimethylamide with MeLi, using Weinreb amides, and using nitriles.

Suggested solutions for Chapter 13

13

Problem 1

In the method for acetylating aromatic amines described in Chapter 8, p. 188 (the Lumière-Barbier method) the amine was dissolved in aqueous HCl. Using 1M amine, pK_a 4.6, and 1M HCl, pK_a –7, what will be the approximate equilibrium constant in the reaction?

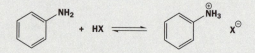

The next step in the reaction is the addition of acetic anhydride and sodium acetate. What will happen to the sodium acetate (NaOAc) in the aqueous solution of HCl? Estimate the equilibrium constant for the reaction between NaOAc and HCl. Will there be enough acid left to keep the amine in solution?

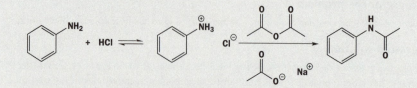

It would be simpler not to add the sodium acetate, keeping the pH low, and thus definitely keep all the amine in solution. Why is it necessary to raise the pH for the second step of the reaction (the reaction with acetic anhydride)?

Purpose of the problem

Simple revision of carbonyl addition from Chapter 6 with a bit of substitution added.

Suggested solution

The answer to the first question is that the difference in pK_a values is very great (11.6 – remember that this means a factor of $10^{11.6}$ as it is a log scale) and there is enough acid to protonate all the amine. The equilibrium constant is enormous.

In the second step, acetate ion (pK_{aH} about 5) and HCl (pK_a –7) set up an equilibrium with acetic acid and chloride ion. The difference in pK_a values is still great and that equilibrium too will be right over on one side. However, the main point is that the pK_a values of acetic acid and $PhNH_3^+$ are about the same so there will be about the same amount of each – the equilibrium constant will be about 1. Most of the amine will still be in solution as the cation but there will be enough free amine to react. If no sodium acetate were added, the amine would all be in solution as the cation and there would be essentially no free amine to act as a nucleophile. It would be simpler but no reaction would occur.

Problem 2

In the comparison of stability of the last intermediates in the carbonyl substitution of acid chlorides on the one hand and anhydrides on the other to make esters we made this statement.

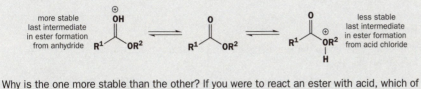

Why is the one more stable than the other? If you were to react an ester with acid, which of the two would you form and why?

Purpose of the problem

Revision of contribution of delocalization to structure, particularly of cations.

Suggested solution

The positive charge on the left-hand cation is delocalized over both oxygen atoms but the positive charge in the right-hand intermediate is localized as there are no π electrons to be moved to the positively charged atom. If an ester is protonated it gives the more stable cation by protonation at carbonyl oxygen but we can use the other oxygen in the mechanism to show the delocalization.

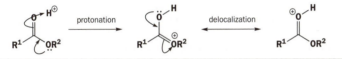

Problem 3

If we carry out an ester exchange reaction on a one molar solution of $MeCO_2R^2$ in an alcohol R^1OH as solvent with catalytic R^1O^- and let the mixture reach equilibrium, what will be the equilibrium composition if the solvent alcohol R^1OH is, say, 25M in itself?

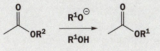

Purpose of the problem

Practice at the estimation of equilibrium constants and assessment of their significance.

Suggested solution

The two esters are in equilibrium by the base-catalysed formation and decomposition of a tetrahedral intermediate. The details of the mechanisms are irrelevant to the equilibrium (as long as there is a good mechanism). The two esters are about as stable as one another so the equilibrium constant will be close enough to 1.

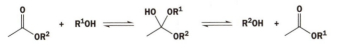

The equilibrium is

$$K = \frac{[R^2OH][MeCO_2R^1]}{[R^1OH][MeCO_2R^2]} \approx 1$$

So, if the available concentrations of $[R^1OH]:[R^2OH]$ are about 25, the ratio of esters must be about 25 the other way with a 25-fold excess of $MeCO_2R^2$. The mixture contains about 96% $MeCO_2R^2$ and about 4% $MeCO_2R^1$. This is quite enough to make ester exchange a practical method.

Problem 4

Write a mechanism for the reaction to give HCl on alumina. You do not need to consider the role of the alumina.

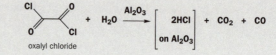

Purpose of the problem

Practice at writing mechanisms of new reactions.

Suggested solution

The first step must surely be attack of water on one of the very reactive acid chlorides.

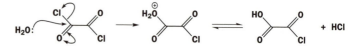

This intermediate, which is half acid and half acid chloride, can now decompose to produce all the products in a step greatly favoured by entropy.

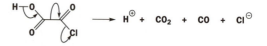

Problem 5

Propose a mechanism for the formation of the diazonium salt referred to in the chapter. The first step is the formation of nitrous acid HONO.

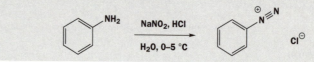

Purpose of the problem

Practice at writing the complicated mechanism of an important reaction that we shall need later.

Suggested solution

The first step is the formation of HONO, we are told, and it is clear that the amine must be a nucleophile as amines can only play that role. HONO must be the electrophile so we had better combine them to form an N–N bond as that is present in the product.

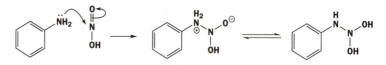

We need to lose two OH groups from this intermediate so we must protonate it on each oxygen and use the lone pair electrons from the nitrogen atom to expel two molecules of water.

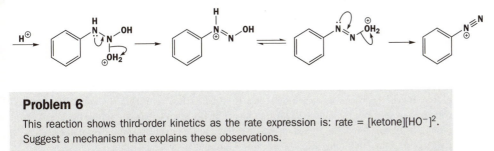

Problem 6

This reaction shows third-order kinetics as the rate expression is: rate = [ketone][HO$^-$]^2. Suggest a mechanism that explains these observations.

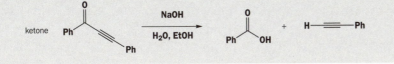

Purpose of the problem

Interpretation of unlikely kinetics in a reaction related to some in the chapter.

Suggested solution

The hydroxide ion must attack the ketone to form a tetrahedral intermediate. The best leaving group from this intermediate is the hydroxide ion that has just come in (pK_{aH} 15) rather than the alkyne anion (pK_{aH} about 35). If we use the second hydroxide ion to deprotonate the intermediate we have only one possible leaving group, though it is a bad one, and the decomposition of the dianion must be the rate-determining step. This mechanism is usually found for nucleophilic substitution at a carbonyl group with a very bad leaving group such as amide hydrolysis.

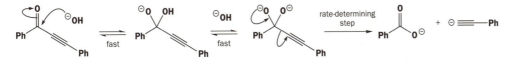

The benzoate ion is the product under the reaction conditions but the alkyne anion collects a proton from a water molecule, regenerating the second hydroxide ion, which therefore is a base catalyst.

Problem 7

Draw an energy profile diagram for this reaction. You will, of course, need to draw a mechanism first. Suggest which step in the mechanism is likely to be the slow step and what kinetics would be observed.

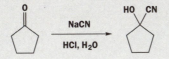

Purpose of the problem

Practice at drawing energy profile diagrams as one way to present the energetics of mechanisms.

Suggested solution

The first thing is to draw the mechanism: this reaction was discussed in Chapter 6.

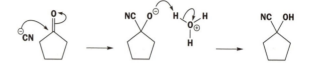

The first step is bimolecular and involves forming a new C–C bond. The second step is just a proton transfer between oxygen atoms and is certainly fast. The first step must be the rate-determining step and the intermediate will have a higher energy than the starting material or the product as it is an anion. In this answer we have used the style of energy profile diagrams used in the chapter but there is nothing sacred about this – any similar diagram is fine.

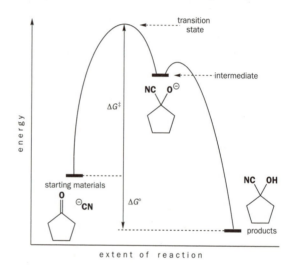

Problem 8

The equilibrium between a carbonyl compound and its hydrate usually favours the aldehyde or ketone. Draw an energy profile to express this and mark the difference in free energy between the two compounds. The hydrate of cyclopropanone is preferred to the ketone. How does the energy profile change for this compound?

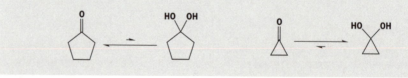

Purpose of the problem

Practice at drawing energy profile diagrams to show the relationship between the same reaction on different compounds.

Suggested solution

This is an equilibrium so the mechanism is irrelevant. The two five-membered rings are both more stable than the two three-membered rings but the important thing to express is that the five-membered cyclic hydrate is *less* stable than cyclopentanone but the three-membered cyclic hydrate is *more* stable than cyclopropanone.

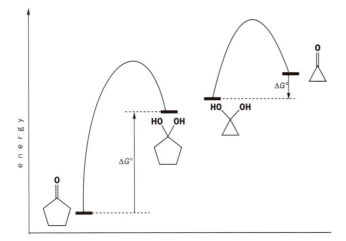

Problem 9

What would be the solvent effects on these reactions? Would they be accelerated or retarded by a change from a nonpolar to a polar solvent?

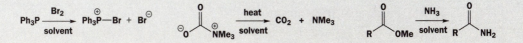

Purpose of the problem

Practice at assessing the likely effect of solvent polarity in terms of the mechanism of reactions.

Suggested solution

It is essential first to draw a mechanism for each reaction and to identify the rate-determining step in each case. The first two reactions are one-step processes so that at least makes life easier.

Now we need to draw the transition state for each reaction so that we can assess whether it is more or less polar than the starting materials. The way to do this is described on p. 318 in the text.

In the first reaction, uncharged starting materials form a partly charged transition state. A polar solvent will stabilize the transition state and speed up the reaction. In the second case, a fully charged (zwitterionic) starting material gives a partly charged transition state. A polar solvent will stabilize both starting material and transition state but it will stabilize the starting materials *more*. The energy gap ($\Delta G^{\ddagger}$) will increase and the reaction will go more slowly.

The third reaction is different because it has more than one step. It is a carbonyl substitution of the kind we met in Chapter 12. The nucleophile (ammonia) attacks the carbonyl group to form a tetrahedral intermediate that decomposes with the loss of the better leaving group.

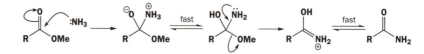

We have marked two steps fast because they are just proton transfers between nitrogen and oxygen atoms. Either of the other two steps might be rate-determining. In this substitution the leaving group is relatively good (compare Problem 6) and the rate-determining step is the first – the usual one for carbonyl substitutions. In this step, neutral starting materials become a dipolar (zwitterionic) intermediate so the transition state is becoming more charged and the reaction is accelerated by more polar solvents.

Problem 10

Comment on the likely effect of acid or base on these equilibria.

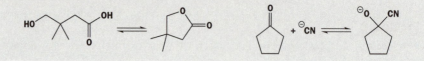

Purpose of the problem

Practice at assessing the likely effect of acid and base on particular examples of equilibria.

Suggested solution

The first example is an esterification (or lactonization as the product is a cyclic ester or lactone) in the forwards direction and an ester hydrolysis in the backwards direction. Ester hydrolysis is catalysed by acid or base but esterification by acid only. In addition, even a weak base will be enough to turn the starting material into the carboxylate anion, which will not cyclize. The equilibrium is to the right in acid solution and to the left in basic solution.

The second example is the familiar one of cyanohydrin formation from a ketone. The reaction is indeed reversible but in basic solution the cyanide anion is more stable than the oxyanion in the product and the carbonyl group is very stable too. In acidic solution (at pHs less than about 12) the oxyanion will be protonated and the reaction driven over to the right.

Problem 11

Elemental sulfur normally exists as an eight membered ring (S_8), but it can also be found in a number of other states. How would entropy and enthalpy affect the equilibrium between sulfur in these two forms?

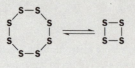

Purpose of the problem

Practice at assessing the likely effect of entropy on equilibria.

Suggested solution

The equilibrium position is determined both by enthalpy and entropy as discussed on p. 312 of the chapter. Entropy favours the larger number of molecules as they have greater randomness so heat will drive the equilibrium to the right. Enthalpy will favour the eight-membered ring as it has no ring strain and so lower temperatures favour the left-hand side.

Problem 12

Draw transition states and intermediates for this reaction and fit each on an energy profile diagram. Be careful to distinguish between transition states and intermediates.

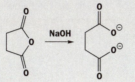

Purpose of the problem

Constructing an energy profile diagram for a multistep process.

Suggested solution

First, a mechanism is essential. This is quite a long job as there are several steps so we must patiently work our way through them. And no short cuts are allowed!

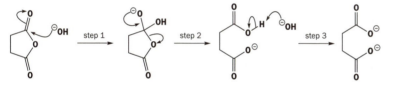

　　Step 3 must be fast as it is just a proton transfer between oxygen atoms. The decomposition of the tetrahedral intermediate is likely to be fast as the leaving group (carboxylate ion, pK_{aH} about 5) is such a good one. The first step, the bimolecular attack of hydroxide on the anhydride, will be the rate-determining step. We need only draw the structures of the transition states and we can construct our diagram.

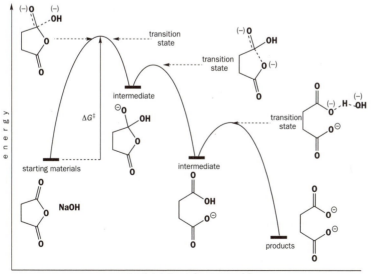

Suggested solutions for Chapter 14

14

Problem 1

In the cyclization of the open-chain form of glucose to form the stable hemiacetal, it may be difficult to work out what has happened. Number the carbon atoms in the open-chain form and put the same numbers of the hemiacetal so that you can see where each carbon atom has gone. Then draw a mechanism for the reaction.

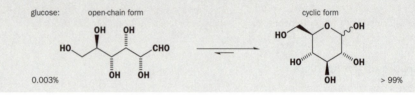

Purpose of the problem

First trial of a simple method to follow complicated looking reactions plus some revision of carbonyl addition reactions.

Suggested solution

Just do what the question says! The starting material is numbered from the aldehyde group. We started numbering the product with C6 as that is the only CH$_2$ group and then followed the carbon chain back to C1 which is now the hemiacetal carbon.

Numbering the carbon atoms makes it clear that the hydroxyl group on C5 has cyclized. To draw the mechanism, we need to redraw the starting material so that the reaction is possible and the easiest way to do that is to draw it like the product. Don't forget to change the thick and thin lines as necessary when you flip a C atom over! The mechanism is just a nucleophilic addition to the aldehyde as described in Chapter 6 and is the first step of the reactions in this chapter.

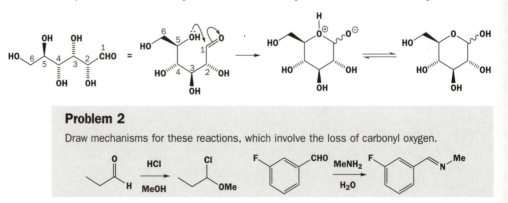

Problem 2

Draw mechanisms for these reactions, which involve the loss of carbonyl oxygen.

Purpose of the problem

To see if you can draw mechanisms for two of the main types of reactions in the chapter.

Suggested solution

As MeOH is the solvent and present in large excess, it probably adds first. This also makes the intermediate in the addition of chloride a more stable oxonium ion. This mechanism is very like that of acetal formation and, if you added chloride first, that is a good mechanism too.

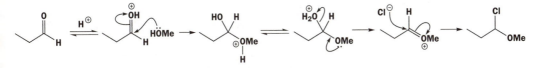

The second example is imine formation – attack by an amine nucleophile and dehydration of the intermediate. Don't forget to protonate the OH group so that it can leave as a water molecule.

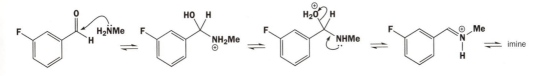

Problem 3

Each of these molecules is an acetal, that is, a compound made from an aldehyde or ketone and two alcohol groups. Which compounds were used to make these acetals?

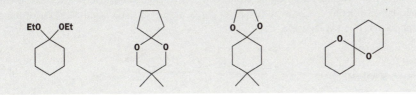

Purpose of the problem

Practice at the recognition of acetals and working out how to make them.

Suggested solution

All we have to do is to identify the hidden carbonyl group by looking for the only carbon atom having two C–O bonds. This atom is marked with a black blob in the answer below. Those are the bonds made during acetal formation so, if we imagine breaking them, we can see the alcohols or diol needed.

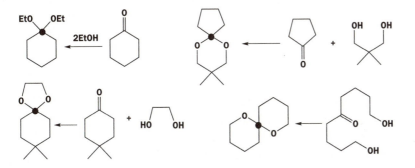

Problem 4

Each of these reactions leads to an acetal or a closely related compound and yet no alcohols are used in the first two reactions and no carbonyl group in the third. How are these acetals formed?

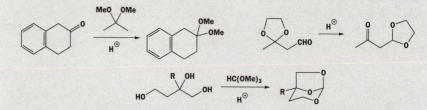

In the first and third of these two reactions, a compound, different in each case, must be distilled from the reaction mixture if the reaction is to go to completion. What are the compounds and why is this necessary? In the second case, why does the reaction go in this direction?

Purpose of the problem

Practice at drawing mechanisms for acetal exchange rather than straightforward acetal synthesis.

Suggested solution

In each case the acetal provided reacts with the acid catalyst to create an oxonium ion and an alcohol. Each has a vital part to play in the rest of the reaction.

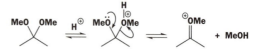

By the usual acetal mechanism the alcohol and the ketone combine to form a hemiacetal, which is dehydrated either by acid or by the oxonium ion. If water is lost, it is captured by the oxonium ion and the end result is the same.

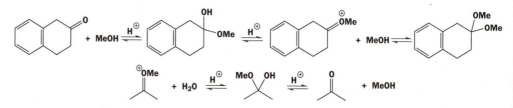

Enough methanol is formed in these reactions to make the final acetal and the water is consumed in this process to give acetone as the other product. So both ketones and both acetals are in equilibrium but the most volatile component is acetone and this is distilled off, driving the equilibrium over to the right.

The third example is essentially the same except that the reagent is one orthoester and the product is another. The volatile by-product is methanol and distillation of this pushes the equilibrium across. The mechanism is very similar to the first example with the formation of methanol and an oxonium ion starting the process off. This time, an oxonium ion must be involved in the capture of each alcohol.

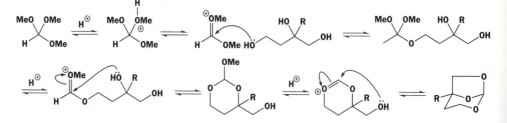

The second example is different because no volatile by-product is removed and the position of equilibrium must represent a genuine difference in stabilities of the two acetals. Some water is needed as the first acetal must be wholly or partially hydrolysed and the water appears when the second acetal is formed. The exact details of this mechanism are uncertain but the whole process is under thermodynamic control: the ketone is more stable than the aldehyde.

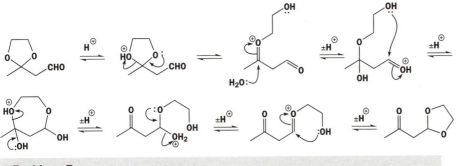

Problem 5

Suggest mechanisms for these two reactions of the smallest aldehyde, formaldehyde (methanal, $CH_2=O$).

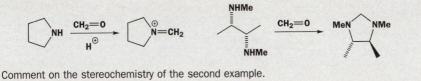

Comment on the stereochemistry of the second example.

Purpose of the problem

Extension of simple acetal chemistry into related reactions with nitrogen.

Suggested solution

Both reactions start in the same way by attack of a nitrogen nucleophile on formaldehyde. Acid catalysis is not necessary for this first step. The first reaction ends with the formation of an iminium ion by acid-catalysed loss of water.

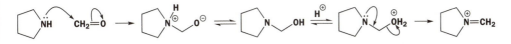

In the second reaction a second amino group is waiting to capture this intermediate by cyclization to form a stable five-membered ring. The stereochemistry does not change but the central bond of the molecule has to be rotated through 180° before the cyclization can occur.

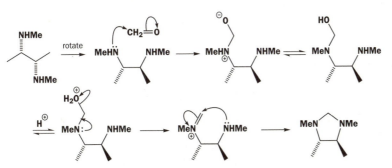

Problem 6

Suggest mechanisms for this reaction. It first appeared in Chapter 3 where we identified the rather unexpected product from its spectra but did not attempt to draw a mechanism for the reaction.

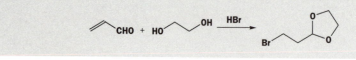

Purpose of the problem

Revision of conjugate addition from Chapter 10 with the drawing of an acetal mechanism and a decision to be made on the order of events.

Suggested solution

Two reactions are required: conjugate addition of bromide and acetal formation. The obvious solution is to take the conjugate addition first while the carbonyl group is intact and make the acetal later.

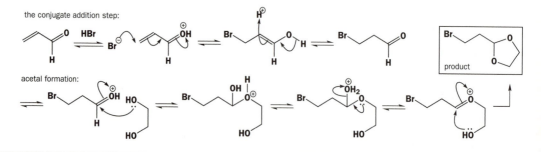

Problem 7

In Chapter 6 we described how the antileprosy drug dapsone could be made soluble by the formation of a 'bisulfite adduct'. Now that you know about the reactions described in Chapter 14, you should be able to draw a mechanism for this reaction. The adduct is described as a 'pro-drug', meaning that it can give dapsone itself in the human body. How might this happen?

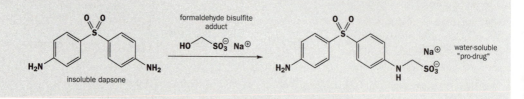

Purpose of the problem

Revision of chemistry from Chapter 6 with a challenging mechanistic problem – did you avoid the trap?

Suggested solution

The trap is to go straight to the product by displacing hydroxide from the formaldehyde bisulfite adduct. Hydroxide is a very bad leaving group and reactions like this never occur.

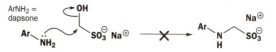

To avoid this trap we must use the chemistry from Chapters 6 and 14 – carbonyl chemistry. Reactions of the kind we need were used in Problem 5 where formaldehyde was also a reagent. First, we must make formaldehyde from its bisulfite adduct and then add it to the amino group of dapsone.

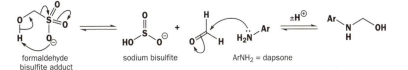

Now we can form an iminium salt and add the bisulfite back into this reactive electrophile to give the final product. This is loss of carbonyl oxygen in an unusual setting as the carbonyl group is not there at the start.

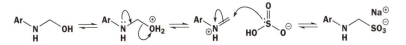

Problem 8

Suggest a detailed mechanism for the acetal exchange used in this chapter to make an acetal of a ketone from an orthoester.

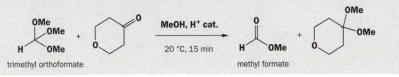

Purpose of the problem

Reminder of chemistry from this chapter – making acetals from ketones is difficult – and drawing a mechanism related to, but different from, one in Problem 4.

Suggested solution

The equilibrium between a ketone and an alcohol on the one hand and the dimethyl acetal on the other is unfavourable and must be driven over by devices such as this. The orthoformate is even more unstable than the acetal (three C–O bonds instead of two at the same carbon atom) and the ester even more stable, because of conjugation, than the ketone. There are various mechanisms you might have drawn, which differ only in details such as the order of the steps. Here is one such.

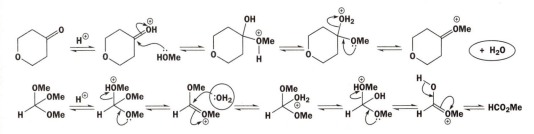

The first line shows the normal mechanism for acetal formation and points out the by-product, a molecule of water. This water is consumed in the hydrolysis of the orthoformate (by an acetal hydrolysis mechanism) to give a stable ordinary ester. The favourable second equilibrium pulls the first across to the right.

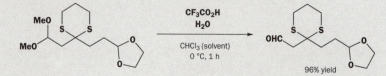

Problem 9

When we introduced cyclic acetals, we showed you this reaction.

What are the two functional groups not affected by this reaction? How would you hydrolyse them?

Purpose of the problem

Revision of the different types of acetals and their relative reactivity.

Suggested solution

This question develops immediately from the last. The noncyclic acetal hydrolyses easily because the equilibrium favours the three molecules (one aldehyde and two alcohols). The two remaining acetals are a cyclic acetal or dioxolane and a cyclic dithioacetal or dithiane. The cyclic acetal just needs more vigorous acidic conditions but the dithiane needs a Lewis acid that likes sulfur. Mercury – Hg(II) – is a good example. The precise reagents and conditions are not important.

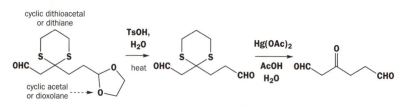

Problem 10

What would actually happen if you tried to make the unprotected Grignard reagent shown here?

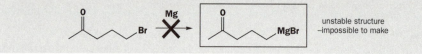

Purpose of the problem

Revision of the danger of mutually destructive functional groups.

Suggested solution

The Grignard reagent would attack the carbonyl group in the same molecule or in another molecule with irreversible formation of a carbon–carbon bond. The intramolecular reaction forms a four-membered ring and so may not be favoured. There are other possibilities too such as radical reactions (Chapter 39) or polymerization (Chapter 52).

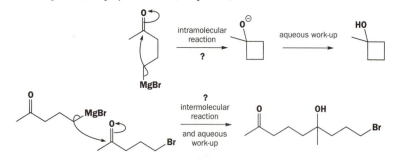

Problem 11

Find the acetals in cellulose (see the answer below for the structure of cellulose).

Purpose of the problem

Reminder that functional groups are the same no matter how complex the structure. Acetals can be stable!

Suggested solution

Just look for the carbon atoms (marked with a blob) having two C–O single bonds.

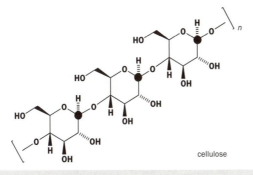

cellulose

Problem 12

A stable product can be isolated from the reaction between benzaldehyde and ammonia discussed in this chapter. Suggest a mechanism for its formation.

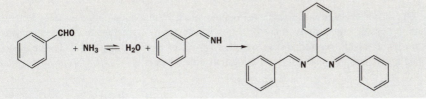

Purpose of the problem

Revision of aminal formation – the all-nitrogen version of acetal formation.

Suggested solution

Imine formation follows its usual course but the imine is unstable and reacts with more benzaldehyde.

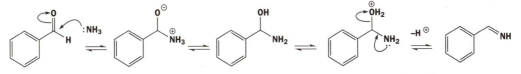

The reaction with benzaldehyde starts normally enough but the dehydration of the first intermediate produces a strange looking cation with two adjacent double bonds to the same nitrogen atom. Since the benzene rings play no part in the reaction, we abbreviate them to 'Ph'.

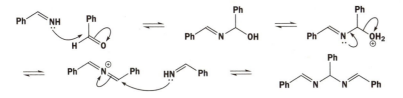

Problem 13

Suggest mechanisms for these reactions.

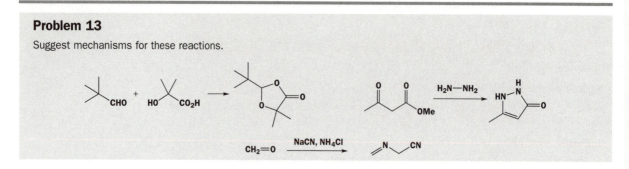

Purpose of the problem

Extension of acetal and aminal formation into examples where the intermediate is trapped by a different nucleophile.

Suggested solution

The first reaction starts with the usual attack of an alcohol on an aldehyde but the second nucleophile is a carboxylic acid. Though a poor nucleophile, it is good enough to react with the oxonium ion, especially when the product is cyclic.

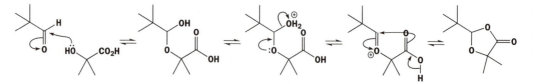

The second reaction starts with nucleophilic attack by the amine on the more electrophilic carbonyl group – the ketone. Imine formation is followed by cyclization and this second step is a normal nucleophilic substitution at the carbonyl group of an ester (Chapter 12). The imine double bond moves into the ring to secure conjugation with the ester.

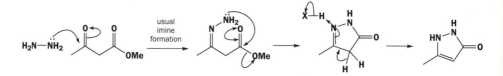

■ B. S. Furniss *et al.*, *Vogel's textbook of organic chemistry* (5th edn), Longmans, Harlow, 1989, p. 753.

The third example uses very simple molecules and again starts with imine formation. Cyanide is the nucleophile that captures the iminium ion and a second imine formation completes the mechanism.

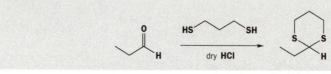

Problem 14

Finally, don't forget the problem at the end of the chapter: suggest a mechanism for this reaction.

Purpose of the problem

Extension of the mechanism for acetal formation into dithioacetal (dithiane) formation.

Suggested solution

The mechanism is a direct analogue of acetal formation. The dehydration step is more difficult because the C=S bond is less stable than the C=O bond because overlap of 2p and 3p orbitals is not as good as overlap of orbitals (for example, two 2ps) of the same size and energy.

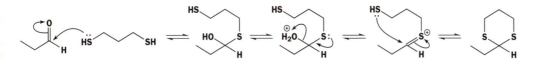

Suggested solutions for Chapter 15 **15**

Problem 1

A compound C_6H_5FO has a broad peak in the infrared at about 3100–3400 cm^{-1} and the following signals in its (proton decoupled) ^{13}C NMR spectrum. Suggest a structure for the compound and interpret the spectra.

δ_C (p.p.m.) 157.38 (doublet, coupling constant 229 Hz), 151.24 (singlet), 116.32 (doublet, coupling constant 7.5 Hz), 116.02 (doublet, coupling constant 23.2 Hz)

Purpose of the problem

Just to remind you that coupling occurs in carbon spectra too and is useful.

Suggested solution

All the signals are in the sp^2 region and two (at > 150 p.p.m.) are attached to electronegative elements. As the formula is C_6, a benzene ring is strongly suggested. The IR peak tells us we have an OH group so the compound is one of these.

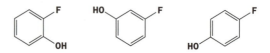

The symmetry of the spectrum (only four signals) suggests the third compound as the other two would have six different carbon atoms in the ring. We can therefore assign the spectrum by saying that the very large coupling (229 Hz) must be a $^2J_{CF}$ and the zero coupling must be the carbon attached to the OH. The other two signals are the carbons in between and we can assume that, the larger the coupling, the nearer the carbon is to fluorine.

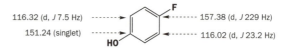

Problem 2

Suggest structures for the products of these reactions.

Compound 2A is $C_7H_{12}O_2$ and has: IR 1725 cm^{-1};
δ_H (p.p.m.) 1.02 (6H, s), 1.66 (2H, t, J 7 Hz), 2.51 (2H, t, J 7 Hz), and 3.9 (2H, s).

Compound 2B has: m/z 149/151 (M$^+$ ratio 3:1); IR 2250 cm^{-1}.
δ_H (p.p.m.): 2.0 (2H, quintet, J 7 Hz), 2.5 (2H, t, J 7 Hz), 2.9 (2H, t, J 7 Hz), and 4.6 (2H, s).

Purpose of the problem

Revision of Chapter 11 and the first example of total structure determination.

Suggested solution

The starting material for 2A is $C_7H_{12}O_3$ and has apparently lost an oxygen atom. As the reagent is NaBH₄ it is more likely to have gained two hydrogen atoms and lost a water molecule. The IR spectrum shows a C=O group and suggests an ester or a strained ketone. The NMR shows two joined CH_2 groups, one (at 2.51 p.p.m.) next to a functional group but not oxygen, so presumably C=O, the intact CMe_2 group, and an isolated CH_2 group next to oxygen at 3.9 p.p.m.. There is only one reasonable structure.

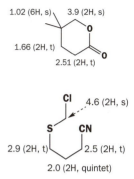

The mass spectrum of compound 2B shows that it has chlorine in it, the IR shows a CN group, and the proton NMR shows eight Hs. If we assume that no C atoms have been lost, the most reasonable formula is C_5H_8ClNS. The compound has lost a water molecule. The proton NMR shows three linked CH_2 groups, a quintet in the middle , and two triplets at the ends. The shifts of the terminal CH_2s (2.5 and 2.9 p.p.m.) show that they are next to functional groups but not Cl (there is no oxygen in the molecule). This must mean that we have a unit: $-SCH_2CH_2CH_2CN$. All that remains is an isolated CH_2 group with a large shift and the chlorine atom so they must be joined to S. The large shift (4.9 p.p.m.) comes from 1.5 + 1 (S) + 2 (Cl) = 4.5 p.p.m. Again one structure emerges.

Problem 3

Two alternative structures are shown for the possible products of the following reactions. Explain in each case how you would decide which product is actually formed. Several pieces of evidence would be required and estimated values are more convincing than general statements.

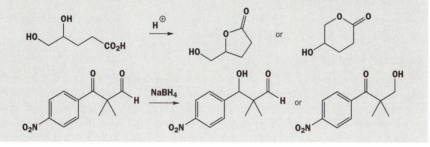

Purpose of the problem

To get you thinking the other way round – from structure to data. What are the important pieces of evidence?

Suggested solution

There are many acceptable ways in which you could answer this question ranging from choosing just one vital statistic for each pair to analysing all the data for each compound. We'll adopt a middle way and point out several important distinctions. In the first example, one main difference is the ring size, seen mainly in the IR. Both are esters (about 1745 cm^{-1}) but we should add 30 cm^{-1} for the five-membered ring. The functional group next to the OCH_2 group is different – an OH group in one case and an ester in the other. There are other differences.

In the second case there are also differences in the IR C=O stretch between the aldehyde (about 1730 cm^{-1}) and the conjugated ketone (about 1680 cm^{-1}) but the main difference is in the proton NMR. The aldehyde proton and the number of protons next to oxygen make a clear distinction. There will also be differences in the ^{1}H and ^{13}C NMR spectra of the benzene rings since one is conjugated with a C=O group and the other is not. The reaction actually gives a mixture.

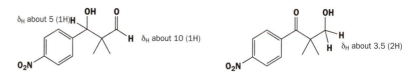

Problem 4

The following products might possibly be formed from the reaction of MeMgBr with the cyclic anhydride shown. How would you tell the difference between these compounds using IR and ^{13}C NMR spectra? With ^{1}H NMR available as well, how would your task be easier?

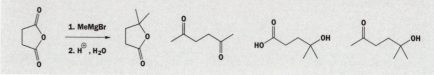

Purpose of the problem

Further thinking the other way round – from structure to data. Contrasting the limitations of IR and ^{13}C NMR with the data from ^{1}H NMR spectra.

Suggested solution

The molecular formula of the compounds varies so mass spectra would also be useful, but in the IR the compounds with an OH group would show a strong broad U-shaped band at above 3000 cm^{-1}. The cyclic ester would have a C=O stretch at about 1775 cm^{-1}, the ketones a band at about 1715 cm^{-1}, and the CO$_2$H group a broad band at about 1715 cm^{-1} as well as a very broad V-shaped band from 2500 to 3500 cm^{-1}. In the ^{13}C NMR the ester and acid would have a carbonyl group at about 170–180 p.p.m. but the ketones would have one at about 200 p.p.m. The number and position of the other signals would also vary.

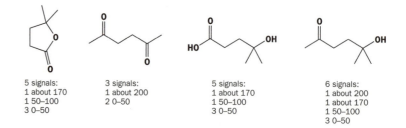

5 signals:	3 signals:	5 signals:	6 signals:
1 about 170	1 about 200	1 about 170	1 about 200
1 50–100	2 0–50	1 50–100	1 about 170
3 0–50		3 0–50	1 50–100
			3 0–50

In the proton NMR, all compounds would show the two linked CH$_2$ groups as a pair of triplets except in the second compound where they have the same chemical shift and do not couple. All have a 6H singlet, for the CMe$_2$ group in all cases except the second where it is the two identical COMe groups. The last has an isolated Me group. The OH and CO$_2$H protons might show up as broad signals at any chemical shift.

Problem 5

The NMR spectra of sodium fluoropyruvate in D_2O are given below. Are these data compatible with the structure shown? If not, suggest how the compound might exist in this solution.

δ_H (p.p.m.): 4.43 (2H, d, J 47 Hz); δ_C (p.p.m.): 83.5 (d, J 22 Hz), 86.1 (d, J 171 Hz), and 176.1 (d, J 2 Hz).

sodium fluoropyruvate

Purpose of the problem

To show how NMR spectra can reveal more than just the identity of a compound.

Suggested solution

The proton NMR is all right as we expect a large shift here: $1.5 + 1$ (C=O) $+ 2$ (F) $= 4.5$ p.p.m. and the coupling to fluorine is fine. The carbon NMR shows the carboxylate carbon at 176 p.p.m. with a small coupling to F as it is far away and the CH_2 carbon at 86.1 p.p.m. with an enormous coupling as it is joined directly to F. But where is the ketone? We should expect it at about 200 p.p.m. but it is at 83.5 p.p.m. with the expected intermediate coupling. It cannot be a carbonyl group at all. So what could have happened in D_2O solution? The obvious answer is that a hydrate is formed (Chapter 6).

Problem 6

An antibiotic isolated from a microorganism crystallized from water and formed (different) crystalline salts on treatment with either acid or base. The spectroscopic data were as follows.

Mass spectrum: 182 (M^+, 9%), 109 (100%), 137 (87%), and 74 (15%).

δ_H (p.p.m.; in D_2O at pH < 1): 3.67 (2H, d,), 4.57 (1H, t,), 8.02 (2H, m), and 8.37 (1H, m).

δ_C (p.p.m.; in D_2O at pH < 1): 33.5, 52.8, 130.1, 130.6, 134.9, 141.3, 155.9, and 170.2.

Suggest a structure for the antibiotic.

Purpose of the problem

Structure determination of a compound with biological activity isolated from a natural source.

Suggested solution

The solubility and salt formation data suggest the presence of acidic and basic groups, probably CO_2H and NH_2 as this is a natural compound. If so, the ^{13}C peak at 170.2 p.p.m. is the CO_2H group. The five carbons in the sp^2 region and protons at 8.0 and 8.4 p.p.m. suggest an aromatic ring. There cannot be just one nitrogen atom as M^+ is an even number and that all looks very like a pyridine ring (typical δ_H about 7–8 p.p.m. but the N atom will be protonated at pH < 1). The two sets of aliphatic protons are coupled and the large shift of the 1H signal at 4.57 p.p.m. suggests a proton between CO_2H and NH_3^+ (pH < 1). We have these fragments.

■ The details of the structure and spectra are in S. Inouye et al., *Chem. Pharm. Bull.*, 1975, **23**, 2669; S. R. Schow et al., *J. Org. Chem.*, 1994, **59**, 6850; and B. Ye and T. R. Burke, *J. Org. Chem.*, 1995, **60**, 2640.

Presumably, the aliphatic part must be X or Y and that leaves just one oxygen atom for a formula of $C_8H_{10}N_2O_3 = 182$. Only six of the ten H atoms show up in the NMR because the OH, CO_2H, and NH_3^+ protons all exchange rapidly at pH < 1 and do not show up. There are various isomers possible and the compound is actually 'azatyrosine' though you cannot be expected to deduce that. The second diagram shows the structure in solution at pH < 1.

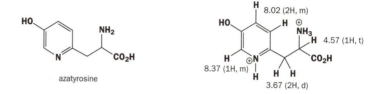

The two carbonyl groups so the [azatyrosine / structures]

Problem 7

Suggest structures for the products of these two reactions.

Compound 7A. m/z: 170 (M⁺, 1%), 84 (77%), and 66 (100%); IR: 1773, 1754 cm⁻¹; δ_H (p.p.m.; CDCl₃): 1.82 (6H, s) and 1.97 (4H, s). δ_C (p.p.m.; CDCl₃): 22, 23, 28, 105, and 169 (the signals at 22 and 105 p.p.m. are weak).

Compound 7B. m/z: 205 (M⁺, 40%), 161 (50%), 160 (35%), 106 (100%), and 77 (42%); IR: 1670, 1720 cm⁻¹; δ_H (p.p.m.; CDCl₃): 2.55 (2H, m), 3.71 (1H, t, *J* 6 Hz), 3.92 (2H, m), 7.21 (2H, d, *J* 8 Hz), 7.35 (1H, t, *J* 8 Hz), and 7.62 (2H, d, *J* 8 Hz); δ_C (p.p.m.; CDCl₃): 21, 47, 48, 121, 127, 130, 138, 170, and 172 p.p.m.

Purpose of the problem

The other important kind of structure determination: compounds isolated from a chemical reaction.

Suggested solution

Compound 7A is much the simpler so we start with that. The two reagents are $C_5H_6O_4$ and $C_5H_8O_2$ which adds up to $C_{10}H_{14}O_6$ or 230 so 60 has been lost. That looks like $C_2H_4O_2$ or less likely C_3H_8O (because it is saturated). If the first is right, 7A is $C_8H_{10}O_4$ which at least fits the proton NMR.

The IR suggests two carbonyl groups, though the ¹³C NMR suggests one only, but the molecule must have some symmetry as there are only five signals for eight carbon atoms. The proton NMR shows a CMe_2 group and four identical protons, which can only be two identical CH_2 groups. The only unsaturation is the two carbonyl groups so the ¹³C signal at 105 p.p.m. is very strange. It, like the signal at 22 p.p.m., is probably a quaternary carbon as it is weak and it must be next to two oxygen atoms to have such a large shift. Either 22 or 105 p.p.m. must be C of CMe_2. The last two degrees of unsaturation must be rings. The cyclopropane would supply one ring and the two linked and symmetrical CH_2 groups and the carbon at 22 p.p.m. would be the atom joining that ring to the two carbonyl groups leaving 105 p.p.m. to be C of CMe_2. So we have:

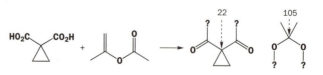

That actually accounts for all the atoms so 7A must be what you get when you join these two fragments together. The carbonyls are arranged rather like those in acyclic anhydride and the two C=O stretches must be the symmetrical and antisymmetrical combinations.

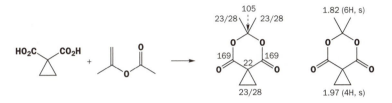

Compound 7B has nitrogen in it (odd MW) and clearly has the benzene ring too from the NMR so we can put down PhN (= 91) as part of the structure. It also has two carbonyl groups (in IR, the one at 1670 cm^{-1} looks like an amide) and they are both acid derivatives (^{13}C NMR). There are three aliphatic carbons, two CH$_2$s and one CH. Adding these together gives us $C_{11}H_{10}NO_2 = 188$ so there is 17 missing which looks like OH. Since we need a second acid derivative and the OH is the only remaining heteroatom, it must be a carboxylic acid. Given that the CH is a triplet, it must be next to one of the CH$_2$s and, as they are both multiplets, they must be joined to each other. There is one more degree of unsaturation so there is a ring. So we have:

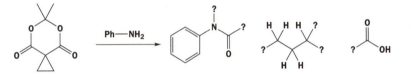

To assemble three fragments into a molecule we must plug the amide into the C$_3$ fragment to make a ring and put the CO$_2$H group on the last free position. We can do this in two ways. Proton NMR distinguishes these. The end CH$_2$ is attached either to the nitrogen atom (estimated 3.2 p.p.m.) or to the carbonyl group (estimated 2.2 p.p.m.) of the amide. The observed value of 3.92 p.p.m. fits the first better. A similar estimate for the CH also suggests the first structure. This is indeed the correct answer.

■ See S. Danishefsky and R. K. Singh, *J. Am. Chem. Soc.*, 1975, **97**, 3239.

■ From now on, having established the general method and approach, we shall simply give the answers with any special points applying to the question under discussion.

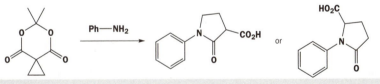

Problem 8

Treatment of the two compounds shown here with base gives an unknown compound with the spectra given here. What is its structure?

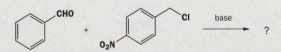

m/z: 241 (M$^+$, 60%), 90 (100%), 89 (62%); δ_H (p.p.m.; CDCl$_3$): 3.89 (1H, d, J 3 Hz), 4.01 (1H, d, J 3 Hz), 7.31 (5H, s), 7.54 (2H, d, J 10 Hz), and 8.29 (2H, d, J 10 Hz); δ_C (p.p.m.; CDCl$_3$): 62, 64, 122, 125, 126, 127, 130, 136, 144, and 148 (the last three are weak).

Purpose of the problem

Further practice at structure determination, adding ideas of the size of the coupling constant.

Suggested solution

The compound is an epoxide: the coupling constants around the ring are small (3 Hz, contrast 10 Hz in the benzene ring) because of ring size and the electronegative oxygen atom. All the Hs in the Ph group happen to come at the same shift but those in the nitrated ring are at lower field and separated by the NO_2 group.

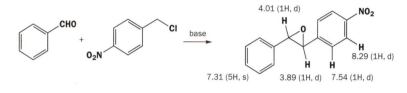

Problem 9

Treatment of this epoxy-ketone gives a compound with the spectra shown below. What is its structure? (*Hint.* You might like to check the comments on pp. 366–7 before deciding on your answer.)

m/z: 138 (M^+, 12%), 109 (56%), 95 (100%), 81 (83%), 82 (64%), and 79 (74%); IR: 3290, 2115, 1710 cm^{-1}; δ_H (p.p.m.; $CDCl_3$): 1.12 (6H, s), 2.02 (1H, t, J 3 Hz), 2.15 (3H, s), 2.28 (2H, d, J 3 Hz), and 2.50 (2H, s); δ_C (p.p.m.; $CDCl_3$): 26, 31, 32, 33, 52, 71, 82, 208.

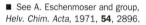

Purpose of the problem

Further practice at structure determination, adding a curious chemical shift.

■ See A. Eschenmoser and group,
Helv. Chim. Acta, 1971, **54**, 2896.

Suggested solution

The compound is an alkyne formed by the Eschenmoser fragmentation (Chapter 38). It is not possible to assign all the ^{13}C NMR with certainty but the alkyne Cs come in the region 70–85 p.p.m.

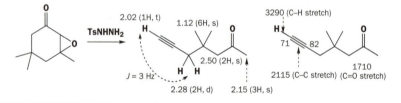

Problem 10

Reaction of the epoxy-alcohol below with LiBr in toluene gave a 92% yield of compound 10A. Suggest a structure for this compound.

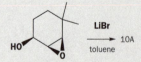

Compound 10A: m/z: $C_8H_{12}O$; v_{max} (cm^{-1}): 1685, 1618; δ_H (p.p.m.): 1.26 (6H, s), 1.83 (2H, t, J 7 Hz), 2.50 (2H, dt, J 2.6, 7 Hz), 6.78 (1H, t, J 2.6 Hz), and 9.82 (1H, s); δ_C (p.p.m.): 189.2, 153.4, 152.7, 43.6, 40.8, 30.3, and 25.9.

Purpose of the problem

Further practice at structure determination including a change in the carbon skeleton – a ring contraction.

Suggested solution

The compound is a simple cyclopentenal. The ^{13}C NMR assignment is not all certain.

■ G. Magnusson and S. Thorén, *J. Org. Chem.*, 1973, **38**, 1380.

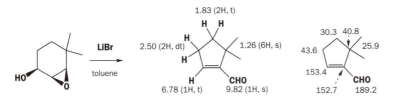

Problem 11

Female boll weevils (a cotton pest) produce two isomeric compounds that aggregate the males for food and sex. A few mg of two isomeric active compounds, grandisol and Z-ochtodenol, were isolated from 4.5 million insects. Suggest structures for these compounds from the spectroscopic data below. Signals marked * exchange with D_2O.

Z-Ochtodenol. m/z: 154 ($C_{10}H_{18}O$), 139, 136, 121, 107, 69 (100%).

ν_{max} (cm^{-1}): 3350, 1660.

δ_H (p.p.m.): 0.89 (6H, s), 1.35–1.70 (4H, broad m), 1.41 (1H, s*), 1.96 (2H, s), 2.06 (2H, t, J 6 Hz), 4.11 (2H, d, J 7 Hz), and 5.48 (1H, t, J 7 Hz).

Grandisol. m/z: 154 ($C_{10}H_{18}O$), 139, 136, 121, 109, 68 (100%).

ν_{max} (cm^{-1}): 3630, 3250–3550, and 1642.

δ_H (p.p.m.): 1.15 (3H, s), 1.42 (1H, dddd, J 1.2, 6.2, 9.4, 13.4 Hz), 1.35–1.45 (1H, m), 1.55–1.67 (2H, m), 1.65 (3H, s), 1.70–1.81 (2H, m), 1.91–1.99 (1H, m), 2.52* (1H, broad t, J 9.0 Hz), 3.63 (1H, ddd, J 5.6, 9.4, 10.2 Hz), 3.66 (1H, ddd, J 6.2, 9.4, 10.2 Hz), 4.62 (1H, broad s), and 4.81 (1H, broad s).

δ_C (p.p.m.): 19.1, 23.1, 28.3, 29.2, 36.8, 41.2, 52.4, 59.8, 109.6, and 145.1.

Purpose of the problem

Further practice at structure determination of natural products with all the data properly presented.

Suggested solution

The structures are shown below.

■ J. M. Tumlinson *et al.*, *Science*, 1969, **166**, 1010, but see K. Mori *et al.*, *Liebig's Annalen*, 1989, 969 for the spectra of ochtodenol and K. Narasaka *et al.*, *Bull. Chem. Soc. Jap.*, 1991, **64**, 1471 for the spectra of grandisol.

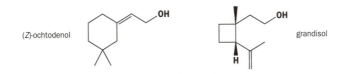

Problem 12

Suggest structures for the products of these reactions.

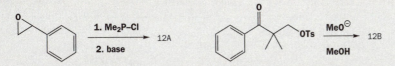

Data for compound 12A: $C_{10}H_{13}OP$, IR (cm^{-1}): 1610, 1235; δ_H (p.p.m.): 6.5–7.5 (5H, m), 6.42 (1H, t, J 17 Hz), 7.47 (1H, dd, J 17, 23 Hz), and 2.43 (6H, d, J 25 Hz).

Data for compound 12B: $C_{12}H_{16}O_2$; IR: CH and fingerprint only; δ_H (p.p.m.): 7.25 (5H, s), 4.28 (1H, d, J 4.8 Hz), 3.91 (1H, d, J 4.8 Hz), 2.96 (3H, s), 1.26 (3H, s), and 0.76 (3H, s).

Purpose of the problem

Structure determination of reaction products by proton NMR alone with extra twists – an element with spin (P) and protons on the same carbon atom that are different in the NMR.

Suggested solution

The coupling constants ($^3J_{PH}$) across the double bond in 12A are very large. Typically *cis* $^3J_{PH}$ is about 20 Hz and *trans* $^3J_{PH}$ is about 40 Hz. Geminal ($^2J_{PH}$) coupling constants are also large but more variable. In 12B there is a stereogenic centre so that the hydrogen atoms and methyl groups in the ring are either on the same side as MeO or the same side as Ph. They are diastereotopic with different chemical shifts and the Hs couple (^2J). We cannot say which is which.

■ See F. Nerdel *et al.*, *Tetrahedron Lett.*, 1968, 5751.

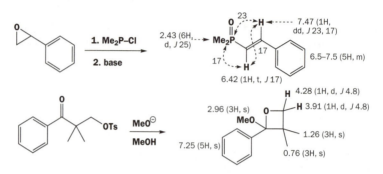

Problem 13

Identify the compounds produced in these reactions. Warning! Do not attempt to deduce the structures from the starting materials but use the data! These molecules are so small that you can identify them from ^{1}H NMR alone.

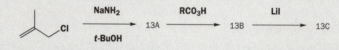

Data for compound 13A (C_4H_6): δ_H (p.p.m.): 5.35 (2H, s) and 1.00 (4H, s)

Data for compound 13B (C_4H_6O): δ_H (p.p.m.): 3.00 (2H, s), 0.90 (2H, d, J 3 Hz), and 0.80 (2H, d, J 3 Hz)

Data for compound 13C (C_4H_6O): δ_H (p.p.m.): 3.02 (4H, d, J 5 Hz) and 1.00 (2H, quintet, J 5 Hz).

Purpose of the problem

Structure determination of reaction products, including cyclopropanes, by proton NMR alone.

Suggested solution

The very small shifts of cyclopropane protons may have troubled you, but they often have δ_H < 1.0 p.p.m. Compounds 13A and C are simple enough but 13B may have amazed you. It is unstable but can be isolated and the two '*spiro*' rings sit at right angles to each other. There is again a CH$_2$ group whose protons are different.

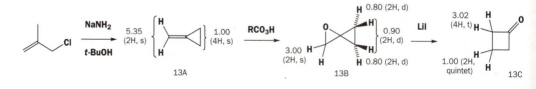

Problem 14

The yellow crystalline antibiotic frustulosin was isolated from a fungus in 1975 and it was suggested that the structure was an equilibrium mixture of 14A and 14B. Apart from the difficulty that the NMR spectrum clearly shows one compound and not an equilibrium mixture of two compounds, what else makes you unsure of this assignment? Suggest a better structure. Signals marked * exchange with D_2O.

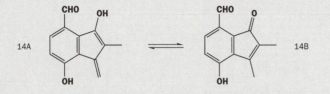

14A   14B

Frustulosin.

m/z: 202 (100%), 187 (20%), 174 (20%); v_{max} (cm^{-1}): 3279, 1645, 1613, and 1522; δ_H (p.p.m.): 2.06 (3H, dd, J 1.0, 1.6 Hz), 5.44 (1H, dq, J 2.0, 1.6 Hz), 5.52 (1H, dq, J 2.0, 1.0 Hz), 4.5* (1H, broad s), 7.16 (1H, d, J 9.0 Hz), 6.88 (1H, dd, J 9.0, 0.4 Hz), 10.31 (1H, d, J 0.4 Hz), and 11.22* (1H, broad s); δ_C (p.p.m.): 22.8, 80.8, 100.6, 110.6, 118.4, 118.7, 112.6, 125.2, 126.1, 151.8, 154.5, and 195.6.

Warning! This is difficult – after all the original authors initially got it wrong! *Hint.* How might the DBEs be achieved without a second ring?

Purpose of the problem

A serious and difficult structure determination of a natural product to end with.

Suggested solution

Structure 14B is definitely wrong because the NMR shows only one methyl group, not two and only one carbonyl group, not two. Structure 14A is unlikely because it looks unstable but that is not evidence! The NMR shows two protons on the end of a double bond (at 5.44 and 5.52 p.p.m.) but they are coupled to a methyl group, presumably by allylic coupling, and the methyl group is too far away in 14B. Also in the ^{13}C NMR, what is the signal at 80.8 p.p.m.? The 'hint' was meant to guide to an alkyne. Most of the alkene and aryl carbons cannot be assigned with confidence.

■ The true structure was later described with the help of NMR as you can read in R. C. Ronald *et al.*, *J. Org. Chem.*, 1982, **47**, 2541 and M. S. Nair and M. Anchel, *Phytochemistry* 1977, **16**, 390 revised from M. S. Nair and M. Anchel, *Tetrahedron Lett.*, 1975, 2641.

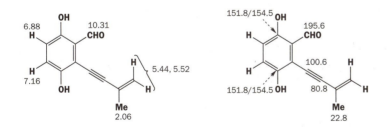

Problem 1

Assign a configuration, *R* or *S*, to each of these compounds.

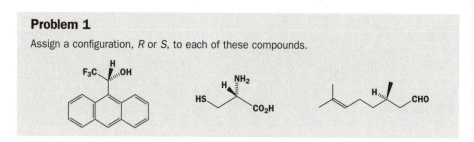

Purpose of the problem

Nomenclature may be the least important of the organic chemists' necessary skills, but giving *R* or *S* designation to simple compounds does matter. These three check your basic knowledge of the rules.

Suggested solution

Carrying out the procedure given in the chapter (p. 387) we label the substituents 1–4 and deduce the configuration. In all these case, '4' is H and goes at the back when we work out *R* or *S*. The first compound is 'Pirkle's reagent' used to check the purity of enantiomers. The second is the natural amino acid cysteine and is *R* because S ranks higher than O. The third is natural citronellal. Though atoms 1, 2, and 3 are all carbon, 1 and 2 are easily distinguished from 3 by the next atom in the chain (C or H) and 1 and 2 are distinguished by the *next* atom in the chain (O or C).

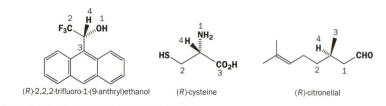

(*R*)-2,2,2-trifluoro-1-(9-anthryl)ethanol   (*R*)-cysteine   (*R*)-citronellal

Problem 2

If a solution of a compound has a rotation of +12, how could you tell if this was actually +12, or really –348, or +372?

Purpose of the problem

Revision of the meaning of rotation and what it depends on.

Suggested solution

Check in the chapter (p. 389) to see the equation that states that the rotation depends on three things: the rotating power of the molecule, the length of the cell used in the polarimeter, and the concentration of the solution. We can't change the first, we may be able to change the second, but the third is easiest to change. If we halve the concentration, the rotation will change to +6, –174, or +186. That is not quite good enough as those last two figures are the same. Any other change in concentration will do the trick.

Problem 3

Cinderella's glass slipper was undoubtedly a chiral object. But would it have rotated the plane of polarized light?

Purpose of the problem

Revision of the cause of rotation and optical activity.

Suggested solution

No. The macroscopic shape of the object is irrelevant. Only the molecular structure matters as light interacts with the electrons in the molecules. Glass is not chiral (usually made up of inorganic borosilicates these days). Only if the slipper had been made of a single enantiomer of a transparent substance would it have rotated polarized light. The molecules in Cinderella's left foot are of the same absolute configuration as those in her right foot despite her feet being macroscopically enantiomeric.

Problem 4

Are these compounds chiral? Draw diagrams to justify your answer.

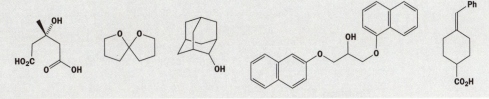

Purpose of the problem

Reinforcement of the very important criterion for chirality. The previous problems were almost childish compared with this one; make sure you understand the answer.

Suggested solution

Only one thing matters – does the molecule have a plane of symmetry? We need to redraw some of them to see if they do. On no account look for chiral centres or carbon atoms with four different groups or whatever. *Just look for a plane of symmetry.*

The first compound has been drawn with the carboxylic acid represented in two different ways and a false chiral centre drawn at the top. The two CO$_2$H groups are, in fact, the same and the molecule has a plane of symmetry.

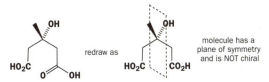

The second molecule is a '*spiro*' compound having two rings joined at a tetrahedral carbon atom. These two rings are orthogonal so there is no plane of symmetry. The third molecule does have a plane of symmetry. It is much easier to see this if you make a model.

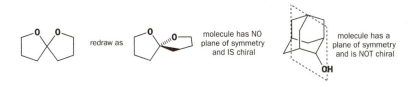

The fourth molecule is a bit of a trick. It needs to be redrawn to see if it has a plane of symmetry but when you did the redrawing you might not have noticed that the two naphthalene rings were joined at different positions. The molecule is chiral.

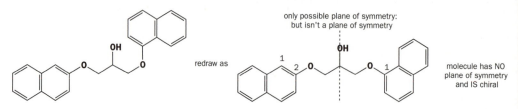

The last molecule is an interesting case. It is chiral but, if you got this one wrong, don't be too disappointed. Again, making a model will help but the vital thing is to realize that the CO_2H group is on a tetrahedral centre so the ring itself is not a plane of symmetry. The alkene puts the phenyl group to one side and a hydrogen atom to the other so the plane at right angles to the ring (dotted line) isn't a plane of symmetry either.

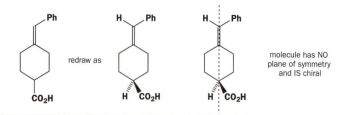

Problem 5

What makes molecules chiral? Give three examples of different types of chirality. State with explanations whether the following compounds are chiral.

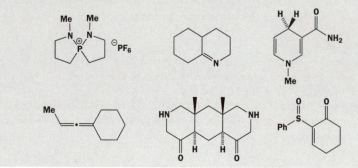

Purpose of the problem

Revision of the criterion for chirality with examples of the main classes of chiral molecules. Exam-style question.

Suggested solution

Molecules are chiral if they have no plane of symmetry. This may arise from a tetrahedral atom with four different substituents or from a molecule that is forced to adopt a shape that lacks a plane of symmetry. Examples include *spiro* compounds, axial chirality in allenes, chiral C, P, S, etc. You should give definite examples in this part of the answer, which are different from those given in the question. Ask someone to check if yours are all right.

The phosphorus compound does not have a chiral phosphorus atom but the *molecule* is chiral because it is a *spiro* compound like the second molecule in the last question. The second molecule is nearly planar but the combination of a double bond and a tetrahedral centre at the other ring junction removes all possibility of a plane of symmetry. This too is chiral.

The third molecule tries to look chiral but it is almost planar because of conjugation, and the hydrogen atom above the plane reflects the hydrogen atom below it. The plane of the ring is a plane of symmetry and the molecule is not chiral. The fourth molecule is an allene with the two alkenes orthogonal to each other. It needs to be drawn more realistically to show that there is a plane of symmetry cutting the cyclohexane ring at right angles and passing through the methyl group on the other end of the allene. Not chiral either.

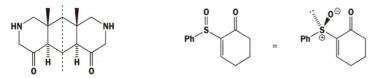

The last two molecules are more straightforward. The tricyclic compound has a plane of symmetry vertically down the middle and is not chiral. The sulfoxide is a simple example of a stereogenic atom other than carbon. Sulfoxides are tetrahedral with the oxygen atom and the lone pair above and below the plane as drawn. This one is chiral.

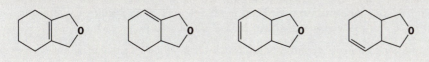

Problem 6

Discuss the stereochemistry of these compounds. (*Hint.* This means saying how many diastereoisomers there are, drawing clear diagrams of each, and saying whether they are chiral or not.)

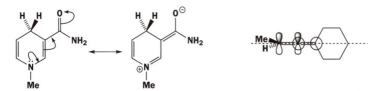

Purpose of the problem

Making sure that you can handle this important approach to the stereochemistry of molecules.

Suggested solution

Just follow the hint given in the question! Diastereoisomers are different compounds so they must be distinguished first. Then it is easy to say if each diastereoisomer is chiral or not. The first two are simple.

one compound
no diastereoisomers
plane of symmetry
not chiral

one compound
no diastereoisomers
no plane of symmetry
chiral

The third structure may exist as two diastereoisomers: one has a plane of symmetry (a *meso* compound) but the other is chiral (it has C_2 symmetry).

The last compound is most complicated as it has no symmetry. Again we can have a *cis* or *trans* ring junction but this time both diastereoisomers are chiral.

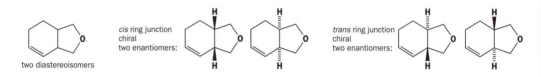

Problem 7

In each case state with explanations whether the products of these reactions are chiral and/or enantiomerically pure.

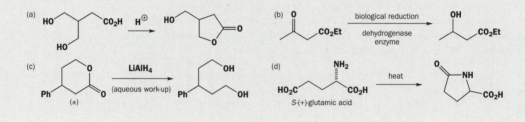

Purpose of the problem

Combining mechanisms with stereochemical analysis for the first time.

Suggested solution

We need a mechanism for each reaction, a stereochemical description for each starting material (for example, chiral?, enantiomerically enriched?), and an analysis of what has happened to the stereochemistry during the reaction. Don't forget: you can't get enantiomers out of nothing – if everything that goes into a reaction is racemic or achiral, so is the product.

(a) The starting material is achiral (the two side chains are of equal length) but the product is chiral. It cannot be enantiomerically enriched because the starting materials and reagents are achiral. The stereochemistry is defined in the first step: if one OH cyclizes you get one enantiomer and, if the other, the other.

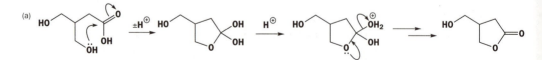

(b) The starting material is achiral and the product is chiral but this time the reagent (the enzyme) is undoubtedly chiral! The actual reducing agent will be a dihydropyridine bound to the enzyme: we shall get one enantiomer of the product, though we cannot deduce which as different dehydrogenase enzymes have different stereoselectivities.

(c) The starting material is chiral but racemic and the reagent is achiral. The product is also chiral but must be racemic. Nothing happens to the chiral centre during the reaction: it was 50:50 in the ester and will be 50:50 in the diol product.

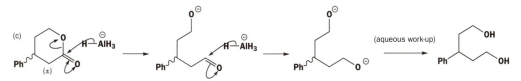

(d) The starting material is a single enantiomer and in the cyclization (intramolecular amide formation) nothing happens to the one chiral centre. The reaction goes with retention of configuration and the product is a single enantiomer too.

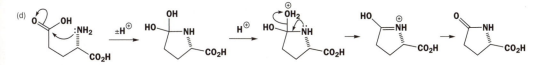

Problem 8

Propose mechanisms for these reactions that explain the stereochemistry of the products. All compounds are enantiomerically pure.

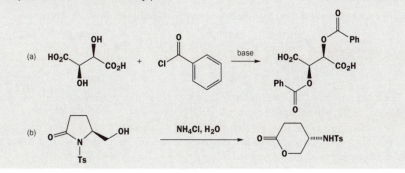

Purpose of the problem

Practice at tracing the stereochemical course of reactions.

Suggested solution

Reaction (a) is ester formation from alcohols and acid chlorides (Chapter 12) and the mechanism is the same for both alcohols. No bonds are made or broken at the two chiral centres so both acylations go with retention.

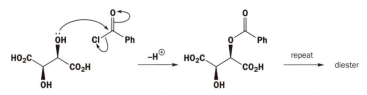

Reaction (b) is more tricky. The two compounds are isomeric and the OH group is part of an ester in the product so it makes sense to attack the carbonyl group with the OH as nucleophile. The Ts group (toluene-*para*-sulfonyl) on nitrogen makes it a better leaving group than an oxyanion so the reaction proceeds. The OH group has to come in from the top and that pushes the NHTs group downwards. In fact, it is another example of no change in stereochemistry.

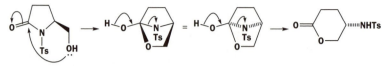

Problem 9

Discuss the stereochemistry of these compounds. The diagrams are deliberately poor ones that are ambiguous about stereochemistry – your answer should use good diagrams that show the stereochemistry clearly.

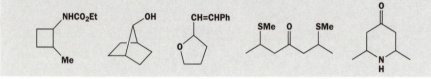

Purpose of the problem

Practice at spotting stereochemistry and unravelling the different possible stereochemical relationships.

Suggested solution

The first compound is simple: two diastereoisomers, *cis* and *trans*; both are chiral.

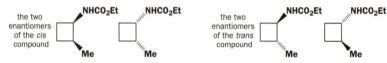

The second is simple too – the molecule has a plane of symmetry passing through the black dots and is not chiral. No diastereoisomers.

The third compound has two stereochemical units: an alkene that can be *Z* (*cis*) or *E* (*trans*) and that provides two different compounds, or diastereoisomers. There is also a chiral centre so each alkene isomer has two enantiomers.

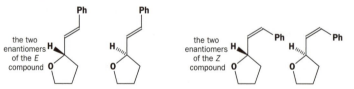

The fourth compound has some symmetry. There are two diastereoisomers with the MeS groups arranged *syn* or *anti* (as drawn). One has a plane of symmetry and is a *meso* compound while the other is chiral.

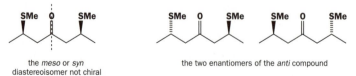

The fifth compound is similar: two diastereoisomers; one is chiral.

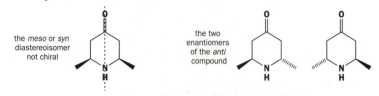

Problem 10

This compound racemizes in base. Why is that?

Purpose of the problem

To draw your attention to the dangers in nearly symmetrical molecules and revision of ester exchange.

Suggested solution

Ester exchange in base goes through a symmetrical tetrahedral intermediate with a plane of symmetry. Loss of the right-hand leaving group gives one enantiomer of the ester and loss of the left-hand leaving group gives the other.

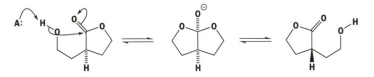

Problem 11

Draw mechanisms for these reactions. Will the products be single stereoisomers?

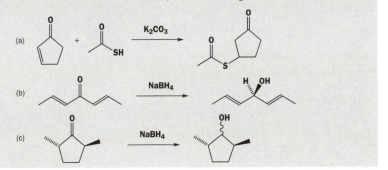

Purpose of the problem

Revision of direct (Chapter 6) conjugate addition (Chapter 10) and some minor stereochemical points.

Suggested solution

Reaction (a) proceeds by a straightforward conjugate addition of a sulfur nucleophile. One stereogenic centre is formed from two achiral reagents and the product must therefore be racemic.

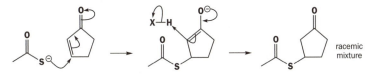

Example (b) creates a centre that is drawn to look chiral but isn't. The molecule has a plane of symmetry. It consists of a single (*E,E*-) geometrical isomer, of course, as the alkenes are not affected by the reaction.

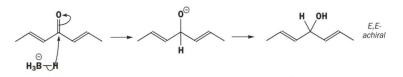

Example (c) might look like the example (b) except that the product has three 'chiral centres' and looks more complicated. In fact, the starting material is C_2 symmetric and the two 'stereoisomers' suggested by the wiggly line to the OH group are identical. Check for yourself. The product is a single diastereoisomer (*anti-*) but there is no meaning in any stereochemical designation of the central (OH) centre.

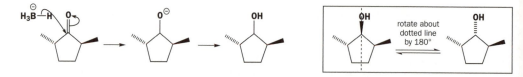

Problem 12

How many diastereoisomers of compound 1 are there? State clearly whether each diastereoisomer is chiral or not. If you had made a random mixture of stereoisomers by a chemical reaction, by what types of methods might they be separated? Which isomer(s) would be expected from the hydrogenation of compound 2?

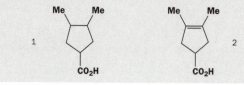

Purpose of the problem

A more difficult example with a chemical reaction.

Suggested solution

The simplest way to do the first part is to say that the two methyl groups can either be on the same or opposite sides of the ring and that in each case the CO_2H group can be up or down. Let's see what we get from that.

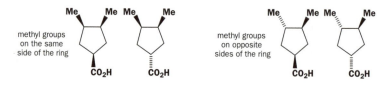

The first two compounds (methyls *cis*) are obviously different diastereoisomers and both have a plane of symmetry so neither is chiral. The second pair (methyls *trans*) are slightly more tricky – they are the same! And this isomer does not have a plane of symmetry and is chiral. In fact this is like the last example in Problem 11. There are three diastereoisomers and only one of them is chiral.

Reduction of a double bond by catalytic hydrogenation puts two hydrogen atoms on the same side of the alkene. This can give only the first two compounds but we might expect hydrogen to add more easily to the side of the alkene opposite the CO_2H group. Our prediction is:

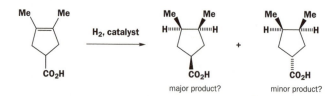

Problem 13

Just for fun, you might like to try and work out just how many diastereoisomers inositol has and how many of them are *meso* compounds.

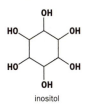

inositol

Purpose of the problem

Fun, it says! There is a more serious purpose in that the relationship between symmetry and stereochemistry is interesting and, in this human brain chemical, important to understand.

Suggested solution

If we start with all the OH groups on one side and gradually move them over, we should get the right answer. If you got too many diastereoisomers, check that some of yours aren't the same as others. There are eight diastereoisomers altogether and, incredibly, all except one are achiral. Some have one, some two, and two of the most symmetrical have many planes of symmetry.

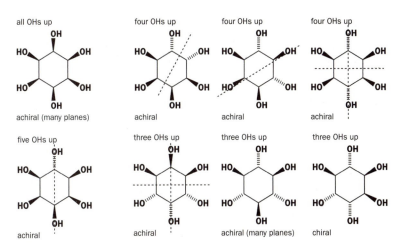

Suggested solutions for Chapter 17

17

Problem 1

Suggest mechanisms for the following reactions, commenting on your choice of S_N1 or S_N2.

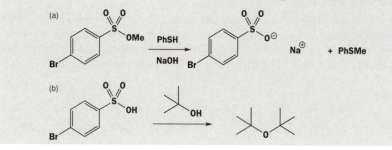

(a)

(b)

Purpose of the problem

Simple example of the two basic mechanisms of Chapter 17: S_N1 and S_N2.

Suggested solution

The electrophile in example (a) is the methyl group of the sulfonate ester and the nucleophile is a sulfur anion formed by deprotonation of PhSH (pK_a about 7) by NaOH (pK_{aH} about 16). The mechanism must be S_N2 with a good nucleophile and a methyl electrophile.

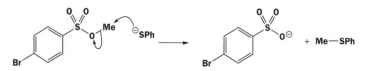

The first reagent in example (b) looks like a reagent in example (a) but, in fact, it is not incorporated into the product and must be a catalyst. It is an aryl sulfonic acid and will be a strong acid. The product contains two t-butyl groups and can be formed only by an S_N1 mechanism.

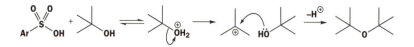

Problem 2

Draw mechanisms for the following reactions. Why were acidic conditions chosen for the first reaction and basic conditions for the second?

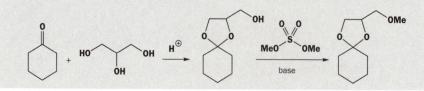

Purpose of the problem

Simple example of an S_N2 mechanism and a reminder that acetal formation is an S_N1 reaction.

Suggested solution

The first step is acetal formation. It doesn't matter which OH group you use first as the five-membered ring is preferred to the six-membered ring. The S_N1 reaction is the loss of water from the intermediate and the capture of the oxonium ion by the second hydroxyl group.

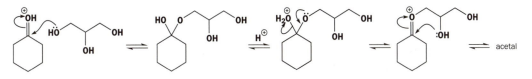

The second reaction is a straightforward S_N2 mechanism at a methyl group. The base need not be strong enough to remove the OH proton before the reaction.

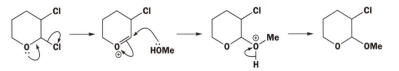

Problem 3

Draw mechanisms for these reactions, explaining why these particular products are formed.

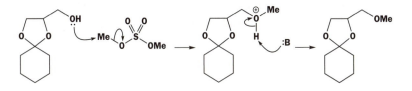

Purpose of the problem

How to choose between the S_N1 or S_N2 mechanism when the choice is more subtle.

Suggested solution

The first compound has two leaving groups – both secondary chlorides. The one that leaves is next to oxygen and that can be explained by an S_N1 mechanism.

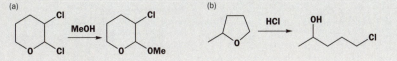

The second reaction starts by protonation at the ether oxygen. Chloride is now the nucleophile and could attack at a primary or a secondary alkyl group. As it chooses the primary centre, the mechanisms must be S_N2.

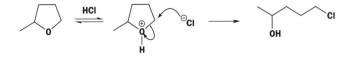

Problem 4

The chemistry shown here is the first step in the manufacture of Pfizer's doxazosin (Cardura), a drug for hypertension. Draw mechanisms for the reactions involved and comment on the bases used.

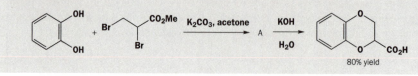

80% yield

Purpose of the problem

How to choose between the S_N1 or S_N2 mechanism when the choice is more subtle.

Suggested solution

Carbonate is a strong enough base to remove protons from phenols (pK_a about 10) but not from alcohols. One oxyanion will be formed at a time and will displace a bromide ion in an S_N2 reaction. It doesn't matter which goes first. Both react by the S_N2 mechanism because one is primary and the other, though secondary, is next to a carbonyl group and the cation would be very unstable while the S_N2 mechanism is excellent.

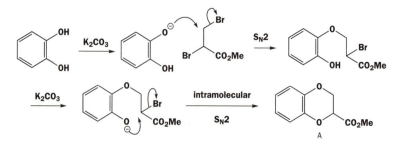

The last reaction is ordinary ester hydrolysis (Chapter 12) in aqueous strong base. The two leaving groups (HO^- and MeO^-) are about equally good but the formation of the carboxylate anion under the reaction conditions drives the equilibrium to the right.

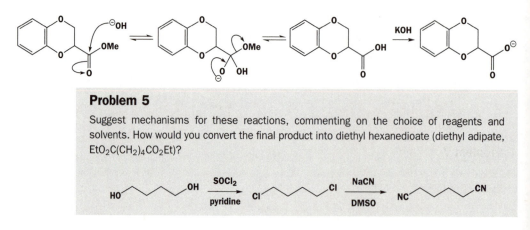

Problem 5

Suggest mechanisms for these reactions, commenting on the choice of reagents and solvents. How would you convert the final product into diethyl hexanedioate (diethyl adipate, $EtO_2C(CH_2)_4CO_2Et$)?

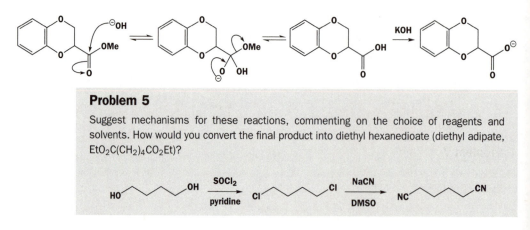

Purpose of the problem

Drawing mechanisms for useful substitutions and extending the sequence by other reactions.

Suggested solution

These are standard S_N2 reactions at primary carbon atoms. In the first case, the OH group has to be turned into a good leaving group. We show one reaction of each kind, though actually everything happens twice. The dinitrile is converted into the diester by treatment with ethanol in acid solution; see Problem 9 in Chapter 12.

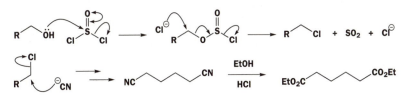

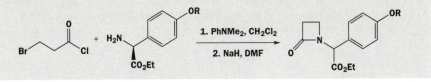

Problem 6

Draw mechanisms for these reactions and describe the stereochemistry of the product.

Purpose of the problem

Drawing mechanisms for two types of substitutions in the same sequence to make a useful molecule.

Suggested solution

The two reactions are an S_N2 at primary carbon and a nucleophilic substitution at the carbonyl group with the amino group as the nucleophile in both cases. Substitution at the carbonyl group probably happens first. Don't worry if you didn't draw the deprotonation of the amide and the S_N2 attack of nitrogen exactly as we did. The stereochemistry of the product is the same as that of the starting material (CO_2Et group up) as no change has occurred at the chiral centre.

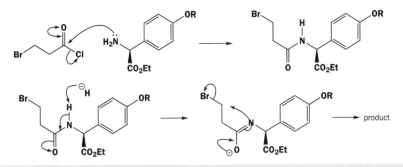

Problem 7

Suggest a mechanism for this reaction. You will find it helpful first of all to draw good diagrams of reagents and products.

$$t\text{-BuNMe}_2 + (\text{MeCO})_2\text{O} \longrightarrow \text{Me}_2\text{NCOMe} + t\text{-BuO}_2\text{CMe}$$

Purpose of the problem

Revision of Chapter 2 and practice at drawing mechanisms of nonstandard reactions.

Suggested solution

First draw good diagrams of the molecules as the question suggests.

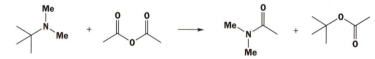

With an unfamiliar reaction, it is best to identify nucleophile and electrophile and see what happens when we unite them. The nitrogen atom is obviously the nucleophile and one of the carbonyl groups of the anhydride must be the electrophile as nitrogen is bonded to an acetyl group in one of the products.

We must lose a *t*-butyl group from this intermediate to give one of the products and unite it with the acetate anion to give the other. This must be an S_N1 rather than an S_N2 reaction at a *t*-butyl group.

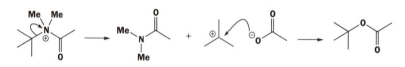

Problem 8

Predict the stereochemistry of these products. Are they single diastereoisomers, enantiomerically pure, or racemic, or something else?

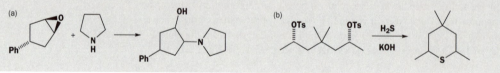

Purpose of the problem

Revision of stereochemistry from Chapter 16 and practice at applying it to substitution reactions.

Suggested solution

The two sides of the epoxide in example (a) are the same as the molecule has a plane of symmetry. Attack at either side by the S_N2 mechanism must occur from the bottom face of the molecule so that inversion occurs. The product is a single diastereoisomer but cannot, of course, be one enantiomer.

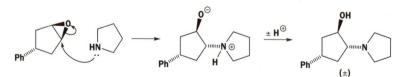

The stereochemistry of the starting material in (b) was discussed in the solution to Problem 9, Chapter 16. The starting material has a plane of symmetry but is a single diastereoisomer and both centres are inverted during the double displacement by sulfur. You must be careful to get the stereochemistry right when twisting the molecule round to draw it as a ring. The product is a single diastereoisomer but is not chiral.

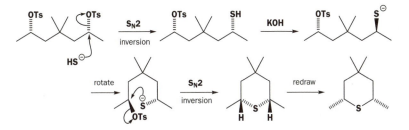

Problem 9

What are the mechanisms of these reactions, and what is the role of the $ZnCl_2$ in the first step and the NaI in the second?

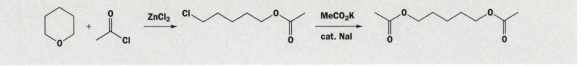

Purpose of the problem

Exploration of the roles of two different kinds of catalysts in substitution reactions.

Suggested solution

The $ZnCl_2$ acts as a Lewis acid and can be used either to remove chloride from MeCOCl or to complex with its carbonyl oxygen atom, in either case making it a better electrophile so that it can attack the unreactive oxygen atom of the cyclic ether. Ring cleavage by chloride follows.

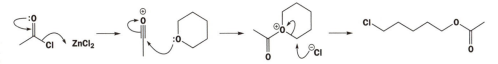

The second reaction is an S_N2 displacement of a reasonable leaving group (chloride) by a rather weak nucleophile (acetate). The reaction is very slow unless catalysed by iodide – a better nucleophile than acetate and a better leaving group than chloride. It is a nucleophilic catalyst.

■ You can read more about this in B. S. Furniss *et al.*, *Vogel's textbook of organic chemistry* (5th edn), Longmans, Harlow, 1989, p. 492.

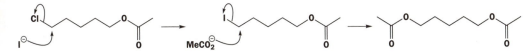

Problem 10

Describe the stereochemistry of the products of these reactions.

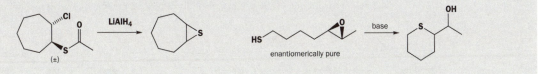

Purpose of the problem

Time for nucleophilic substitution and stereochemistry again with a few extra twists this time!

Suggested solution

The ester in the first example is removed by reduction, leaving an oxyanion that cyclizes by intramolecular S_N2 reaction with inversion of configuration giving a *cis*-fused product.

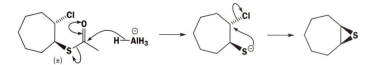

The second case involves an intramolecular S_N2 reaction by a sulfur anion on one end of the epoxide. The reaction occurs stereospecifically with inversion and so one enantiomer of one diastereoisomer of the product is formed. Some redrawing is needed and we have kept the epoxide of the starting material in its original position to avoid mistakes.

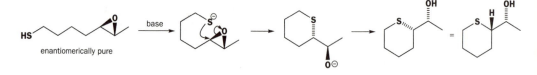

enantiomerically pure

Problem 11

Identify the intermediates in these syntheses and give mechanisms for the reactions.

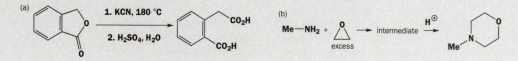

Purpose of the problem

Trying out substitution reactions in synthesis.

Suggested solution

Sequence (a) starts with a displacement of the carboxylate anion followed by acid-catalysed hydrolysis of the nitrile to give the second carboxylic acid. Full details of the second step are in Chapter 12 (p. 294).

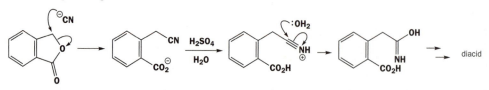

Reaction (b) starts with two consecutive S_N2 reactions of the amine on the epoxide. At that point there are no more NH protons to be replaced. The diol intermediate cyclizes by an intramolecular S_N2 reaction with one of the primary alcohols displacing the other after protonation.

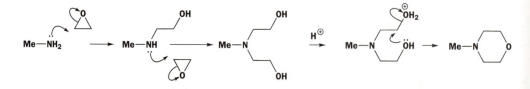

Problem 12

State with reasons whether these reactions will be either S$_N$1 or S$_N$2.

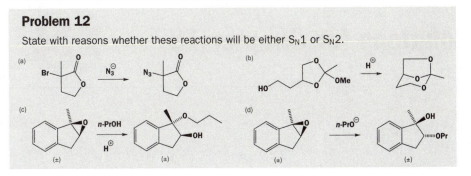

Purpose of the problem

Taxing examples of choice between our two main mechanisms with (c) and (d) differing only in conditions.

Suggested solution

Example (a) offers a choice between an S$_N$2 reaction at a tertiary carbon atom or an S$_N$1 reaction next to a carbonyl group. In fact, these are about the only known examples of S$_N$2 reactions at tertiary carbon and work because the carbonyl group accelerates S$_N$2 reactions so much. The diagram of the transition state shows how the p orbitals on the carbonyl group are parallel to the p orbital in the S$_N$2 transition state. Azide is also an excellent nucleophile being sharp and narrow and not affected much by steric hindrance. We know the reaction is S$_N$2 because we find inversion of configuration.

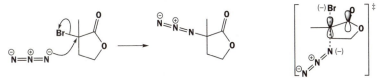

Example (b) is like acetal formation though it is strictly orthoester exchange and should be compared with the chemistry described in Chapter 14 and particularly with Problems 4 and 8 at the end of that chapter. The replacement of OMe by the primary OH is a nucleophilic substitution at saturated carbon and it goes by the S$_N$1 route because of the cation-stabilizing effect of the other two oxygen atoms.

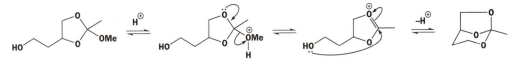

Examples (c) and (d) add the same group (PrO) to the same starting material (an epoxide) under different conditions. We can tell that (c) must be S$_N$1 because the nucleophile has added to the *tertiary* rather than the *secondary* end of the epoxide and that (d) must be S$_N$2 because of the reverse regioselectivity. Acid catalysis makes S$_N$1 better by improving the leaving group and base catalysis makes S$_N$2 better by improving the nucleophile. Notice that inversion of configuration occurs in both reactions. This is expected in S$_N$2 and may be the case in S$_N$1 because the underside of the cation is not blocked by the OH group. If you said that inversion implied S$_N$2 with a loose cationic transition state, you may well be right as this is also a valid explanation.

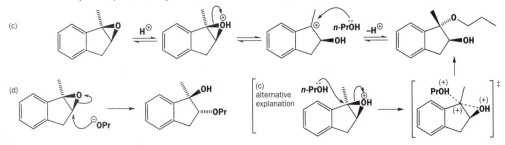

Suggested solutions for Chapter 18

18

Problem 1

Identify the chair or boat six-membered rings in the following structures and say why that particular shape is adopted.

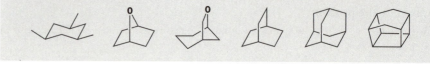

Purpose of the problem

Simple examples of chair and boat forms.

Suggested solution

The first three are relatively simple.

 simple chair with all substituents equatorial

 boat in cage molecule with no choice: chair impossible

 chair in cage molecule with some choice: C* can move

The next two have several rings each, all boat in the first. We'll link that to the sixth molecule as it also has a boat and neither of these cage structures has any choice.

 three boats cage compound has no choice

 one boat cage compound has no choice

The rings are all chairs in the fifth molecule adamantane – a tiny fragment of a diamond molecule. The rings don't all look very chair-like in these diagrams – making a model of adamantane is the only way to appreciate this beautiful and symmetrical structure and to see all the chairs.

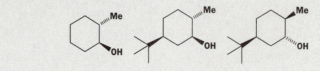 adamantane: four chairs cage compound has no choice

Problem 2

Draw clear conformational drawings for these molecules, labelling each substituent as axial or equatorial.

Purpose of the problem

Simple practice at drawing chair cyclohexanes with axial and equatorial substituents.

Suggested solution

Your drawings may look different from ours but make sure the rings have parallel sides and don't 'climb upstairs'. Make sure that the axial bonds are vertical and the equatorial bonds parallel to the next-ring-bond-but-one. The first molecule has a free choice so it puts both substituents equatorial. The last two molecules are dominated by the *t*-butyl groups, which insist on being equatorial.

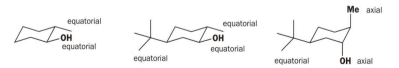

Problem 3

Would the substituents in these molecules be axial or equatorial or a mixture of the two?

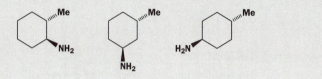

Purpose of the problem

Simple practice at drawing chair cyclohexanes and deciding whether the substituents are axial or equatorial. Remember to decide by drawing and not by some memory trick like '*trans* means diequatorial' or any such nonsense.

Suggested solution

All three molecules have a free choice as the substitutents aren't large and are about the same size. Notice that *trans* means diequatorial in two cases and axial/equatorial in the third.

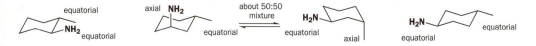

Problem 4

Why is it difficult for cyclohexyl bromide to undergo an E2 reaction? When it is treated with base, it does undergo an E2 reaction to give cyclohexene. What conformational changes must occur during this reaction?

Purpose of the problem

Simple exploration of the relationship between conformation and mechanism.

Suggested solution

Cyclohexyl bromide has the chair conformation with the bromine atom equatorial. It cannot do an E2 reaction (Chapter 19) in this conformation as E2 requires the reacting C–H and C–Br bonds to be antiperiplanar. This can be achieved if the molecule first flips to an unfavourable axial conformation.

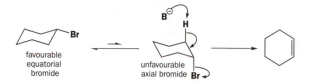

favourable
equatorial
bromide

unfavourable
axial bromide **Br**

Problem 5

Treatment of this diketoalcohol with base causes an elimination reaction. What is the mechanism, and which conformation must the molecule adopt for the elimination to occur?

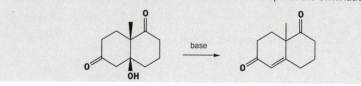

Purpose of the problem

Exploration of the relationship between conformation and mechanism in a less simple example.

Suggested solution

Since this is a multistep mechanism, it is best to draw the mechanism first before considering conformation.

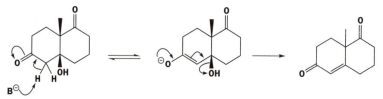

In the deprotonation step, the axial proton will be lost, though this doesn't affect the conformation of the molecule. In the elimination step the OH group must be axial so that the C–O bond is parallel to the p orbitals in the π system. The molecule is a *cis*-decalin so it has a choice of conformations. Don't be disheartened if you got the wrong conformation first time – we took several tries before getting it right.

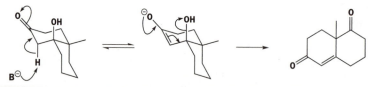

Problem 6

Which of these two compounds would form an epoxide on treatment with base?

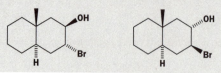

Purpose of the problem

Exploration of the relationship between conformation of important *trans*-decalins and mechanism.

Suggested solution

The mechanism is easy and the conformations of *trans*-decalins are fixed, so we can start with the conformation.

Only the first compound can get the necessary 'attack from the back' angle of 180° between nucleophile (O⁻), carbon atom, and leaving group (Br) for the intramolecular S_N2 reaction to make the epoxide.

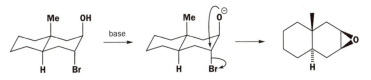

Problem 7

Draw conformational diagrams for these compounds. State in each case why the substituents have the positions you state. To what extent could you confirm your predictions experimentally?

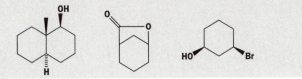

Purpose of the problem

Further exploration of conformation and establishing the link with NMR spectra.

■ This problem uses material from Chapter 19. If you find it too difficult, we suggest you come back to it when you have read that chapter.

Suggested solution

The first two molecules have no choice about their conformation but the third does.

Confirming the conformations experimentally means measuring coupling constants in the proton NMR so we need to look at the vital protons (marked 'H' on the diagrams below) and consider whether they can be seen in the spectrum. Fortunately, the interesting protons are all next to functional groups so they can be seen. We probably can't see the axial proton at the ring junction in the first example but *trans*-decalins must have axial protons there, so that is not so important.

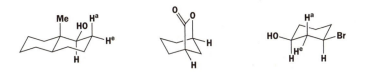

In the first molecule, proton H has two neighbours, one axial (H^a) and one equatorial (H^e) and will appear as a double doublet with characteristic axial/axial and axial/equatorial coupling constants. By contrast, the two equatorial Hs in the second compound have each got two axial and two equatorial neighbours (not shown) and all the coupling constants will be the same. They will both appear as narrow triplets of triplets. The two axial protons in the third example have each got two axial and two equatorial neighbours (one pair shown) and will appear again as approximate tts, but this time one triplet will have a large axial/axial coupling constant.

Problem 8

It is more difficult to form an acetal of compound **8A** than of **8B**. Why is this?

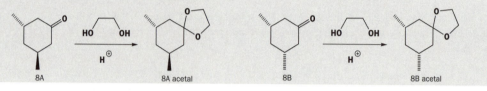

Purpose of the problem

Exploration of the effect of conformation on equilibria in thermodynamically controlled reactions.

Suggested solution

The mechanism of these reaction is normal acetal formation as described in Chapter 14 and is irrelevant to this question as we are dealing only with stabilities in this thermodynamically controlled reaction. We need to look at the conformations of the molecules.

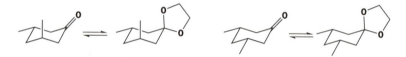

Axial groups in the ketones are not very important as there is no 1,3-diaxial interaction with the ketone itself. But 1,3-diaxial interactions with the acetal are bad because one of the the acetal oxygen atoms has to be axial. Though the ketone **8A** is slightly less stable than **8B**, the acetal of **8A** is much less stable than the acetal of **8B**.

Problem 9

Predict which products would be formed on opening these epoxides with nucleophiles, say, cyanide ion.

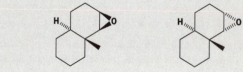

Purpose of the problem

Practice at choosing the correct regiochemistry of a conformationally controlled epoxide opening.

Suggested solution

The opening of cyclohexene epoxides is controlled by the need to get *trans* diaxial products. This is the reverse of the reaction discussed in Problem 6 and is described in full in the text on pp. 468–70.

To get the right answer we need merely to draw the only possible *trans* diaxial product from each of these conformationally fixed *trans*-decalins. Cyanide must, of course, open the epoxide with inversion so the OH group in the products is on the same side as the original epoxide.

■ We deduced the structure of the product by conformational arguments but eventually draw its structure in configurational style like that of the starting material.

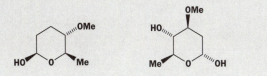

Problem 10

These two sugar analogues are part of the structure of two compounds used to treat poultry diseases. Which conformations would they prefer?

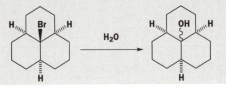

Purpose of the problem

Exploration of conformations in natural products with many substituents.

Suggested solution

Trial and error gives the conformations with the most equatorial substituents. The first compound can have all its groups equatorial and the second can have three equatorial and only one axial.

■ This OH group actually prefers to be axial; see text, p. 1129.

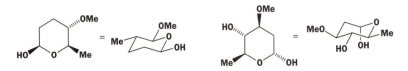

Problem 11

Hydrolysis of the tricyclic bromide shown here in water gives an alcohol. What is the conformation of the bromide and what will be the stereochemistry of the alcohol?

Purpose of the problem

To show the conformation of a tricyclic system and that conformation can control stereochemistry in S_N1 reactions.

Suggested solution

The mechanism of hydrolysis of this tertiary bromide is obviously S_N1 so the water molecule can approach from either side of the planar cation intermediate. In fact, there is a unique conformation of the (achiral) starting material with all three rings in a chair conformation and this will be much preferred in the product. The reaction goes with retention.

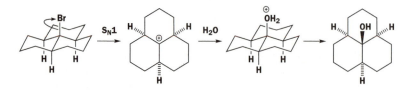

Problem 12

Treatment of the triol 12A with benzaldehyde in acid solution produces one diastereoisomer of the acetal 12B and none of the alternative acetal. Why is this acetal preferred? (*Hint*. What controls acetal formation?) What is the stereochemistry of the undefined centre in 12B?

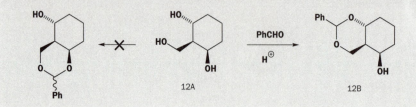

Purpose of the problem

More revision of acetal formation with extra selectivity controlled by conformation.

Suggested solution

Acetal formation is thermodynamically controlled so we need look only for the most stable possible acetal. The one that is not formed is a *cis*-decalin – significantly less stable than the *trans*-decalin (12B) that is formed. The equatorial conformation of the phenyl group will decide its configuration as all the acetals are in equilibrium.

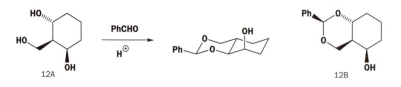

Suggested solutions for Chapter 19

19

Problem 1

Draw mechanisms for these reactions, which were listed in the chapter.

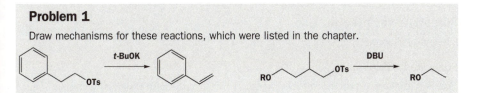

Purpose of the problem

Simple examples of elimination mechanisms.

Suggested solution

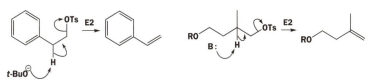

■ The structure of DBU appears on p. 482 of the textbook. We shall use 'B' for this and other bases.

Problem 2

Give a mechanism for the elimination reaction in the formation of tamoxifen from p. 489 and comment on the fact that it gives a mixture of *cis* and *trans* alkenes.

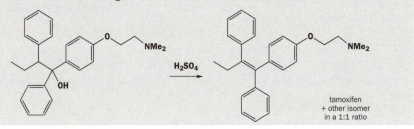

tamoxifen
+ other isomer
in a 1:1 ratio

The leaving group is on a primary carbon atom in both cases so these must be E2 reactions.

Purpose of the problem

Letting you switch from E2 to E1 with a bit of evidence to help you.

Suggested solution

The tertiary alcohol leaving group, the acid catalyst, and the mixture of alkenes all suggest E1 rather than E2. There is only one proton that can be lost and, depending on the conformation of the molecule at that moment, we may get the *cis* or *trans* alkene.

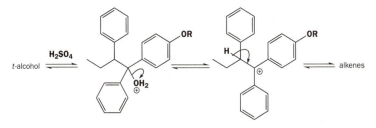

Problem 3

Account for the contrasting results in these two reactions.

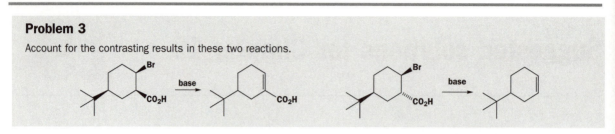

Purpose of the problem

Revision of conformational analysis in the context of elimination reactions.

Suggested solution

The only difference between the two molecules is stereochemistry and, as these are *t*-butyl cyclohexanes, the conformation of each molecule is determined by putting the *t*-butyl group equatorial. In each case it is the group anti-periplanar to the bromine that is lost.

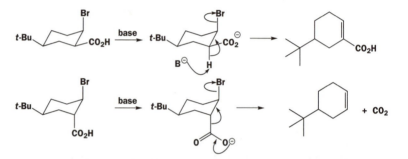

Problem 4

Draw mechanisms for these two elimination reactions of epoxides.

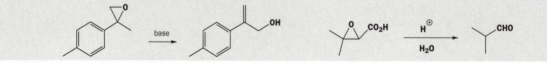

Purpose of the problem

Finding the mechanism of less obvious elimination reactions.

Suggested solution

The first mechanism is not too difficult to find – just lose a proton from the methyl group in an E2 reaction.

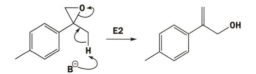

The second mechanism also involves opening an epoxide, but in acid solution the more stable cation is formed and we can see from the structure of the product that CO_2 is lost, as in Problem 3.

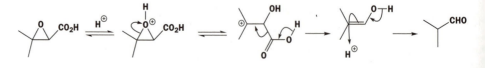

Problem 5

Only one of the diastereomeric bromides shown here eliminates to give alkene A. Why?
Neither bromide gives alkene B. Why not?

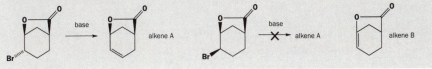

Purpose of the problem

Helping to appreciate that cage molecules often have restricted opportunities for eliminations.

Suggested solution

The first molecule has one anti-periplanar H and Br so elimination can occur. The second has no
hydrogens in the right place. Alkene B is a bridgehead alkene and could not exist.

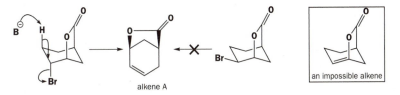

Problem 6

Suggest mechanisms for these reactions, paying particular attention to the elimination steps.

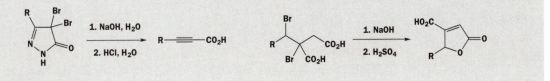

Purpose of the problem

Exploiting multiple eliminations.

Suggested solution

The first compound is an amide and hydrolyses in base to a nitrogen anion that can lose one
bromide. A second elimination then expels the remaining bromide and a molecule of nitrogen.

■ Read about this in B. S. Furniss
et al., *Vogel's textbook of organic
chemistry* (5th edn), Longmans,
Harlow, 1989, p. 744.

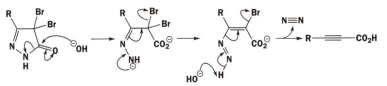

The second example includes a cyclization and an elimination. They could occur in either order
but starting with the cyclization removes any difficulty about the geometry of the alkene. In
whichever order you do the reactions, the elimination should be by the E1cB mechanism.

■ Read about this in B. S. Furniss
et al., *Vogel's textbook of organic
chemistry* (5th edn), Longmans,
Harlow, 1989, p. 805.

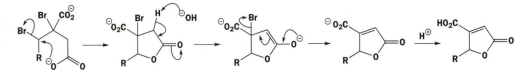

Problem 7

Suggest mechanisms for these two eliminations. Why does the first one give a mixture and the second a single product?

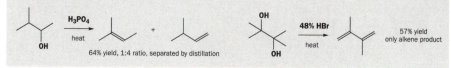

64% yield, 1:4 ratio, separated by distillation

57% yield
only alkene product

Purpose of the problem

Learning about the regioselectivity of E1 reactions.

Suggested solution

Whether the first reaction is E1 or E2, there are two (sets of) hydrogens that can be lost in the alkene formation.

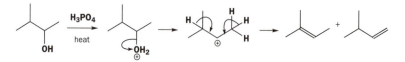

■ See B. S. Furniss *et al.*, *Vogel's textbook of organic chemistry* (5th edn), Longmans, Harlow, 1989, p. 490. Another product from this reaction is discussed in Chapter 37.

The second reaction produces more stable tertiary cations from which any of six hydrogens can be lost, but all give the same alkene.

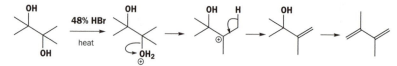

Problem 8

Explain the position of the double bond in the products of these reactions. The starting materials are enantiomerically pure. Are the products also enantiomerically pure?

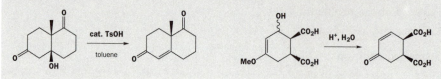

Purpose of the problem

Examples of E1cB reactions in the context of absolute stereochemistry.

Suggested solution

The first reaction is an E1cB elimination of a hydroxy-ketone. The product is still chiral although it has lost one stereogenic centre. The other (quaternary!) centre is not affected by the reaction and the product is enantiomerically pure.

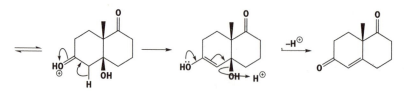

The second example already has the electron-rich alkene present in the starting material so this is more of an E1 than an E1cB reaction. The starting material is an enol ether and the intermediate oxonium ion hydrolyses to the ketone by an extension of acetal hydrolysis (see Chapter 14). The product has two chiral centres unaffected by the reaction and all intermediates are chiral so this product also is a single enantiomer.

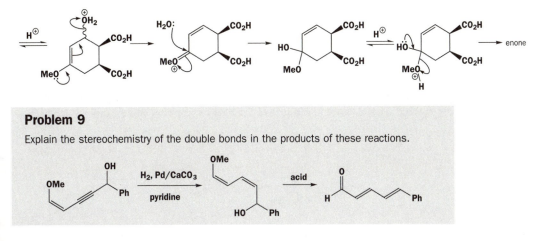

Problem 9

Explain the stereochemistry of the double bonds in the products of these reactions.

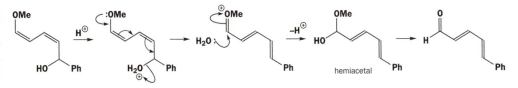

Purpose of the problem

Display your skill in a deceptive example of control of alkene geometry by elimination.

Suggested solution

The first reaction is a stereospecific *cis* addition of hydrogen to the alkyne to give a *cis*-alkene. The intermediate is therefore a *cis, cis*-diene and it seems remarkable that it should turn into a *trans, trans*-diene on elimination. However, when we draw the mechanism for the elimination we see that there need be no relationship between the stereochemistry of the two compounds as this is an E1 reaction, and the stable and therefore long-lived intermediate cation can rotate into the most stable shape before conversion to the ketone.

■ Even *after* conversion to the dienal, reversible conjugate addition of water would allow *cis/trans* interconversion.

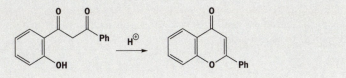

Problem 10

Why is elimination preferred to hemiacetal formation in the acid-catalysed cyclization of this ketone?

Purpose of the problem

Exploring what else might happen on the way to an elimination.

Suggested solution

The strict answer is that it isn't! Hemiacetal formation is part of the mechanism to form the product. The real question is: why doesn't it stop there? Why does elimination occur under the same conditions as hemiacetal formation?

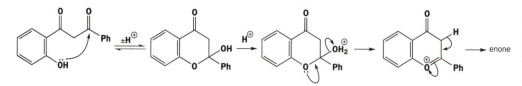

The enone is a fully conjugated cyclic system with six electrons in it – in fact, it is aromatic. This is clearer when we delocalize a lone pair from the ether oxygen atom into the carbonyl group.

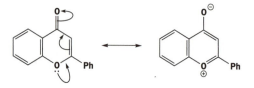

Problem 11

Comment on the position taken up by the alkene in these elimination reactions.

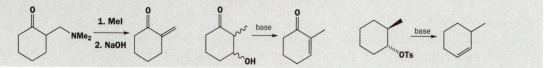

Purpose of the problem

Further exploration of the regioselectivity of alkene formation in elimination reactions (E1cB and E2).

Suggested solution

The first is an E1cB reaction after methylation at nitrogen and the double bond has to go where the leaving group was.

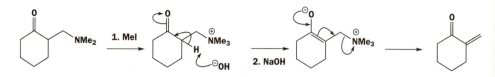

The second reaction is also E1cB and so the alkene must end up conjugated with the carbonyl group. The stereochemistry is irrelevant as this is not an E2 reaction and there is no requirement for H and OH to be *anti*-planar.

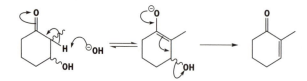

The third example is E2 and so there *is* a requirement for H and OTs to be *anti*-planar. There is no hydrogen on the tertiary centre *anti* to OTs so elimination must go the other way and an initial flip to the diaxial conformer is also needed (revision of Chapter 18).

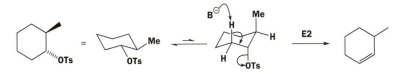

Problem 12

Give mechanisms for these reactions drawn from Chapter 19 and comment on the stereochemistry.

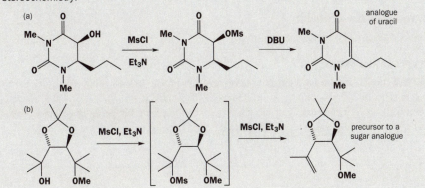

Purpose of the problem

Extending simple examples of elimination mechanisms to mesylation and elimination of alcohols.

Suggested solution

The mechanism of mesylation involves an E1cB elimination.

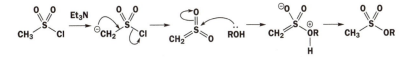

The first elimination uses DBU, a stronger base than Et₃N, but the second elimination occurs under the conditions of the mesylation with Et₃N as base, possibly because any of six easily accessible Hs could be removed.

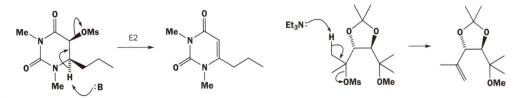

Suggested solutions for Chapter 20 20

Problem 1

Predict the orientation in HCl addition to these alkenes.

Purpose of the problem

Simple examples of addition mechanisms.

Suggested solution

The first and last alkenes have different *numbers* of substituents at each end and will give the more highly substituted carbocation on protonation. The middle one has the same number of substituents at each end (one) but they are very different in *kind*. The secondary benzylic cation is preferred to the simple secondary alkyl cation.

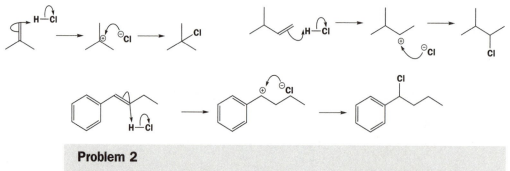

Problem 2

Suggest mechanisms and products for these reactions.

Purpose of the problem

Checking that you understand the bromonium ion mechanism.

Suggested solution

The question of what product is formed is easily answered as we know we get *trans* addition to each alkene. These products are, of course, racemic (all reagents achiral) and the diagrams show relative stereochemistry only.

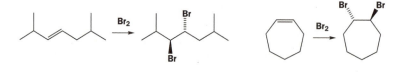

The mechanisms involve bromine attack on the centre of each alkene to form a bromonium ion, which is captured (with inversion) by bromide ion to give the *trans* dibromide. Drawing this for one alkene only:

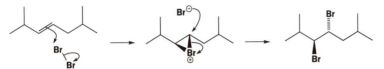

Problem 3

What will be the products of addition of bromine water to these alkenes?

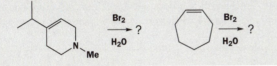

Purpose of the problem

Checking that you understand the bromonium ion mechanism with an external nucleophile.

Suggested solution

The bromonium ion is formed again but now water attacks as the nucleophile (it is present in large excess as the solvent). When the alkene is unsymmetrical, water attacks the more substituted end of the bromonium ion and in any case it does so with inversion.

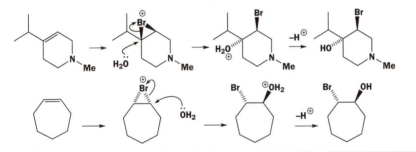

Problem 4

By working at low temperature with one equivalent of a buffered solution of a peroxy-acid, it is possible to prepare the monoepoxide of cyclopentadiene. Why are the precautions necessary and why does the epoxidation not occur again?

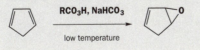

Purpose of the problem

A more complicated electrophilic addition with questions of stability and selectivity added.

Suggested solution

One of the double bonds in the diene reacts in the usual way first of all to give the monoepoxide. Two questions then arise. If the reaction can be made to stop there, the monoalkene must be less nucleophilic than the diene. This is so because the HOMO reacts. The HOMO of an alkene is just the populated π orbital. The HOMO of a diene is Ψ^2 resulting from the antibonding addition of the two separate π orbitals and is less stable than either π orbital.

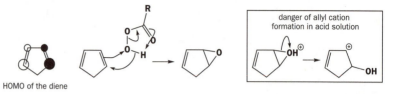

HOMO of the diene

The other question is why work at low temperature and buffer the solution? The low temperature favours the kinetic product and discourages the slower second epoxidation. A by-product from epoxidation is the acid RCO_2H and there is a danger that this may catalyse opening of the monoepoxide to give the allyl cation in the frame. A buffer prevents the solution from becoming too acidic.

Problem 5

The synthesis of a tranquillizer uses this step. Give mechanisms for the reactions.

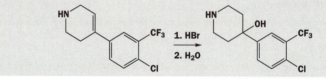

Purpose of the problem

An electrophilic addition followed by a substitution (revision of Chapter 17) as part of a useful application.

Suggested solution

HBr adds to the unsymmetrical alkene to give the tertiary benzylic cation and then the tertiary bromide.

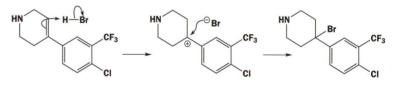

The bromide hydrolyses by an S_N1 mechanism to the alcohol. The same cation is an intermediate in the addition of HBr to the alkene and in the S_N1 substitution.

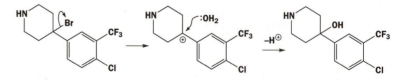

Problem 6

Explain this result.

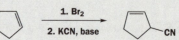

Purpose of the problem

An electrophilic addition followed by an elimination (revision of Chapter 19) and a substitution (revision of Chapter 17).

Suggested solution

Addition of bromine occurs first to give the *trans* dibromide in the usual way (see Problem 2). Base then eliminates one of the bromides in an E2 reaction using the only available *trans* H atom. This gives a reactive allylic bromide (Chapter 17), which reacts with cyanide by a favourable S$_N$2 mechanism to give the product.

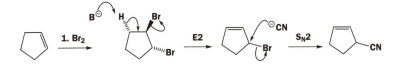

Problem 7

Bromination of this alkene in water gives a single product in good yield. What is the structure and stereochemistry of this product?

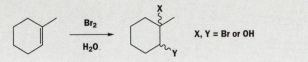

Purpose of the problem

Analysis of the regioselectivity of addition of electrophile and nucleophile with stereochemistry.

Suggested solution

The bromonium ion mechanism leads to *trans* addition of Br and OH with water attacking the bromonium ion at the more substituted carbon atom. This is an S$_N$2 reaction with a loose transition state and some positive charge at the carbon atom under attack (pp. 511–13).

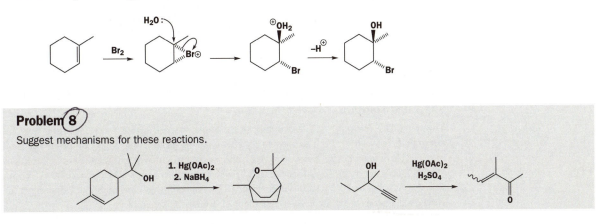

Problem 8

Suggest mechanisms for these reactions.

Purpose of the problem

Exploration of electrophilic addition to alkenes with Hg(II) as the electrophile.

Suggested solution

Hg(II) is a good soft electrophile for alkenes and alkynes and attacks the first alkene to give the usual three-membered ring intermediate, which is in turn captured by the OH group at the more highly substituted centre to give the cage ether. Borohydride reduction removes the mercury (actually by a radical mechanism; see Chapter 39).

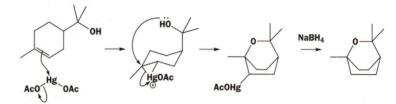

The hydration of the alkyne in the second example is described on p. 519 of the text, but it is possible to write several reasonable mechanisms for the dehydration. The hydroxy-ketone could be formed first and dehydrated but the OH group could get involved earlier and that makes the dehydration easier.

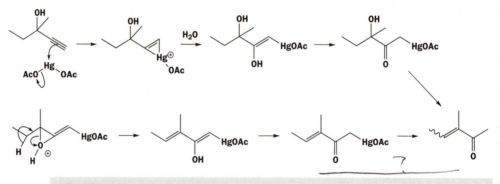

Problem 9

Comment on the formation of a single diastereoisomer in this reaction.

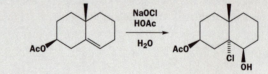

Purpose of the problem

Revision of conformational analysis from Chapter 18 and a mechanism for a new electrophile.

Suggested solution

The anion ClO$^-$ cannot be an electrophile: one protonation makes it neutral HOCl but it becomes electrophilic enough to attack an alkene only after a second protonation. Attack of the chlorine end of ClOH$_2^+$ evidently occurs from underneath the alkene, presumably because the two substituents (Me and AcO) are on the top surface. Attack by water must then be from the opposite (top) face but why does it occur at the *less* substituted carbon? This is a conformational question: we must get the *trans* diaxial product for the correct alignment of orbitals (Chapter 18).

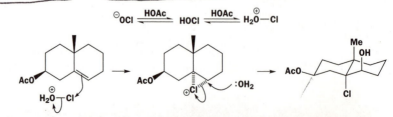

Problem 10

Chlorination of this triarylethylene leads to a chloro-alkene rather than a dichloroalkane. Suggest a mechanism and an explanation.

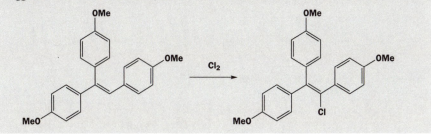

Purpose of the problem

Can you draw a mechanism for a new reaction? Electrophilic addition and then...?

Suggested solution

The first reaction must surely be electrophilic addition of chlorine. One possibility is that the dichloride is formed but then undergoes E1 elimination via the very stable tertiary and doubly benzylic cation. Another possibility is that the addition to this very reactive and unsymmetrical alkene occurs in a stepwise manner to give the very stable cation direct (see diagram in frame in the margin) and not via a three-membered ring intermediate.

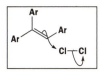

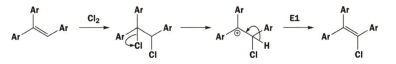

Problem 11

Revision problem. Give mechanisms for each step in this synthesis and explain any regio- and stereochemistry.

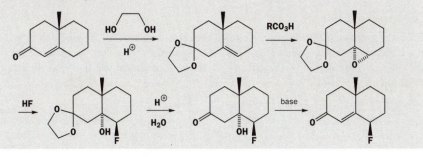

Purpose of the problem

Revision of Chapter 14 (acetal formation and hydrolysis) and elimination (Chapter 19) together with two reactions from this chapter: epoxidation and conformationally controlled opening of an epoxide on a six-membered ring (see also Chapter 18).

Suggested solution

The full mechanisms for acetal formation (step 1) and hydrolysis (step 4) and for epoxidation (step 2) have appeared so often that we shall not repeat them. See Chapter 14 and this chapter if you have some doubts. The epoxidation occurs from the bottom face of the ring because the axial methyl group blocks the top face of the alkene (see also Problem 9, this chapter). Ring opening with HF gives the trans diaxial product. The elimination to give an enone is by the E1cB mechanism. The OH group is axial and so is easily lost.

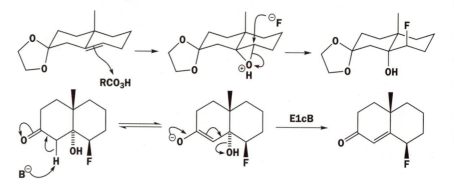

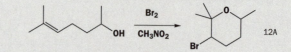

Problem 12

Suggest a mechanism for the following reaction. What is the stereochemistry and conformation of the product?

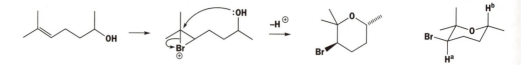

12A has these signals in its NMR spectrum: δ_H 3.9 (IH, ddq, J 12,4,7) and δ_H 4.3 (IH, dd, J 11,3).

Purpose of the problem

Drawing a mechanism for a bromination with an internal nucleophile and revision of NMR.

Suggested solution

The mechanism is formation of the bromonium ion and nucleophilic attack by the OH group at the more substituted carbon atom. The NMR spectrum shows that the protons next to Br (H^a) and oxygen (H^b) are both axial: the Br atom and the methyl group must therefore be equatorial.

Problem 13

Give a mechanism for this reaction and show clearly the stereochemistry of the product.

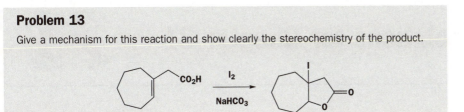

Purpose of the problem

Drawing a mechanism for iodolactonization, discussing selectivity, and working out the stereochemistry.

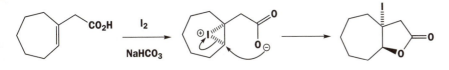

Suggested solution

Iodine attacks the alkene to form an iodonium ion, which is attacked by the carboxylate anion. Attack at the usual more substituted centre would give a four-membered ring so attack at the other end is preferred.

Problem 1

Draw all the possible enol forms of these carbonyl compounds and comment on the stability of the various enols.

Purpose of the problem

Simple exercise in drawing enols with an extra twist.

Suggested solution

There is only one enol of the first compound and it might be quite stable as it is aromatic.

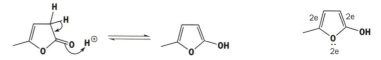

The second compound has more possibilities, one of which is very stable indeed as it has a benzene ring (delocalization shown). We haven't drawn the mechanism for enolization this time but notice the different 'reaction' arrows used for tautomerism (equilibrium arrows) and delocalization (double-headed arrow).

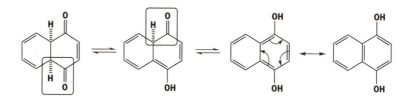

Problem 2

The proportions of enol in a neat sample of the two ketones below are shown. Why are they so different?

Purpose of the problem

Simple exercise in drawing enols with an extra twist.

Suggested solution

The first compound is just an ordinary ketone with its strong C=O bond and so the enol, with its weaker C=C bond, is present in only a tiny amount. The second compound has the special 1,3-relationship between two carbonyl groups that gives a very stable conjugated enol. The stability arises from (a) conjugation of C=C and C=O, and (b) intramolecular hydrogen bonding.

Problem 3

Draw mechanisms for these reactions using just enolization and its reverse.

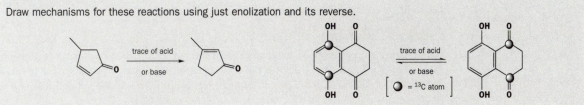

Purpose of the problem

Exercise in using enolization to carry out simple reactions.

Suggested solution

Two enols are possible from the first compound: one (in the margin) leads back only to starting material but the other leads on to product. The whole system is in equilibrium favouring the enone with the more highly substituted alkene.

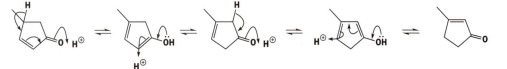

The second example is a bit of a trick. Of course, the ^{13}C label hasn't moved. The molecule just keeps enolizing and going back to a ketone until the functional groups in the two rings have changed places. Here we leave out the mechanism and just show which ketones or enols are tautomerizing with a frame.

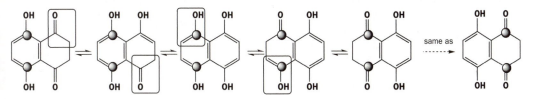

Problem 4

The NMR spectrum of this dimethyl ether is complicated – the two MeO groups are different as are all the hydrogen atoms on the rings. However, the diphenol has a very simple NMR – there are only two types of protons (marked a and b) on the rings. Explain.

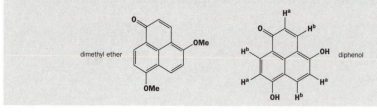

Purpose of the problem

Exploring the relationship between tautomerism and equivalence.

Suggested solution

The protons in the ether are obviously all different as it has no symmetry. Tautomerization between the carbonyl and one of the phenols in the diphenol makes any of the phenols into a ketone and all structures are equivalent. If this proton transfer (note that it is *not* a delocalization!) is fast on the NMR time-scale (p. 258), all H^{a}s will appear the same as will all H^{b}s.

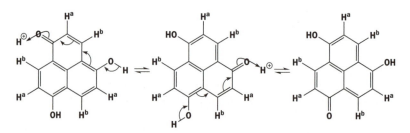

Problem 5

Suggest mechanisms for these reactions.

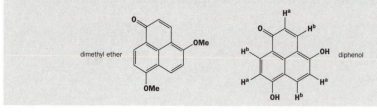

Purpose of the problem

- Revision of Chapters 9 and 14 and a start on mechanisms for chemical reactions involving enolization.

Suggested solution

Reaction (i) starts with an acetal hydrolysis (see Chapter 14 – not repeated here) but the product would be a nonconjugated enone. One enolization under the acidic conditions allows the alkene to move into conjugation.

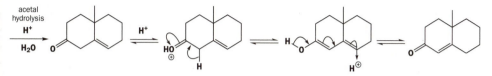

Example (ii) starts with a Grignard addition to a nitrile and hydrolysis of the imine to a ketone (p. 301). Bromination in acid solution at the only enolizable position gives the product.

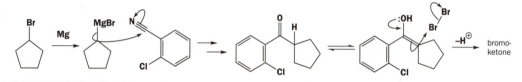

Problem 6

Treatment of this ketone with basic D_2O leads to rapid replacement of two hydrogen atoms by deuterium. Then, more slowly, all the other nonaromatic hydrogens *except* the one marked 'H' are replaced. How is this possible?

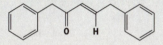

Purpose of the problem

Working through the various ways in which enols and enolates can exchange hydrogen atoms.

Suggested solution

The protons on the CH_2 group next to the ketone exchange by simple enolization and reversion to the ketone. Repetition of this process replaces both H atoms by D. Since D_2O is in large excess, the equilibrium favours D_2-ketone.

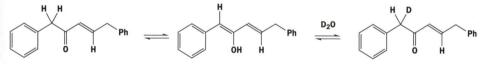

Next, enolization can occur at the other end of the molecule to form a dienol. This leads to replacement of the other CH_2 group with a CD_2 group after two exchanges.

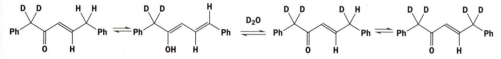

The same dienol can lead to exchange of the remaining proton by a more complicated series of reactions. Deuteration at the carbon atom next to the ketone and then loss of the remaining proton back to the dienol is all that is needed.

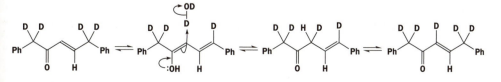

Problem 7

A red alga growing in sea water produces an array of bromine-containing compounds including $CHBr_3$, CBr_4, and $Br_2C=CHCO_2H$. The brominating agent is believed to be derived by the oxidation of bromide ion (Br^-) and can be represented as Br–OX. Suggest mechanistic details for the proposed biosynthesis of $CHBr_3$ in the alga.

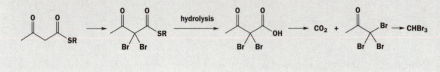

Purpose of the problem

Pushing simple halogenation of ketones into more complex molecules of biological importance.

Suggested solution

The first two brominations occur at the most easily enolized position between the two carbonyl groups. These steps and all the following ones will, of course, be catalysed by enzymes. We are showing the uncatalysed reactions – the enzyme would help by removing or adding protons at the right places and holding the molecule in the right shape.

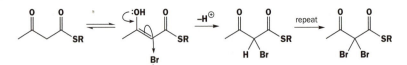

The hydrolysis of the thiolester occurs by simple substitution at the carbonyl group (Chapter 12). Since a thiol is a good leaving group (pK_a RSH about 7) no strong acid or base is needed. The two bromine atoms will also accelerate nucleophilic attack both because they are electron-withdrawing and because they prevent enolization.

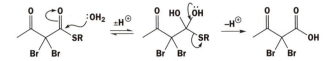

Decarboxylation occurs by the cyclic mechanism explained on p. 678 to give the enol of the product, which is brominated again by the biological brominating agent.

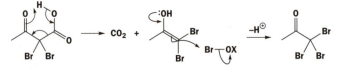

Finally the tribromo ketone adds water and CBr_3^- is a just good enough leaving group to leave the tetrahedral intermediate with C–C bond cleavage (cf. the bromination reaction, p. 537). The anion is protonated by water as it goes.

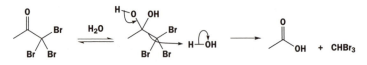

Problem 8

Suggest mechanisms for these reactions and explanations as to why these products are formed.

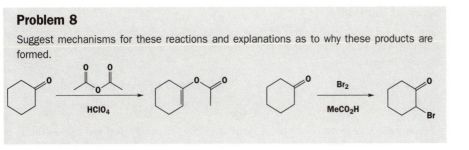

Purpose of the problem

Making sure that you understand the reasons why enols sometimes attack through oxygen and sometimes through carbon.

Suggested solution

The same ketone reacts in similar (acidic) conditions with different electrophiles and with different selectivity. The selectivity must come from the nature of the electrophile. First, we should draw mechanisms. Note that these reactions must occur through the enol and not the enolate as the catalysts are acids.

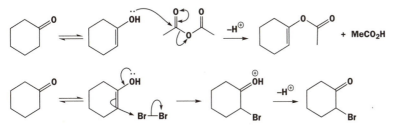

Acid anhydrides, being carbonyl electrophiles with polarized C=O bonds, respond to charge density (they are 'hard' electrophiles) and react well with oxygen nucleophiles. Bromine, by contrast, is uncharged and unpolarized (it is a 'soft' electrophile) and reacts well with neutral nucleophiles such as alkenes (Chapter 20). Each electrophile reacts regioselectively with the part of the enol that suits it best.

Problem 9

1,3-Dicarbonyl compounds such as A are usually mostly enolized. Why is this? Draw the enols available to compounds B-E and explain why B is 100% enol but C, D, and E are 100% ketone.

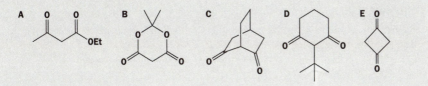

Purpose of the problem

Exploring enols in different kinds of 1,3-dicarbonyl compounds – an important class of enolizable compounds.

Suggested solution

Compound A is mostly enol because the enol is delocalized over five atoms. A minor reason for the stability of this particular compound is that the enol has an internal hydrogen bond. The main reason is conjugation.

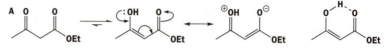

You may also have pointed out that there is an alternative and equally good enol that has the other carbonyl group enolized. These two structures are tautomers of each other and of the keto-ester.

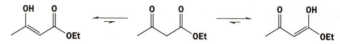

That compound B is totally enolized shows that the conjugation is much more important than the internal hydrogen bond, which is impossible with B. The two enols are the same and have extra conjugation from the lone pair electrons on the other oxygen atoms.

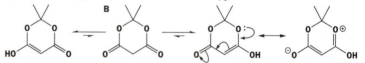

The rest of the compounds have special problems with enolization. Compound C cannot form an enol between the two carbonyl groups at all as the alkene would be a bridgehead alkene unable to get its p orbitals parallel (p. 484). It can, of course, form an ordinary enol on the other side of the ketone but this has no special stabilization.

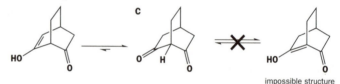

impossible structure

Compounds D and E can form enols but the *t*-butyl group in D has to move into the plane of all the other atoms and there is a bad steric clash with the adjacent oxygen atoms. The *t*-butyl group is out of the plane in the keto form. Compound E is strained but the enol is more strained as another sp^2 atom has to fit into the four-membered ring. The dienol is worse – even more strained and anti-aromatic (four electrons) to boot.

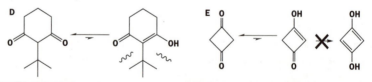

Problem 10

Bromination of ketones can be carried out with molecular bromine in a carboxylic acid solution. Give a mechanism for the reaction.

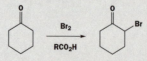

The rate of the reaction is *not* proportional to the concentration of bromine [Br_2]. Suggest an explanation. Why is the bromination of ketones carried out in acidic and not in basic solution?

Purpose of the problem

Extension of the material of Chapter 13 to enol chemistry.

Suggested solution

We need to draw the mechanism out in full with no short cuts. It has four steps (A–D). Steps A and D are merely proton transfers between oxygen atoms and cannot be slow. The rate-determining step must be either B or C. If the rate is not proportional to the concentration of bromine, then the rate-determining step must be B – a step *before* bromine gets involved. Step B is also just a proton transfer, but the proton is being removed from a carbon atom and this can be slow.

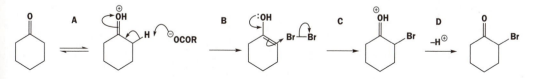

The bromination is carried out in acidic solution to prevent multiple bromination. There is a full analysis of this problem in the chapter (p. 536).

Suggested solutions for Chapter 22

22

Problem 1

All you have to do is to spot the aromatic rings in these compounds. It may not be as easy as you think and you should state some reasons for your choice!

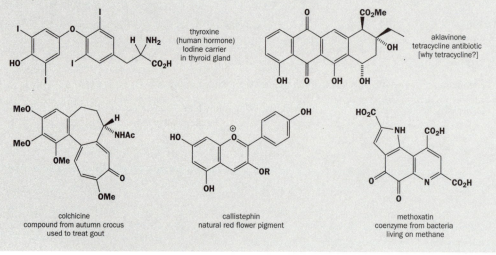

thyroxine
(human hormone)
Iodine carrier
in thyroid gland

aklavinone
tetracycline antibiotic
[why tetracycline?]

colchicine
compound from autumn crocus
used to treat gout

callistephin
natural red flower pigment

methoxatin
coenzyme from bacteria
living on methane

Purpose of the problem

Simple exercise in counting electrons with a few hidden tricks.

Suggested solution

Truly aromatic rings are marked with thick lines. Thyroxine has two benzene rings, which are aromatic and that is that. Aklavinone has again two aromatic rings, one definitely nonaromatic ring (D), and one (B) that we might argue about. However, try as you may you can't get six electrons into ring B (one extreme delocalized version is shown). 'Tetracycline' because of the four rings.

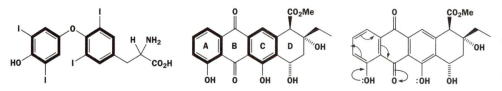

Colchicine has one benzene ring and one aromatic seven-membered ring with six electrons (don't count the electrons in the C=O bond) as the delocalized structure makes clear. Callistephin has a benzene ring and a two-ring oxygen-based cation, which is like a naphthalene. You can count it as one ten-electron system or as two fused six-electron systems sharing one C=C bond, whichever you prefer.

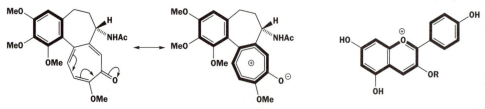

Methoxatin has an aromatic pyridine ring (don't count the lone pair as it is in an sp^2 orbital) and an aromatic pyrrole ring (do count the lone pair as it is in a p orbital) but the middle ring cannot be given six electrons even if you try delocalization (example given).

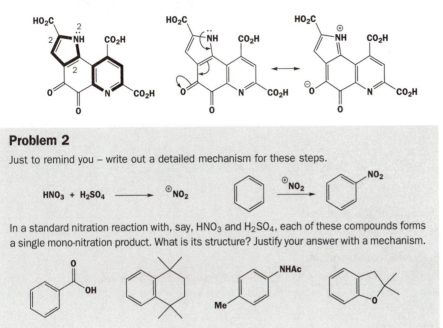

Problem 2

Just to remind you – write out a detailed mechanism for these steps.

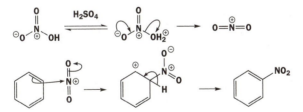

In a standard nitration reaction with, say, HNO$_3$ and H$_2$SO$_4$, each of these compounds forms a single mono-nitration product. What is its structure? Justify your answer with a mechanism.

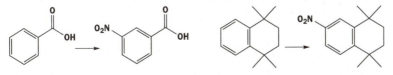

Purpose of the problem

Revision of the basic nitration mechanism and extension to compounds where selectivity is an issue.

Suggested solution

The basic mechanism involves the formation of the active nitrating species, NO$_2^+$, and adding it to benzene. Don't forget to draw in the hydrogen atom at the point of substitution in the intermediate.

The first compound has one electron-withdrawing substituent (CO$_2$H), which is *meta*-directing and deactivating. The second has two identical *ortho*, *para*-directing substituents (alkyl groups), which activate all positions. Steric hindrance decides where the nitro group will go.

The remaining two compounds have two competing *ortho*, *para*-directing substituents, but in each case the one with lone pair electrons (NHAc or OR) beats the simple alkyl group. In the first case, that alone defines the position of nitration – *ortho* to NHAc. In the second, steric hindrance makes the position *para* to OR the more attractive.

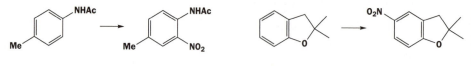

Problem 3

Write mechanisms for these reactions, justifying the position of substitution.

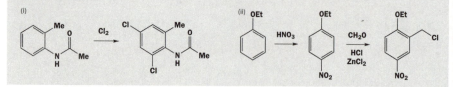

Purpose of the problem

Making you draw mechanisms for two other important reactions and discuss selectivity further.

Suggested solution

Reaction (i) is like the last one in Problem 2: NHAc dominates Me because it has a lone pair of electrons. Chlorine is small and goes in twice: it doesn't matter whether you substitute *ortho* or *para* first.

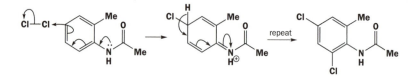

The nitration in reaction (ii) goes *para* because of the large electron-donating OEt group. The details of the second stage – chloromethylation – are on p. 575 of the chapter so we give just the key step.

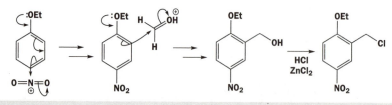

Problem 4

How reactive are the different sites in toluene? Nitration of toluene produces the three possible products in the ratios shown. What would be the ratio of products if all sites were equally reactive? What is the actual relative reactivity of the three sites? (You could express this as *x:y:*1 or as *a:b:c* where $a + b + c = 100$.) Comment on the ratio you deduce.

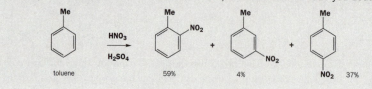

Purpose of the problem

A more quantitative assessment of the relative reactivities of the various positions in an aromatic ring.

Suggested solution

There are two *ortho* and two *meta* sites but only one *para* site so the ratio would be 2:2:1 *o:m:p* if they were equally reactive. It is actually 30:2:37 or 15:1:18 or 43:3:54 depending on how you expressed it. The *ortho* and *para* positions are roughly equally reactive because the alkyl group is electron-donating. The *para* is slightly more reactive than the *ortho* because of steric hindrance. The *meta* is an order of magnitude less reactive because the intermediate cation is not so stabilized by σ-conjugation from the methyl group.

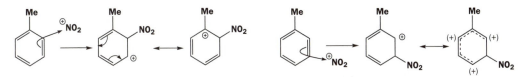

Problem 5

Revision problem. The local anaesthetic proparacaine is made by this sequence of reactions. Deduce a structure for each product. Draw a mechanism for each step and explain why it gives that particular product.

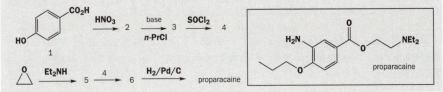

Purpose of the problem

Revision of Chapters 12 and 17 with a bit of aromatic substitution to start with.

Suggested solution

The answer is set out below. The nitration occurs *ortho* to the OH group, which, being strongly electron-donating, dominates the CO_2H group. The rest is revision. The steps to make 4 and 6 are nucleophilic substitution at C=O.

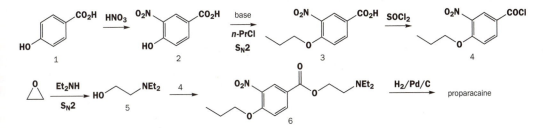

Problem 6

In the chapter, we established that electron-withdrawing groups direct *meta*. Among such reactions is the nitration of trifluoromethyl benzene. Draw out the detailed mechanism for this reaction and also for a reaction that does not happen – the nitration of the same compound in the *para* position. Draw all the delocalized structures of the intermediates and convince yourself that the intermediate for *para* substitution is destabilized by the CF₃ group while that for *meta* substitution is not.

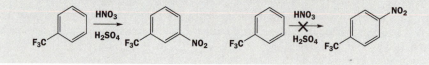

Purpose of the problem

Making sure that you really do understand why groups like CF_3 are *meta*-directing.

Suggested solution

Just do what the question says! You need to do this once to convince yourself. In the mechanism for *meta* substitution, the positive charge in the intermediate is *not* delocalized to the carbon atom carrying the CF₃ group.

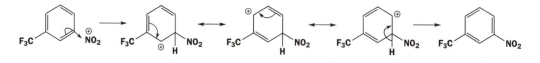

In the mechanism for *para* substitution, the positive charge in the intermediate *is* delocalized to the carbon atom carrying the electron-withdrawing CF_3 group. This intermediate is unstable and is not formed.

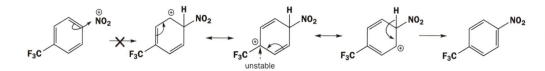

Problem 7

Draw mechanisms for the following reactions and explain the position(s) of substitution.

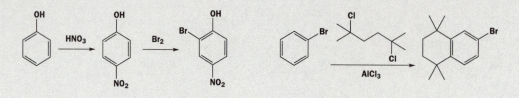

Purpose of the problem

Further exercises in explaining the position of substitution.

Suggested solution

The OH group has a lone pair of electrons and dominates both reactions. Steric hindrance favours the *para* product in the first reaction. *Ortho* substitution has to occur in the second because the *para* position is blocked.

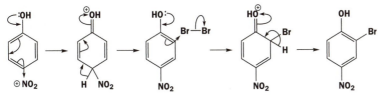

The second example has two Friedel-Crafts alkylations with *tertiary* alkyl halides. The first occurs *para* to Br, a deactivating but *ortho,para*-directing group (pp. 566–8 of the chapter), because of steric hindrance and the second has to occur *ortho* to the first as the new ring cannot stretch any further.

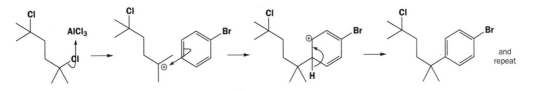

Problem 8

Nitration of these compounds gives products with the proton NMR spectra shown. Deduce the structures of the products from the NMR and explain the position of substitution.

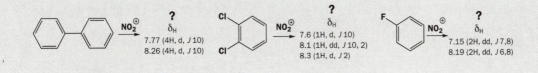

Purpose of the problem

Revision of the relationship between substitution pattern and proton NMR spectra.

Suggested solution

The first product has only eight hydrogen atoms and is symmetrical so two nitro groups must have been added. Each benzene ring acts as an *ortho, para*-directing group on the other so that the intermediate cation can be delocalized round both rings. Steric hindrance favours the *para* product in both rings.

■ Strictly, of course, you can't tell this by NMR!

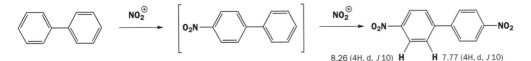

The next two are mono-nitro compounds. The position of the nitro group in the first is obvious from the one hydrogen (at 8.3) which is *ortho* to the nitro group and has no coupling to an *ortho* hydrogen. The two chlorine atoms are *ortho, para*-directing and steric hindrance favours the 1,2,4

pattern rather than the 1,2,3. The symmetry of the NMR shows that the fluoro compound gives the *para* product. Fluorine is also *ortho, para*-directing. The 8 Hz coupling is between neighbouring Hs and the 6 and 7 Hz couplings are to the fluorine atom.

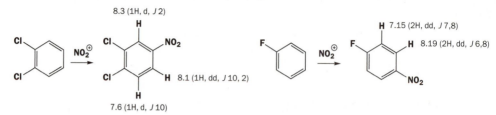

Problem 9

Attempted Friedel–Crafts acylation of benzene with *t*-BuCOCl gives some of the expected ketone, as a minor product, and also some *t*-butyl benzene, but the major product is the disubstituted compound C. Explain how these compounds are formed and suggest the order in which the two substituents are added to form compound C.

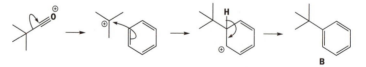

Purpose of the problem

Detailed analysis of a revealing example of a Friedel–Crafts acylation.

Suggested solution

The expected reaction to give A is a simple Friedel–Crafts acylation with the usual acylium ion (text, p. 554) as the reactive intermediate.

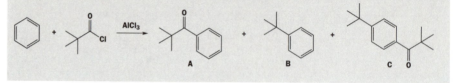

Product B must arise from a Friedel–Crafts alkylation with the *t*-butyl cation as intermediate. This comes from the loss of carbon monoxide from the acylium ion. Such a reaction happens only when the simple carbocation is stable.

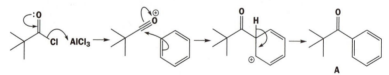

The main product, C, comes from the addition of both these electrophiles, but which adds first? The ketone in A is *meta*-directing but the *t*-butyl group in B is *para*-directing. Product C has a *para* relationship and must come from Friedel–Crafts acylation of B with the acylium cation.

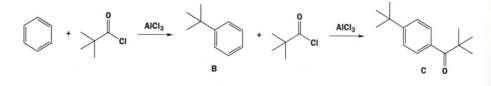

We have now answered the question but you might like to go a stage further. Both A and C are formed with alkylation of benzene as the first step. The decomposition of the acylium ion is evidently *faster* than the acylation of benzene. However, when B reacts further, it is mainly acylated (only a small amount of di-*t*-butyl benzene is formed). Evidently, the decomposition of the acylium ion is *slower* than the acylation of B! This is not, in fact, unreasonable as the *t*-Bu group in B accelerates electrophilic attack – it is just a dramatic demonstration of that acceleration.

Problem 10

Draw mechanisms for the following reactions.

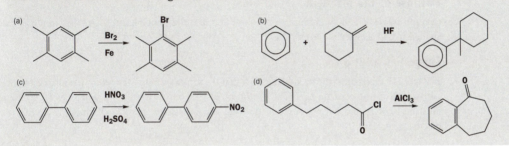

Purpose of the problem

Drawing mechanisms for a variety of electrophilic substitutions.

Suggested solution

Example (d) is the same as the first part of Problem 9 and the orientation was explained there. Parts (a) and (b) raise no orientation questions and *ortho* attack in (d) is simply because the chain cannot reach any further round the benzene ring.

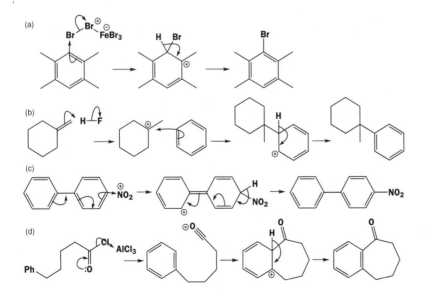

Problem 11

Nitration of this aromatic heterocycle with the usual mixture of HNO_3 and H_2SO_4 gives a product whose NMR spectrum is given. Though you have not yet met heterocycles you should be able to deduce the structure of the product and explain why it is formed.

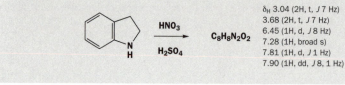

$$C_8H_8N_2O_2$$

δ_H 3.04 (2H, t, J 7 Hz)
3.68 (2H, t, J 7 Hz)
6.45 (1H, d, J 8 Hz)
7.28 (1H, broad s)
7.81 (1H, d, J 1 Hz)
7.90 (1H, dd, J 8, 1 Hz)

Purpose of the problem

Revision of NMR and an attempt to convince you that the principles of Chapter 22 can easily be applied to molecules you haven't met before.

Suggested solution

The two 2H triplets and the broad NH show that the saturated ring is intact so one nitro group has been added to the benzene ring. The proton at 7.81 p.p.m. with only one small (*meta*) coupling must be between the nitro group and the ring and is marked in each structure. There are two possible structures.

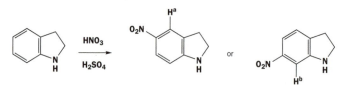

Though both NH and CH_2 are *ortho, para*-directing, we should expect NH to win as it has a lone pair of electrons. This would favour the first structure, but is not evidence. Calculation of the expected chemical shifts of protons in the two structures gives a more reliable answer. For example, the two marked hydrogen atoms are predicted to come at:

Proton	*ortho-*	*ortho-*	*meta-*	Predicted δ_H (p.p.m.)
H^a at 7.27+	$NO_2 = +0.95$	$CH_2 = -0.14$	$NH = -0.25$	7.73
H^b at 7.27+	$NO_2 = +0.95$	$NH = -0.75$	$CH_2 = -0.06$	7.31

The observed value of 7.81 p.p.m. clearly fits H^a better than H^b. The main difference between the two comes from the NH group and the result favours the compound predicted from the expected mechanism. This one is correct.

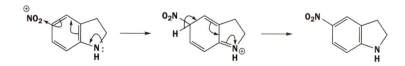

Problem 12

Explain the position of substitution in the following reactions and predict the structure of the final product. Why is a Lewis acid necessary for the second bromination but not for the first?

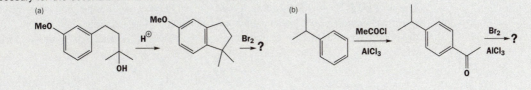

Purpose of the problem

Further exploration of important synthetic reactions: Friedel–Crafts and bromination.

Suggested solution

Reaction (a) is an intramolecular Friedel–Crafts alkylation at an activated position.

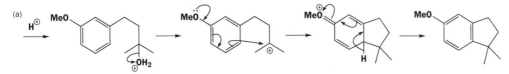

Reaction (b) starts with a Friedel–Crafts acylation *para* to the large *ortho,para*-directing isopropyl group.

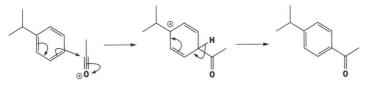

Bromination in part (a) will also be directed by the delocalized electrons of the OMe group and will choose the less hindered of the two positions *ortho* to OMe. In the second part the activating alkyl group and the deactivating acyl group both direct bromination to the same position: *ortho* to *i*-Pr and *meta* to acyl. No Lewis acid is needed in (a) as the ring is activated and the reaction is intramolecular, whereas in (b) the ring is weakly activated by *i*-Pr, more strongly deactivated by the acyl group, and the reaction is intermolecular.

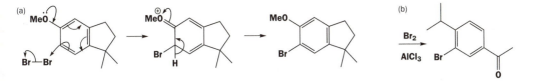

Problem 13

Suggest mechanisms for the methylation step at the end of the synthesis which concludes the chapter. Why is it necessary to go to these lengths rather than just react with MeI?

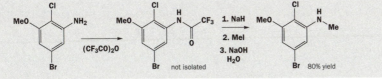

Purpose of the problem

Revision of Chapters 12 and 17.

Suggested solution

The acylation is a straightforward nucleophilic substitution at the carbonyl group and the CF_3CO group is chosen to make the anion at nitrogen more stable. Alkylation can occur only once whereas direct alkylation with MeI could occur several times. Hydrolysis of the amide is also easier with the CF_3CO group.

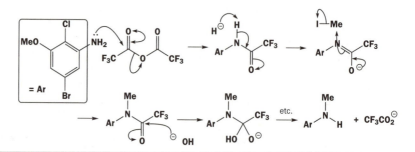

Problem 14

So what happens if we force phenol to react again with bromine? Will reaction then occur in the *meta* positions? It is possible to brominate 2,4,6-tribromophenol if we use bromine in acetic acid. Account for the formation of the product.

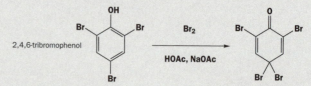

This product can be used for bromination as in the monobromination of this amine. Suggest a mechanism and explain the selectivity.

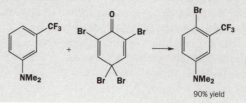

Purpose of the problem

Revision of Chapters 12 and 17.

Suggested solution

The first step is surprising but simple. Electrophilic attack occurs in the *para* position to the activating OH group. The intermediate cannot lose a proton from the tetrahedral centre so it loses the OH proton instead. The phenol would rather lose its aromaticity than react in the *meta* position.

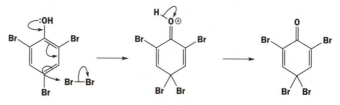

The second reaction uses the compound we have just made to brominate another organic molecule. One of the bromine atoms in the *para* position is used – a case of last in, first out. The large and strongly activating NMe$_2$ group directs the reaction both electronically and sterically. Notice that in the intermediate the two Me groups on nitrogen are fixed in the plane of the ring and would clash sterically with an *ortho* substituent.

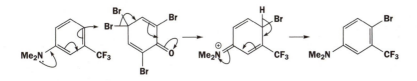

Suggested solutions for Chapter 23

23

Problem 1

What is the structure of the product of this reaction and how is it formed?

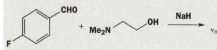

δ_C(p.p.m.) 191, 164, 132, 130, 115, 64, 41, 29
δ_H(p.p.m.) 2.32 (6H, s), 3.05 (2H, t, J 6 Hz),
4.20 (2H, t, J 6 Hz), 6.97 (2H, d, J 7 Hz),
7.82 (2H, d, J 7 Hz), 9.97 (1H, s)

Purpose of the problem

Revision of NMR with an exercise in nucleophilic aromatic substitution.

Suggested solution

This is the answer. You should, of course, work out the structure from the spectra in the style we used in the solutions for Chapters 11 and 15.

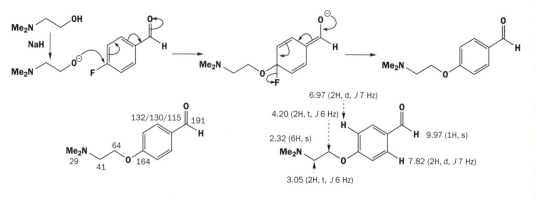

Problem 2

Draw a detailed mechanism for this reaction. Note that no base is added to the mixture. Why is base unnecessary?

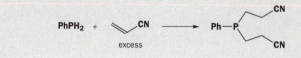

Purpose of the problem

Simple example of conjugate addition with a nucleophile from the second row of the periodic table.

Suggested solution

The phosphine is a good soft nucleophile with a high-energy lone pair, well able to add in a conjugate fashion without help. In particular, the neutral phosphine is a good nucleophile and it does not need to be converted into the anion first. The proton transfer occurs after the conjugate addition and need not be intramolecular.

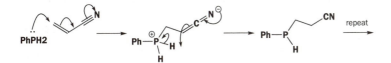

Problem 3

Which of the two routes suggested here would actually lead to the product? What might happen in the other sequence?

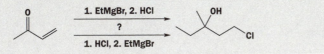

Purpose of the problem

Do you understand the essentials of conjugate addition and can you say when it *won't* happen?

Suggested solution

The chloride must add in a conjugate fashion and the ethyl Grignard in a direct fashion that removes the carbonyl group. Conjugate addition can happen only if the carbonyl group is intact so it must happen first and the lower suggestion will work.

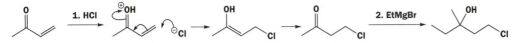

In the other sequence, EtMgBr is likely to add to the carbonyl direct and the further addition of HCl may either substitute on the allylic alcohol or add the 'wrong way round' to the alkene.

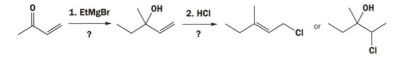

Problem 4

Suggest reasons for the different outcome of each of these reactions. Your answer must, of course, include a mechanism for each reaction.

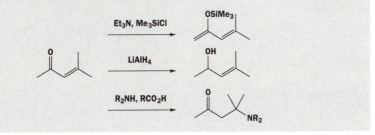

Purpose of the problem

A reminder of the reactions possible with enones.

Suggested solution

The three reactions are: enolization and trapping with silicon (Chapter 21); direct addition with a hard irreversible nucleophile; and conjugate addition by a softer reversible nucleophile.

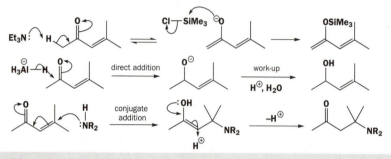

Problem 5

Suggest mechanisms for these reactions. You should explain why one of the cyanides is lost but not the other.

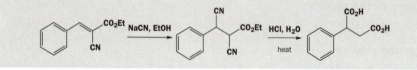

Purpose of the problem

Revision of conjugate addition (Chapter 10 and this chapter), nucleophilic substitution at the carbonyl group (Chapter 14), and decarboxylation of malonates (Chapter 21).

Suggested solution

The first step is conjugate addition promoted by two electron-accepting groups at one end of the alkene. We show the involvement of only one of them in the mechanism.

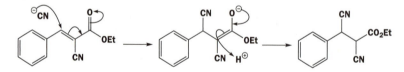

Next, the ester and both nitriles (cyanides) are hydrolysed. The mechanisms for these reactions are in Chapter 14 – make sure you can draw them. Finally, the triacid loses a molecule of CO_2 by the cyclic mechanism introduced in Chapter 21.

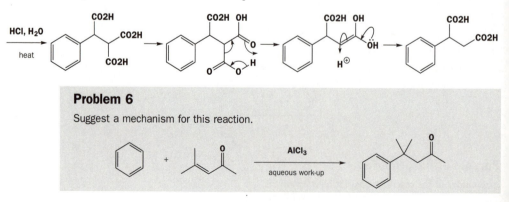

Problem 6

Suggest a mechanism for this reaction.

Purpose of the problem

Combination of conjugate addition and electrophilic aromatic substitution (Chapter 22).

Suggested solution

The weakly nucleophilic benzene ring has evidently added in conjugate fashion to the enone in a kind of Friedel–Crafts reaction and we can use the Lewis acid to make the electrophile into the necessary cation.

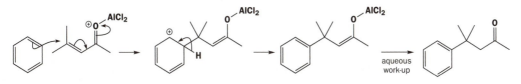

Problem 7

Suggest a mechanism for this reaction explaining the selectivity.

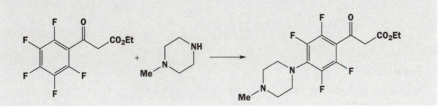

Purpose of the problem

Introduction to the mechanism and selectivity of nucleophilic aromatic substitution.

Suggested solution

The amine attacks in the *para* position so that the intermediate anion is stabilized by conjugation with the ketone and by the five electronegative fluorine atoms.

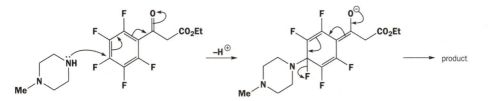

Problem 8

Suggest mechanisms for all of the steps in this synthesis of 2,4-dinitrophenylhydrazine given in the chapter.

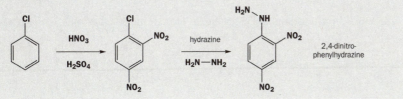

Purpose of the problem

Revision of electrophilic substitution and a problem in nucleophilic aromatic substitution.

Suggested solution

The first reaction is standard nitration of benzene with an *ortho, para*-directing group (Chapter 22). The first nitration can go either 2- or 4- to the chlorine and the second goes *meta* to the first.

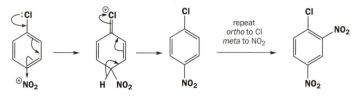

The reaction with hydrazine is a nucleophilic substitution activated by two nitro groups.

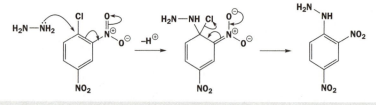

Problem 9

Pyridine is a six-electron aromatic system like benzene. You have not yet been taught anything systematic about pyridine but see if you can work out why 2- and 4-chloropyridines react with nucleophiles but 3-chloropyridine does not.

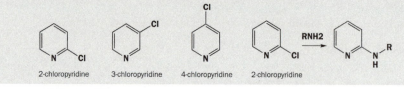

2-chloropyridine   3-chloropyridine   4-chloropyridine   2-chloropyridine

Purpose of the problem

Extension of the ideas on nucleophilic aromatic substitution into new compounds.

Suggested solution

The main need is to find somewhere to park the negative charge in the intermediate. The only place we have in pyridine is the nitrogen atom. We can put the negative charge on the nitrogen when we attack 2- or 4-chloropyridine, but not when we do the same thing with 3-chloropyridine. Try for yourself.

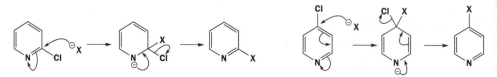

The amine formation given in the question is just one important example of this reaction.

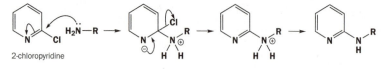

2-chloropyridine

Problem 10

Draw detailed mechanisms for the last two steps in the ranitidine synthesis that involve conjugate substitution. Why is it possible to replace one MeS group at a time?

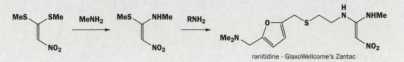

ranitidine - GlaxoWellcome's Zantac

Purpose of the problem

A chance to explore the details of a rare but important conjugate substitution.

Suggested solution

Conjugate addition of the nucleophile is possible because of the nitro group. After the first substitution the compound gains extra stabilization from conjugation of the amino and the nitro groups (in frame in margin). This conjugation is interrupted during the second substitution so it is slower but follows the same mechanism.

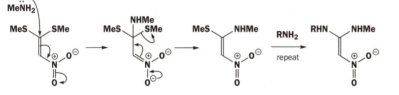

Problem 11

How would you convert this aromatic compound into the two derivatives shown?

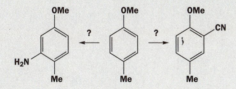

Purpose of the problem

Application of nucleophilic aromatic substitution to the synthesis of aromatic compounds.

Suggested solution

We want to introduce both CN and NH$_2$ as nucleophiles but the ring is unactivated so we must use special methods. The most obvious are a benzyne route for the amine and a diazonium route for the nitrile. In the second example the amide anion attacks the benzyne so that the negative charge ends up *ortho* to the OMe group (p. 603).

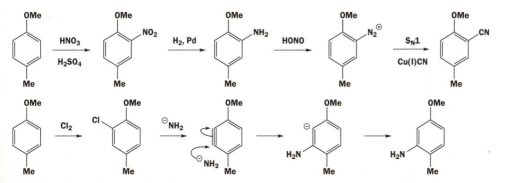

Problem 12

Comment on the selectivity shown in these reactions.

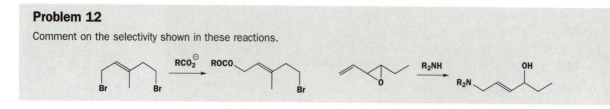

Purpose of the problem

Examples of allylic substitution – can you select the more electrophilic centre?

Suggested solution

Each electrophile has three sites: we need to assess the number of substituents and whether they are allylic.

In both cases, the nucleophile selects the less substituted allylic site and ignores an equally substituted site that is not allylic. Notice that the more highly substituted alkene is formed in both cases.

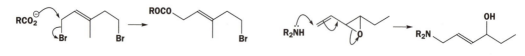

Problem 13

Suggest what products might be formed from the unsaturated lactone and the various reagents given and comment on your choice.

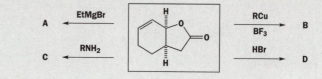

Purpose of the problem

Your chance to rehearse the methods of control in allylic substitution. Prediction is more difficult than explanation.

Suggested solution

The Grignard reagent is likely to attack the carbonyl group direct and will react again with the first formed ketone. The copper compound is expected to react in an S_N2' manner with retention of configuration.

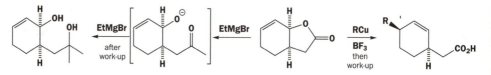

The amine is expected to attack the carbonyl group directly, but it might possibly do an allylic substitution. HBr will protonate the carbonyl group and then will do a nucleophilic substitution either S_N2 (shown) or S_N2'.

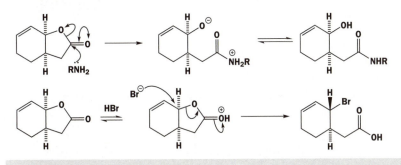

Problem 14

Suggest mechanisms for these reactions, pointing out what guided you to choose these pathways.

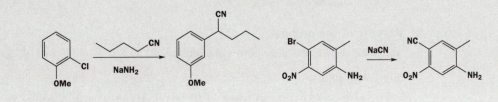

Purpose of the problem

Studies in selectivity in nucleophilic aromatic substitutions by different mechanisms.

Suggested solution

The nucleophile adds to the 'wrong' position in the first example so a benzyne mechanism is indicated. The second example is a straightforward addition-elimination mechanism activated by a nitro group.

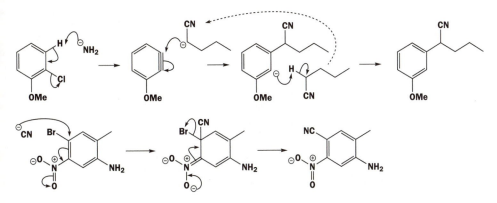

Suggested solutions for Chapter 24

24

■ Several problems in this chapter ask you to suggest ways to carry out conversions of one molecule into another. We give one possible answer and sometimes comment on alternatives but you should realize that there are many possible right answers to this sort of question. Make sure you have understood the principles behind each question and, if your answer is different from ours, check it with someone who knows.

Problem 1

How would you convert this bromoaldehyde chemoselectively into the two products shown?

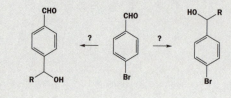

Purpose of the problem

A simple exercise in chemoselectivity and protection.

Suggested solution

You would like to add a Grignard reagent RMgBr (or RLi) to go to the right and you would like to convert the bromoaldehyde into a Grignard reagent for addition to RCHO to go to the left. This second reaction is not possible as the Grignard reagent will react with itself. The aldehyde must be protected first and an acetal of some kind is the obvious choice. We have used a dimethyl acetal; you may very well have used a cyclic acetal. Both are fine.

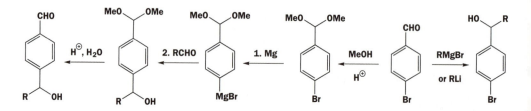

Problem 2

Explain the chemoselectivity of these reactions. What is the role of the Me$_3$SiCN?

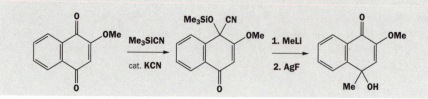

Purpose of the problem

A combination of rather delicate chemoselectivity and protection.

Suggested solution

We want to add MeLi to the less reactive of the two carbonyl groups. Why less reactive? Because it is conjugated with the MeO group.

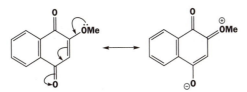

The more reactive carbonyl group needs to be protected, and of the various methods tried, this rather unusual cyanohydrin formation (Chapter 6) was the best. The Me₃SiCN provides the protecting group and F⁻ removes it.

■ You can read more about this in D. A. Evans *et al.*, *J. Am. Chem. Soc.*, 1973, **95**, 5822.

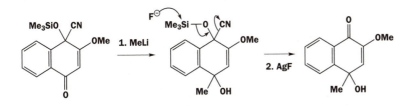

Problem 3

How would you convert this lactone selectively either into the hydroxy-acid or into the unfunctionalized acid?

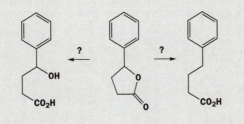

Purpose of the problem

A combination of chemoselectivity and selective reduction.

Suggested solution

The conversion to the hydroxy-acid is just hydrolysis and can be carried out in aqueous base. Reduction to the unfunctionalized acid demands attack at the secondary benzylic centre and possibilities include catalytic hydrogenation or HBr followed by reduction.

■ You can read about an application of this chemistry in S. Torii *et al.*, *Bull. Chem. Soc. Japan*, 1978, **51**, 3590.

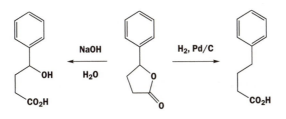

Problem 4

Predict the products of Birch reduction of these aromatic compounds:

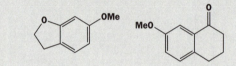

Purpose of the problem

Establishing the principles of Birch reduction.

Suggested solution

In each case a unique product puts electron-donating groups on the alkenes and keeps electron-withdrawing groups away from the alkenes.

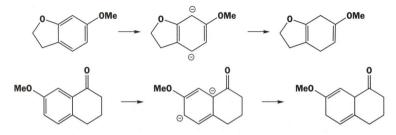

Problem 5

How would you carry out these reactions? In some cases more than one step may be needed.

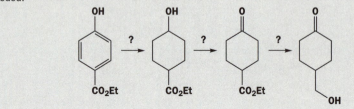

Purpose of the problem

Reduction, selectivity, and protection in the same sequence.

■ The product was used to make an analogue of a thromboxane, a human blood clotting agent by M. Hayashi and group, *Tetrahedron Lett.*, 1979, 3661.

Suggested solution

Everything is straightforward except the final reduction where a less reactive ester must be reduced in the presence of a more reactive ketone. Protection is the answer and an acetal is suitable.

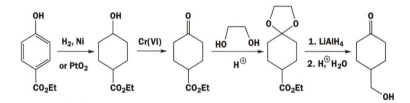

Problem 6

How would you convert this nitro compound into the two products shown? Explain the order of events with special regard to reduction steps.

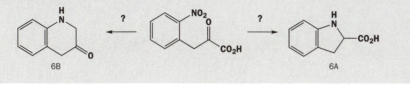

6B   6A

Purpose of the problem

Reduction, selectivity, and protection in the same sequence.

Suggested solution

The nitro group must be reduced to an amino group and cyclized with the ketone (reductively) to give 6A and with the acid to give 6B. The five-membered ring will be formed faster so only the formation of 6B needs special conditions.

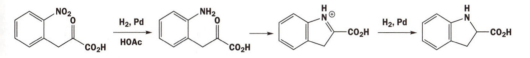

The acid must be converted into a derivative (ester, acid chloride) that will react with the amino group, and the ketone must be protected during the nitro group reduction, the cyclization, and during the reduction of the amide. Again, an acetal is ideal. There are many possible good answers to this problem.

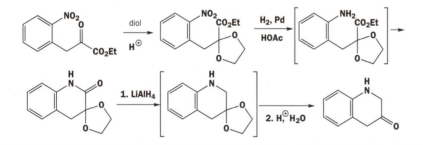

Problem 7

What kinds of selectivity are operating in these reactions and how do they work?

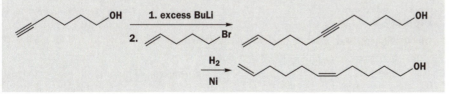

Purpose of the problem

Showing you how to combine selectivity, reduction, stereochemistry, and a dianion reaction.

Suggested solution

■ The product was used at the US Department of Agriculture to make the pheromone of the Douglas Fir tussock moth (G. D. Daves and group, *J. Org. Chem.*, 1978, **43**, 2361).

The very strong base BuLi will first remove the OH proton and then the alkyne proton. The alkyne anion is more reactive (last formed, first to react) and is alkylated. Hydrogenation goes for the alkyne first and adds a molecule of hydrogen to the same side of the triple bond to give a *cis* alkene.

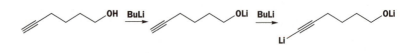

Problem 8

These two Wittig reactions (Chapter 14) give very different results. The first gives a single alkene in high yield (which?). The second gives a mixture from which one alkene can be separated with difficulty and in low yield. Why are they so different?

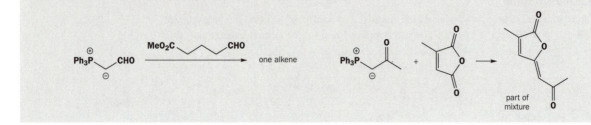

Purpose of the problem

A demonstration of good and bad selectivity for you to unravel.

■ The question of alkene geometry from the Wittig reaction is discussed in Chapter 31.

■ This product was used to make a human leukotriene at Smith, Kline, and French in Philadelphia (J. G. Gleason *et al.*, *Tetrahedron Lett.*, 1980, **21**, 1129).

■ The problem was solved by a group in Adelaide who made the natural product freelingyne in this way (C. F. Ingham *et al.*, *Aust. J. Chem.*, 1974, **27**, 1477).

Suggested solution

Both Wittig reagents have carbanions that are further stabilized by carbonyl groups and should give *E*-alkenes selectively. The first example reacts with the aldehyde rather than with the ester to give the *trans* enal in good yield.

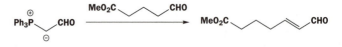

In the second example, the electrophile is an anhydride with two reactive carbonyl groups that are nearly but not quite the same. This is a recipe for bad selectivity – reaction will occur at both carbonyls. In addition, stereoselectivity will not be so good as the alkenes are trisubstituted and there is not as much difference in stability between *cis* and *trans* as there was in the disubstituted enal.

Problem 9

Why is this particular amine formed by reductive amination?

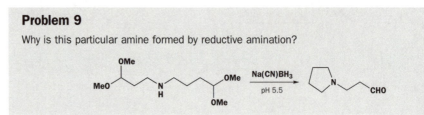

Purpose of the problem

Extending the concept of reductive amination by combining it with deprotection and cyclization.

Suggested solution

The two acetals will be hydrolysed at pH 5.5 and give the amine a choice between cyclization to one or other of two aldehydes.

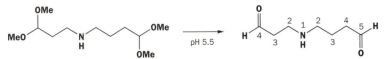

Cyclization to a five-membered ring is preferred to cyclization to a four-membered ring so reductive amination occurs to the right and not to the left (as drawn). Cyanoborohydride is stable under the weakly acidic conditions necessary and does not reduce the remaining aldehyde.

■ This problem was based on work by G. W. Gribble and R. M. Soll, *J. Org. Chem.*, 1981, **46**, 2433.

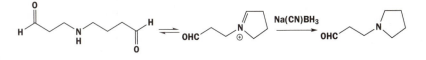

Problem 10

Account for the chemoselectivity of the first reaction and the stereoselectivity of the second. A conformational drawing of the intermediate is essential.

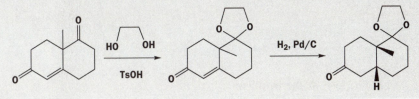

Purpose of the problem

Extending chemoselectivity into more subtle distinctions and conformational analysis.

Suggested solution

The two ketones are different only because one is conjugated. Since acetal formation is thermodynamically controlled, the one that is formed retains the conjugation of the other ketone. A conformational diagram of the intermediate shows that there is an inevitable axial oxygen atom belonging to the acetal preventing the bottom face of the alkene from getting close to the catalyst. The axial methyl group is further away and more in the alkene plane.

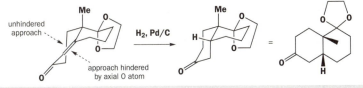

Problem 11

Account for the chemoselectivity of these reductions.

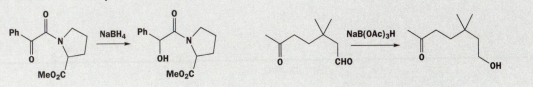

Purpose of the problem

Simpler examples of chemoselectivity between different carbonyl groups.

Suggested solution

■ These examples come from work by K. Soai *et al.*, *J. Chem. Soc., Chem. Commun.*, 1982, 1282 and by C. F. Nutaitis and G. W. Gribble, *Tetrahedron Lett.*, 1983, **24**, 4287.

There are three carbonyl groups in the first compound: a ketone, and ester, and an amide. Reactivity is in the same order (ketone > ester > amide) but the ketone and amide groups are on adjacent carbon atoms (α-keto-amide) so both are more reactive than normal. The ketone is the most reactive: esters and amides are not usually reduced by $NaBH_4$. The second example has a ketone and an aldehyde: the aldehyde is more reactive. If we used $NaBH_4$ as the reagent it would probably reduce both, so the less reactive modified borohydride $NaB(OAc)_3H$, sodium triacetoxyborohydride, is used instead.

Problem 12

How would you carry out the following conversions? More than one step may be needed and you should comment on any chemoselective steps.

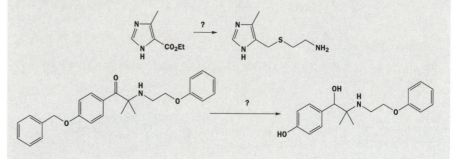

Purpose of the problem

Further examples of chemoselectivity between different functional groups with danger of overreaction.

Suggested solution

We need to reduce the first ester, convert the OH group to an SH, couple it with a two-carbon electrophilic fragment such as an epoxide, and convert OH to NH_2. The plan is something on these lines.

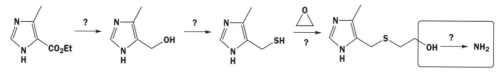

■ This is part of the synthesis of the anti-ulcer drug cimetidine.

The reduction is simple ($LiAlH_4$ will reduce the ester and won't affect the aromatic ring) but the conversion of OH to SH requires care to stop the sulfide forming. We suggest one possibility. Similarly, the conversion of OH to NH_2 requires a nitrogen nucleophile that will react once only (Chapter 17, p. 437) and again we offer one suggestion.

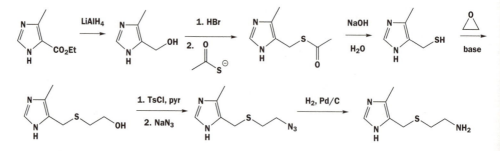

The second example requires reduction of the ketone – no problem – and the chemoselective cleavage of one of two ethers – chemoselectivity problem!

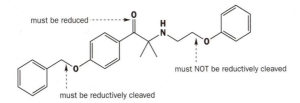

Fortunately, the ether that must be cleaved is a benzyl ether and the C–O bond can be reduced by catalytic hydrogenation using hydrogen and palladium (p. 621 of text). The ketone can be reduced with $NaBH_4$.

Suggested solutions for Chapter 25

■ Several problems in this chapter ask you to suggest ways to carry out conversions of one molecule into another. We give one possible answer and sometimes comment on alternatives but you should realize that there are many possible right answers to this sort of question. Make sure you have understood the principles behind each question and, if your answer is different from ours, check it with someone who knows.

Problem 1

Suggest two different syntheses for these ethers and say which you prefer (and why!).

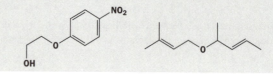

Purpose of the problem

An exercise in choosing good routes to simple compounds.

Suggested solution

You will be making C–O bonds in making these ethers and the choice hinges on which C–O bond you think can be made more easily or without selectivity problems. In the first case either route will do: the aromatic substitution is all right because of the *para*-nitro group and an epoxide can be used as the electrophile in the alternative.

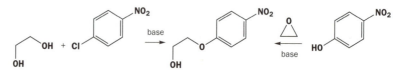

Whichever way you make the second ether, you will have to react an allylic alcohol with an allylic halide. There is no problem with the allylic alcohol, but we need to be sure that the allylic halide will react at the right end. Compound A has two sites for attack marked with a black blob. Attack at either leads to the same product (work it out for yourselves!). In compound B they are different but attack at the primary centre will be much preferred. Either route is fine. In both cases we now know that we can choose either route because we have analysed them mechanistically.

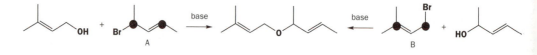

Problem 2

Suggest syntheses for these esters. The starting materials might also need to be made.

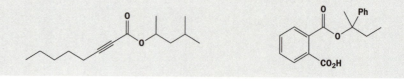

Purpose of the problem

First steps in making C–C bonds as a preliminary to joining molecules together.

Suggested solution

Esters are made from acid derivatives and alcohols so we need first to look at those starting materials.

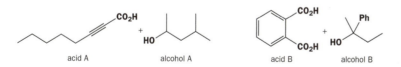

acid A   alcohol A   acid B   alcohol B

Acid A can be made by alkyne chemistry, adding the alkyl group first and CO_2 second. Alcohol A is available. Acid B is available phthalic acid and alcohol B can be made by PhLi or PhMgBr addition to an available ketone. Acid A can be joined to alcohol A via the acid chloride but acid B is best used as the cyclic anhydride to get the monoester.

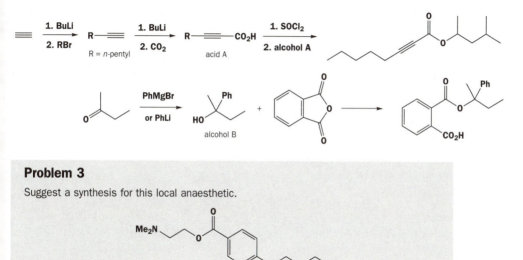

Problem 3

Suggest a synthesis for this local anaesthetic.

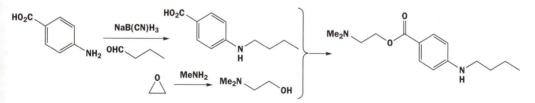

Purpose of the problem

Another exercise in making C–O and C–N bonds but with selectivity required.

Suggested solution

The ester will be made from an alcohol (made in turn from an epoxide and an amine) and an acid. In the *para* position, the amine can be made from either an amide or an imine by reduction. There is a slight problem in doing the reactions in the right order as we must avoid a reduction step that would reduce too many groups. Coupling the alcohol and the acid in acidic solution will avoid unwanted reactions at the amino group as the nitrogen atom will be protonated.

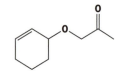

Problem 4

Suggest syntheses for this simple compound. What selectivity problems must be overcome?

Purpose of the problem

A simple synthesis with many solutions.

Suggested solution

This is another ether and there are two obvious ways to make it, each involving the formation of one of the C–O bonds.

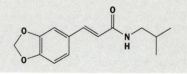

In both cases we must consider the danger of enolate formation from the ketone. In the first case the alcohol might displace the bromide or attack the ketone and in the second the allylic bromide might be attacked at the alkene though this makes no difference as the allylic system is symmetrical. The first approach is easier as the bromoketone is easily made from acetone (Chapter 21) and the allylic halide in the second approach would probably be made from the alcohol used in the first synthesis.

Problem 5

Suggest a synthesis for this compound. Justify your choice of methods and reagents.

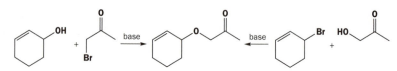

Purpose of the problem

More advanced synthesis planning with more steps in the synthesis.

Suggested solution

The amide must be made from the amine and an acid derivative and any of the methods described in the chapter could be used though we must be sure that the amine will attack the carbonyl group and not do conjugate addition. The acid can be made by a Wittig reaction (Chapter 14) from the available aldehyde 'piperonal'. If the amine does conjugate addition to the ester, change to the acid chloride by hydrolysis to the acid and treatment with $SOCl_2$ first.

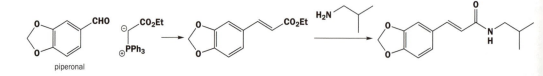

piperonal

Problem 6

Suggest how these amines might be synthesized.

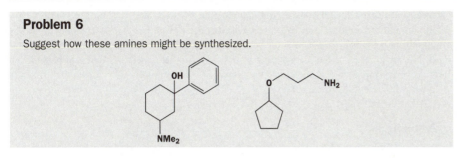

Purpose of the problem

First steps into counting the relationship between functional groups

Suggested solution

The tertiary alcohol comes from addition of PhLi or PhMgX to a ketone, and the relationship between the ketone carbonyl and the amine is just right for conjugate addition (Chapter 10).

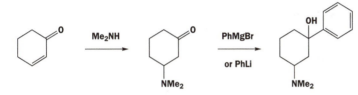

The second compound arises in much the same way except that cyclopentanol can be added to acrylonitrile in conjugate fashion and the primary amine derived by reduction of the nitrile.

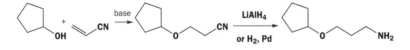

Problem 7

This hexa-alcohol can be deprotected, one OH at a time, by the sequence of reagents shown below. Explain how each reagent works, stating, of course, which protecting group it removes! Would any other order of events be successful?

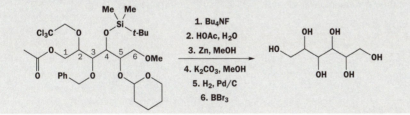

1. Bu$_4$NF
2. HOAc, H$_2$O
3. Zn, MeOH
4. K$_2$CO$_3$, MeOH
5. H$_2$, Pd/C
6. BBr$_3$

Purpose of the problem

A slightly artificial exercise in matching conditions to deprotection.

Suggested solution

Step 1: Fluoride attacks silicon and releases the OH group on C4. Step 2: Mild acidic conditions hydrolyse the only acetal, a THP derivative at C5. Step 3: Zinc is a two-electron donor that attacks

■ Details of all these reactions appear in the chapter and are discussed in E. J. Corey and A. Venkateswarlu, *J. Am. Chem. Soc.*, 1972, **94**, 6191.

halogens and releases the OH on C2 by an elimination reaction. Step 4: These are the only basic conditions and are ideal for ester hydrolysis, releasing the OH on C1. Step 5: The only protecting group susceptible to hydrogenolysis is the benzyl ether at C3. Step 6: The most robust group and the most difficult to cleave is the methyl ether at C6. Several other orders of events are possible but step 6 is very vigorous and would cleave the acetal at C5 for example.

Problem 8

Suggest a synthesis of the starting material and give mechanisms for the reactions. Why does the last step go under such unusual conditions?

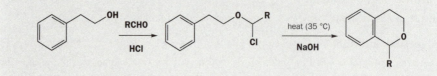

Purpose of the problem

Learning to make C–C bonds and revision of Chapters 12 and 22.

Suggested solution

The CH$_2$CH$_2$OH group in the starting material suggests the addition of PhLi or PhMgBr to ethylene oxide. The next reaction is like acetal formation with the intermediate oxonium ion captured by chloride.

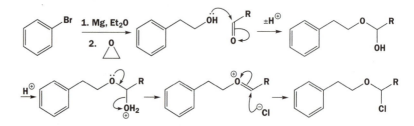

The last step is an electrophilic aromatic substitution with base! Presumably, the easily formed cation (the same one that was involved in the previous reaction) cyclizes rapidly and reversibly, and it is the deprotonation of the intermediate that makes the reaction go.

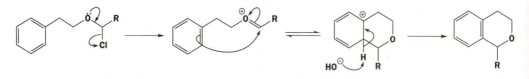

Problem 9

Esters are normally made from alcohols and activated acids. This one is made by a completely different method. Why?

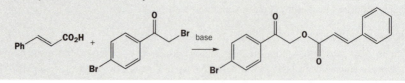

Purpose of the problem

Learning to put mechanistic thinking into the solution to problems in synthesis.

Suggested solution

This ester is clearly made by an S_N2 reaction (Chapter 17) rather than nucleophilic attack by an alcohol on an acid derivative. The base creates the carboxylate anion and the S_N2 reaction next to the ketone carbonyl is such a good reaction that even this weak nucleophile is enough. The bromide is easy to make from the ketone by bromination of the enol and the alcohol would be made in turn from the bromide. So why bother with the usual method?

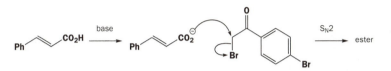

Problem 10

Suggest a synthesis of this non-protein peptide, emphasizing the choice of protecting groups.

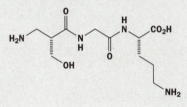

Purpose of the problem

Making sure you understand the approach to peptide synthesis explained in the chapter.

Suggested solution

The three amino acids (serine, glycine, and lysine) need to be coupled together, and the other functional groups (ringed) need to be protected. In principle, you can start at either end and there are many good answers to this problem but we shall give the method used in the chapter.

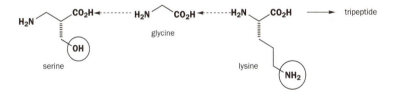

    Choosing a Cbz protecting group for the terminal amino group and any protecting group you choose from the list on p. 657, we convert serine into an aryl ester (Ar = p-nitrophenyl or 2,4,6-trichlorophenyl) and react the aryl ester with glycine protected as its t-butyl ester. Simple acid treatment removes the t-butyl group and the process is repeated, this time with lysine having its side-chain amino group protected by another Cbz group or any method you please from the list on p. 657. Deprotection gives the tripeptide. There are other protecting groups you could have chosen.

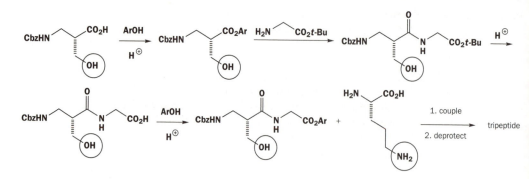

Problem 11

The β-iodoethoxycarbonyl group has been suggested as a protecting group for amines. It is removed with zinc in methanol. How would you add this protecting group to an amine and how does the deprotection occur? What other functional groups might survive the deprotection?

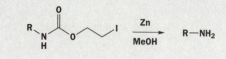

Purpose of the problem

An exploration of a different kind of protecting group.

Suggested solution

This group is very similar to the Cbz and t-BOC groups and might be added in the same way using a chloroformate. The removal is like that of the trichloroethyl group and involves addition of zinc to the C–I bond. Any group normally removed in acid, such as an acetal or a t-butyl ester, or any group removed by hydrogenation, such as a Cbz or benzyl ether, should survive these conditions. However, base might lead to ester hydrolysis and the trichloroethyl group would obviously not survive the treatment with zinc.

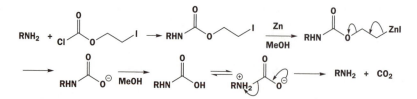

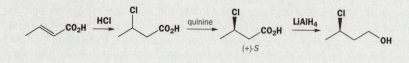

Problem 12

Revision of Chapters 10 and 16. Give mechanisms for this synthesis and suggest why this route was followed.

Purpose of the problem

Revision of Chapters 10 and 16 to show that stereochemical as well as mechanistic considerations matter in synthesis.

Suggested solution

The carboxylic acid group is there both to ensure that conjugate addition is possible and to provide a point of attachment for a resolving agent. Quinine is a naturally occurring base and forms a salt with the chloro-acid that can be separated into diastereoisomers. Acidification of the salt releases the enantiomerically pure acid.

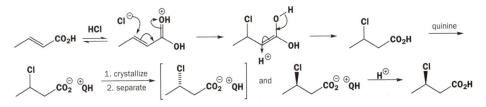

The reduction involves addition to the carbonyl group and loss of the OH group, probably as an aluminium compound, to give the aldehyde, which is further reduced by simple addition to the carbonyl group.

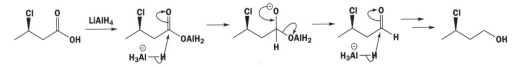

Problem 13

Revision of Chapters 9 and 19. Draw the structures of the intermediates in this synthesis of a diene and comment on the selectivity of the last step.

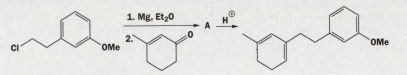

Purpose of the problem

Revision of Chapters 9 and 19 to show that regioselectivity as well as mechanistic considerations matter in synthesis.

Suggested solution

The formation of a Grignard reagent and its direct addition to the enone are expected. The thermodynamically controlled elimination (E1 via the stable tertiary allylic carbocation) gives the more highly substituted conjugated alkene inside the six-membered ring.

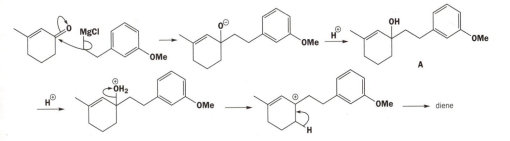

Problem 14

Suggest ways to make these compounds.

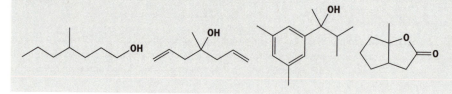

Purpose of the problem

Straightforward exam-style synthesis question.

Suggested solution

One possible synthesis of each compound follows: there are many more correct answers.

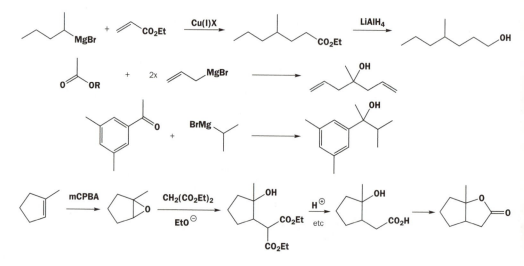

Suggested solutions for Chapter 26

26

■ Several problems in this chapter ask you to suggest ways to carry out conversions of one molecule into another or just to make a molecule. We give one possible answer and sometimes comment on alternatives but you should realize that there are many possible right answers to this sort of question. Make sure you have understood the principles behind each question and, if your answer is different from ours, check it with someone who knows.

Problem 1

Suggest how the following compounds might be made by the alkylation of an enol or enolate.

Purpose of the problem

An exercise in choosing good routes to simple compounds.

Suggested solution

You can see the enolates inside these molecules so it is obvious which electrophiles to use. In both cases they are symmetrical allylic halides. These are good electrophiles for an S$_N$2 reaction. We need to use the electrophile twice in the first case and to use an enol or enolate equivalent that will not self-condense in the second. You might have chosen a 1,3-dicarbonyl compound, a silyl enol ether, or an enamine.

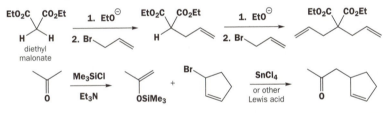

Problem 2

And how might these compounds be made using alkylation of an enol or enolate as one step in the synthesis?

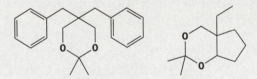

Purpose of the problem

An exercise in using enolate chemistry to make carbonyl compounds disguised as acetals.

Suggested solution

The only functional group in both molecules is an acetal. Cyclic acetals are made from diols and carbonyl compounds so we need to have a look at the 'deprotected' molecules before taking any further decision.

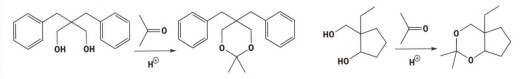

In both cases the carbonyl compound is acetone, but we have to make the diol. Each diol has a 1,3-relationship between the two OH groups so we could make the diols by reduction of 1,3-dicarbonyl compounds and use stable enolates in any alkylation steps. We can abbreviate the benzene rings for clarity.

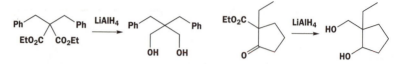

Now the alkylation steps are clear: we need to add two benzyl groups to diethyl malonate in the first case and an ethyl group to the available five-ring keto-ester in the second. In both cases using ethoxide as base avoids substitution problems.

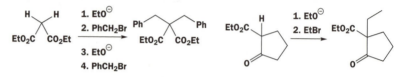

Problem 3

And further, how might these amines be synthesized using alkylation reactions of the enolate style as part of the synthesis?

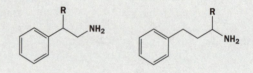

Purpose of the problem

An exercise in using enolate chemistry to make carbonyl compounds disguised as amines.

Suggested solution

The first amine could be made by reduction of a nitrile and that could be made by alkylation of benzyl cyanide (PhCH$_2$CN).

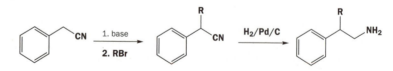

The second amine can be made by reductive amination of a ketone so we need to think how the ketone might be made by enolate alkylation. It is ideal for alkylation of an enol or enolate with a benzyl electrophile. You could have chosen a number of specific enol equivalents for this, we use the enamine.

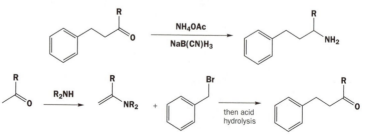

Problem 4

This attempted enolate alkylation does not give the required product. What goes wrong? What products would be expected from the reaction?

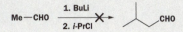

Purpose of the problem

An exercise in troubleshooting – we have had few of these so far but they are important in helping understanding.

Suggested solution

The idea is obviously to make the lithium enolate and alkylate it with *i*-PrCl but BuLi will attack the aldehyde as a nucleophile. Even if it did make some of the enolate, that would self-condense rather than react with the unreactive branched electrophile with a poor leaving group. The aza-enolate or silyl enol ether would be a better bet.

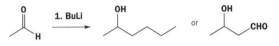

Problem 5

Draw mechanisms for the formation of this enamine, its reaction with the alkyl halide shown, and the hydrolysis of the product.

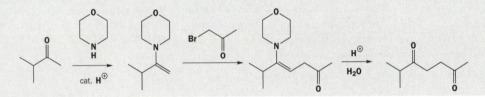

Purpose of the problem

An exploration of the details of enamine formation and alkylation. These are often misunderstood.

Suggested solution

The enamine is formed by nucleophilic attack on the ketone and elimination of water. The less substituted enamine is favoured in enamines formed from cyclic amines because the planarity of alkene and amine required for enamine conjugation would lead to a steric clash between the alkyl groups and the ring in the other enamine (margin).

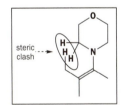

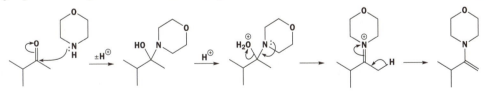

The reaction of the enamine with the alkyl halide is as expected – these very good S$_N$2 electrophiles work particularly well with enamines but the true product under the reaction conditions is a new enamine.

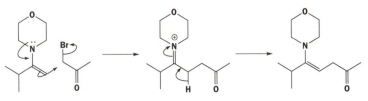

Finally, the enamine is hydrolysed by reprotonation to the same iminium ion and then attack of water. These steps are the exact opposite of what happens in enamine formation.

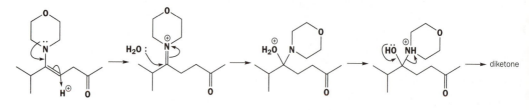

Problem 6

How would you produce specific enol equivalents at the points marked with the arrows (not necessarily starting from the simple carbonyl compound shown)?

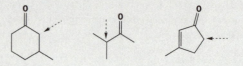

Purpose of the problem

First steps in making enol(ate)s with regiochemical control.

Suggested solution

Ketones 2 and 3 have two different α positions so there is a good chance that we can control enolate formation directly from the ketone. Ketone 1 is different and enolate formation can be controlled only by conjugate addition. The second enolate is thermodynamic and the third kinetic – possible methods are shown.

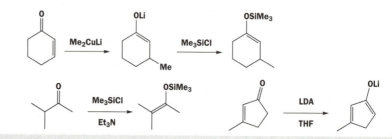

Problem 7

How would the reagents you have suggested in Problem 6 react with: (a) Br_2; (b) a primary alkyl halide RCH_2Br?

Purpose of the problem

Moving on from enol(ate) formation to reactions.

Suggested solution

Two are silyl enol ethers and will react well with bromine to put a bromine atom in the enol position. The third is a lithium enolate and will certainly react though probably rather vigorously. Preliminary conversion to a silyl enol ether would solve that problem.

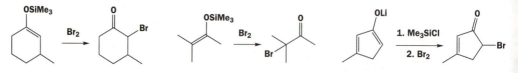

In the reaction with the primary alkyl halide RCH₂Br the boot is on the other foot as there will be a good reaction with the lithium enolate but primary cations are unstable so, unless R is aryl, reaction with RCH₂Br and a Lewis acid will not work well. Preliminary conversion to a lithium enolate with MeLi or a 'naked' enolate with fluoride will do the job.

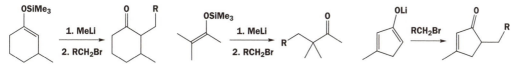

Problem 8

Draw a mechanism for the formation of the imine from cyclohexylamine and the following aldehyde.

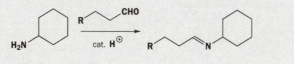

Purposed of the problem

Revision of the standard, but often forgotten, mechanism for imine formation from Chapter 14.

Suggested solution

Simple nucleophilic attack followed by acid-catalysed dehydration of the tetrahedral intermediate.

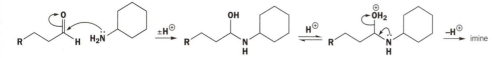

Problem 9

How would the imine from Problem 8 react with LDA followed by n-BuBr? Draw mechanisms for each step: reaction with LDA, reaction of the product with n-BuBr, and the work-up.

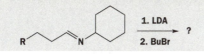

Purpose of the problem

Checking you know how to make and use an aza-enolate.

Suggested solution

LDA makes the aza-enolate, which is alkylated in the α position. After an aqueous acidic work-up, the product is the aldehyde. This is one of the best ways to alkylate aldehydes cleanly.

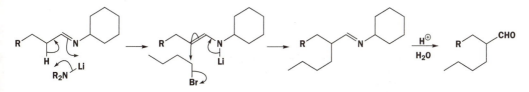

Problem 10

What would happen if this short cut for the reaction in Problems 8 and 9 were tried?

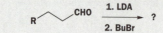

Purpose of the problem

Reminder of the problems with lithium enolates of aldehydes.

Suggested solution

Some aldehydes can be converted into their lithium enolates but it is not generally very successful because the rate of reaction of the lithium enolate with the very electrophilic aldehyde is too great. At least some aldol reaction will occur.

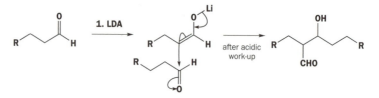

Problem 11

Suggest mechanisms for these reactions.

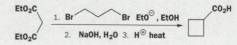

Purpose of the problem

A more complicated-looking sequence which is quite easy when worked out.

Suggested solution

Double alkylation of the malonate enolate to give a four-membered ring is followed by ester hydrolysis and decarboxylation.

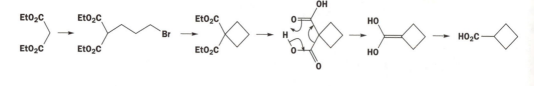

Problem 12

How does this method of making cyclopropyl ketones work? Give mechanisms for all the reactions.

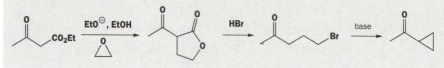

Purpose of the problem

Enols and enolates used in an unlikely looking sequence, which you can work out if you persist.

Suggested solution

Alkylation of the enolate with the epoxide gives an alkoxide anion that cyclizes to the lactone. S_N2 opening of the lactone ring by the soft nucleophile and decarboxylation gives the γ-bromoketone, which cyclizes through its enolate. Three-membered rings are favoured kinetically.

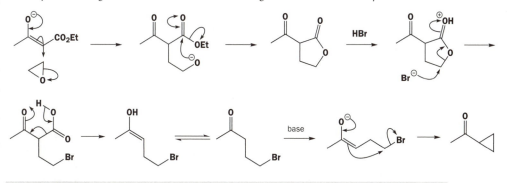

Problem 13

Give the structures of the intermediates in the following reaction sequence and mechanisms for the reactions. Comment on the formation of this particular product.

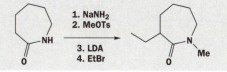

Purpose of the problem

A reminder that enolate-like intermediates can be formed at nitrogen as well as at carbon.

Suggested solution

The first molecule of base removes the NH proton to make an enolate-like intermediate that reacts at nitrogen. Now that the NH is blocked, the second molecule of base makes the amide enolate, which is alkylated at carbon.

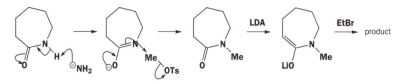

Problem 14

Suggest how the following products might be made using enol or enolate alkylation as at least one step. Explain your choice of specific enol equivalents.

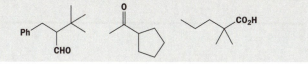

Purpose of the problem

A simple exercise to revise the material of this chapter in the context of synthesis.

Suggested solution

The *t*-butyl group can be added only through the silyl enol ether while the benzyl group can be added in almost any way. The second molecule must be made by a cyclization. The third molecule requires the creation of a quaternary centre which limits the option. Here is one possible solution.

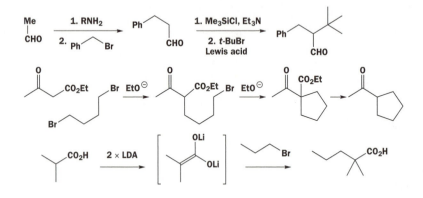

Suggested solutions for Chapter 27   27

Problem 1

Propose mechanisms for the 'aldol' and dehydration steps in the termite defence compound synthesis presented in the chapter.

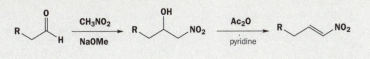

■ Several problems in this chapter ask you to suggest ways to make molecules by aldol reactions. We give one possible answer and sometimes comment on alternatives but you should realize that there are many possible right answers to this sort of question. Make sure you have understood the principles behind each question and, if your answer is different from ours, check it with someone who knows.

Purpose of the problem

Revision of elimination reactions (Chapter 19) and the mechanism for an 'aldol reaction that can't go wrong'.

Suggested solution

One nitro group is worth two carbonyl groups so it will be the nitromethane that forms an 'enolate' anion. The aldehyde is more electrophilic than the nitro compound. Elimination is by the E1cB mechanism after a second 'enolization'.

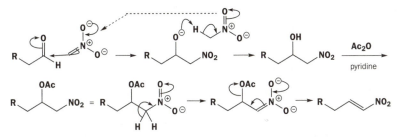

Problem 2

The aldehyde and ketone below are self-condensed with aqueous NaOH so that an unsaturated carbonyl compound is the product. Give a structure for each product and explain why you think this product is formed.

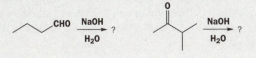

Purpose of the problem

Drawing mechanisms for the simplest aldols – self-condensations of aldehydes and ketones.

Suggested solution

Only one compound can form an enolate and only one compound (the same) can be the electrophile. The ketone can lose a proton from either side but it can produce an enone only if the enolate is formed into the methyl group. The elimination is again by the E1cB mechanism.

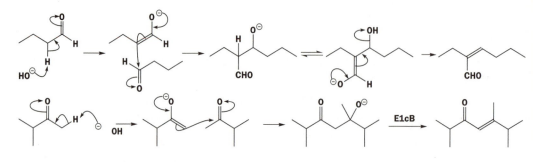

Problem 3

How would you synthesize the following compounds?

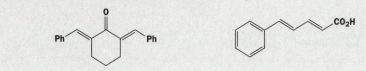

Purpose of the problem

Extending the last problem so that you look at the aldol reaction as a way to make unsaturated carbonyl compounds.

Suggested solution

Just find the conjugated alkene and 'see' the enolate there. In the first case, cyclohexanone provides two enolates to react with two molecules of unenolizable benzaldehyde.

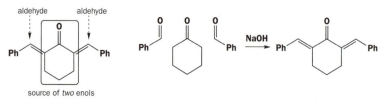

In the second case, two aldol-style reactions must be used, one after the other. The enolate required for the first aldol is an aldehyde but, for the second, an acid derivative must be used and the best is probably malonic acid so that no hydrolysis is needed and decarboxylation occurs under the conditions of the reaction.

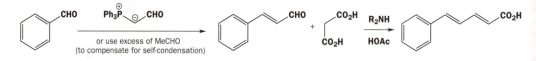

Problem 4

How would you use a silyl enol ether to make this aldol product? Why is it necessary to use this particular intermediate? What would the products be if the two carbonyl compounds were simply mixed and treated with base?

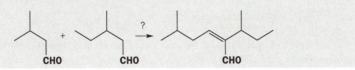

Purpose of the problem

Extending the last problem so that you are forced to use silyl enol ethers and appreciate their effectiveness.

Suggested solution

This is about the most difficult type of aldol reaction to do: two slightly different aldehydes, both enolizable, both electrophilic, both capable of self-condensation. The only solution is to couple the silyl enol ether of one with the other aldehyde under the influence of a Lewis acid. This gives the aldol product which can be dehydrated to the enal.

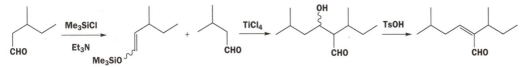

Without this control, each aldehyde would self-condense and each would react with the other. These would be the products.

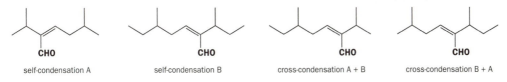

| self-condensation A | self-condensation B | cross-condensation A + B | cross-condensation B + A |

Problem 5

In what way does this reaction resemble an aldol reaction? How could the same product be made without using phosphorus chemistry? Comment on the choice of base.

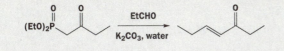

Purpose of the problem

Showing that there are reactions closely related to the aldol that produce similar products.

Suggested solution

The phosphonate ester acts rather like the extra CO_2Et group in malonate to stabilize the enolate. The addition of the enolate to EtCHO is very like the first step of an aldol reaction and the second step is the loss of phosphate instead of water. The very weak base (pK_a about 10) shows how easy it is to make the enolate.

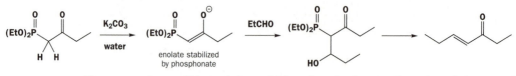

The same product could be made by an aldol condensation between the enolate of a ketone and the same aldehyde (EtCHO) but some other method would have to be used to ensure enolization of the ketone on the right side and to prevent self-condensation of the aldehyde. A silyl enol ether produced via the kinetic enolate would do the trick.

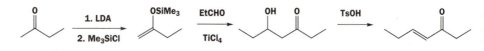

Problem 6

Suggest a mechanism for this attempted aldol reaction. How could the aldol product actually be made?

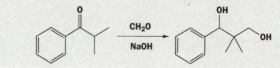

Purpose of the problem

A demonstration of one reason why aldol reactions with formaldehyde (methanal) can fail.

Suggested solution

The aldol reaction appears to have taken place but there has been reduction of the ketone in the product. The only possible reducing agent is more formaldehyde and reduction is by the Cannizzaro reaction (p. 1081). The aldol can be successful if a weaker base (Na$_2$CO$_3$ will do) is used as the Cannizzaro reaction requires a dianion.

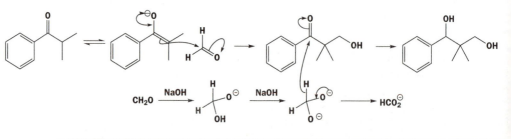

Problem 7

What are the structures of the intermediates and the mechanisms of the reactions leading to this simple cyclohexenone?

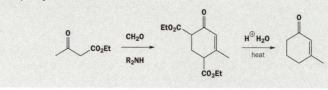

Purpose of the problem

Another demonstration of the problems with formaldehyde but this time they are put to good use.

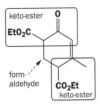

Suggested solution

Obviously, two molecules of the keto-ester have combined with one molecule of formaldehyde to give the first product (margin). Initially, a Mannich product must be formed, which loses Me_2NH by an E1cB elimination, and conjugate addition of the second keto-ester follows. Finally, an intramolecular aldol gives the six-membered ring.

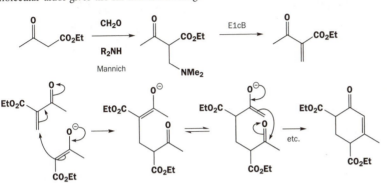

The removal of the two ester groups is by standard ester hydrolysis, decarboxylation of the β-keto-acid by the usual mechanism, and decarboxylation of the γ-keto-acid by an extended version of the same thing.

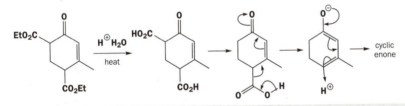

Problem 8

How would you convert the product of that last reaction into these two products?

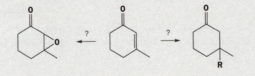

Purpose of the problem

An extension of aldol chemistry to epoxide synthesis and conjugate addition.

Suggested solution

The epoxidation of alkenes is normally carried out with electrophilic peroxy-acids such as *m*-CPBA, but this alkene is itself electrophilic and a better reagent is the nucleophilic anion from hydrogen peroxide itself. Conjugate addition is followed by cleavage of the weak O–O bond to give the epoxide.

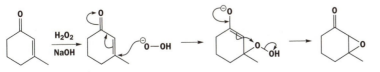

The conjugate addition of 'R' is best accomplished with an organocopper reagent to ensur
conjugate rather than direct addition. You might choose RMgBr with catalytic Cu(I) or R_2CuLi a
the reagent. These methods are described in Chapter 23.

Problem 9

Comment on the selectivity shown in these two cyclizations.

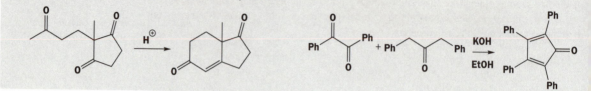

Purpose of the problem

An exploration of selectivity in the context of intramolecular aldol chemistry.

Suggested solution

The first example is the completion of the Robinson annelation (p. 761). There are four possible
sites of enolization (1–4) though two are the same (3 and 4). Enolization at C2 can lead only to am
unstable four-membered ring and is reversible.

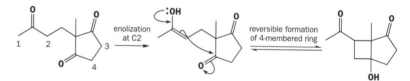

Enolization at C3 or C4 leads to a six-membered ring, but it is bridged and cannot dehydrate to
give a stable enone. This too is reversible.

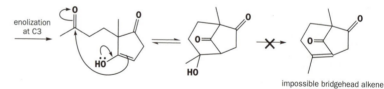

Finally, enolization at C1 and cyclization to either of the ketones in the ring gives a stable six-
membered ring that can dehydrate to give an even more stable conjugated enone and this is the
product.

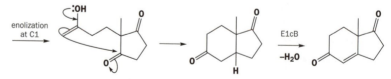

The second case must start with an intermolecular aldol reaction. Only one ketone can enolize
and the 1,2-diketone is more electrophilic because each carbonyl group makes the other more
electrophilic. The first reaction is unambiguous.

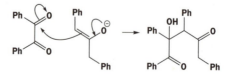

Three more steps are needed – another aldol reaction and two dehydrations. The second aldol reaction will be faster than the first as it is intramolecular and makes a stable five-membered ring. The eliminations are by the E1cB mechanism (Chapter 19) also via an enolate.

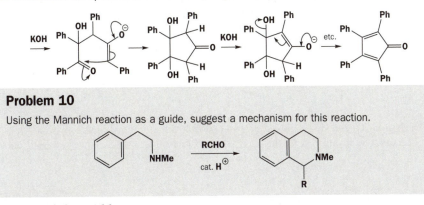

Problem 10

Using the Mannich reaction as a guide, suggest a mechanism for this reaction.

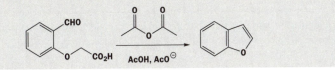

Purpose of the problem

An extension of Mannich chemistry into the synthesis of a new ring system. It is important that you appreciate the relationship between reactions when they have similar mechanisms even if they look rather different.

Suggested solution

The aldehyde and the amine combine in the usual way to form an imine. As the amine is secondary, this will actually be an iminium salt.

The electrophilic iminium salt is perfectly positioned for an intramolecular aromatic substitution

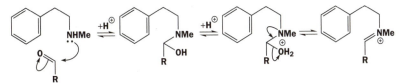

in the style of a Friedel–Crafts reaction (Chapter 22) except that the electrophile is a Mannich-style iminium salt rather than an acylium ion. The position of substitution has to be *ortho* as the side chain can reach no further.

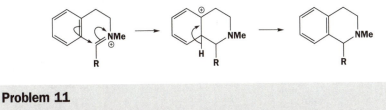

Problem 11

Suggest mechanisms for this reaction. One of the by-products is carbon dioxide.

Purpose of the problem

Further exploration of the aldol reaction: one with surprising consequences.

Suggested solution

The structure of the product shows that no new atoms are added to the starting material: indeed, CO_2 and H_2O are lost. An intramolecular aldol reaction between the enol of the acid (or perhaps the enol of an anhydride formed between the acid and acetic anhydride, see box) and the aromatic aldehyde must start things rolling.

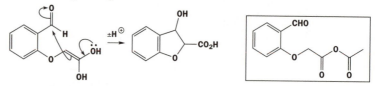

The loss of CO_2 and H_2O may be concerted or else dehydration may be followed by loss of CO_2. In either case one driving force for the reaction is the formation of an all-conjugated aromatic system.

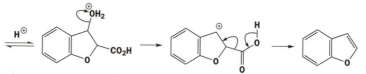

Problem 12

Treatment of this keto-aldehyde with KOH gives a compound $C_7H_{10}O$ with the spectroscopic data shown. What is its structure and how is it formed? You should, of course, assign the NMR spectrum and give a mechanism for the reaction.

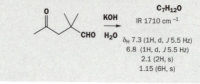

Purpose of the problem

Revision of compound identification, especially NMR, in the context of the aldol reaction.

Suggested solution

The IR suggests a simple ketone, or possibly an aldehyde, but there is no aldehyde proton at about 9–10 p.p.m. so it must be a ketone. The NMR shows an alkene with large chemical shifts (7.3 and 8.8) suggesting that it is conjugated with the ketone. The coupling constant (5.5 Hz) is very small – far too small for a *trans* alkene and small even for a *cis* alkene so it must be in a ring. The CMe_2 group is still there and the 2H singlet at 2.1 must be the CH_2 next to the ketone. We can draw only one structure.

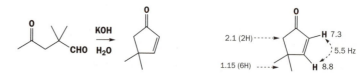

For once, you could have worked out the structure from the mechanism, though this is not generally a good idea. Only the ketone can enolize and it must react with the aldehyde in an intramolecular aldol reaction followed by dehydration.

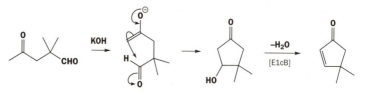

Problem 13

Predict which enone product would be formed in this intramolecular aldol reaction.

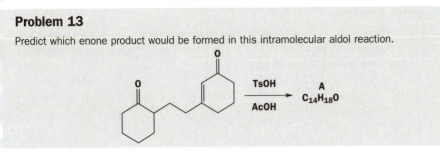

Purpose of the problem

An exercise in considering all the enols from two carbonyl groups and making a reasoned choice between the possibilities.

Suggested solution

The various enols are given below. The first three are straightforward, but the last may have surprised you: the proton removed to make this extended enol came from position C4 in the enone ring.

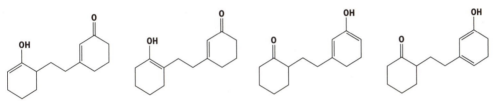

Each enol can cyclize, at least in theory, but we must not be left with a *trans* alkene in a six-membered ring. The most promising cyclization is perhaps from the last enol.

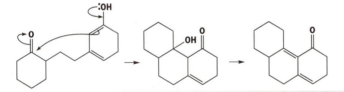

Problem 14

The unstable liquid diketone 'biacetyl' deposits crystals of a dimer slowly on standing or more quickly with traces of base. On longer standing the solution deposits crystals of a trimer. Suggest mechanisms for the formation of the dimer and the trimer. Why are they more stable than the monomer?

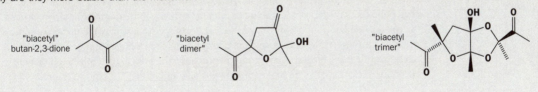

Purpose of the problem

Convincing yourself that the aldol reaction can be applied quite simply to complicated looking molecules.

Suggested solution

'Biacetyl' (butan-2,3-dione) is unstable because the two carbonyl groups are both electron-withdrawing and destabilize each other. One way in which they can escape from this situation is to form an enol from one of the carbonyl groups.

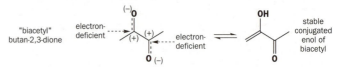

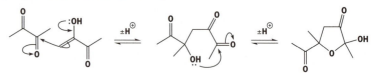

■ Biacetyl dimer and trimer were so named when they were first discovered and their detailed structures were unknown.

This enol can, of course, react with another molecule of biacetyl in a simple aldol reaction. The immediate product once again has the unstable 1,2-diketone structure but it can escape in a different way: it can form a hemiacetal and this is 'biacetyl dimer'. You might have discovered the structure of the intermediate in a different way if you had noticed that biacetyl dimer is a hemiacetal and unravelled it to discover the keto-alcohol.

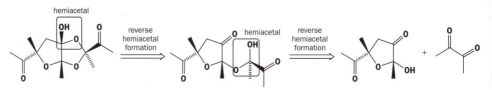

The trimer is also a hemiacetal and, if we take that functional group apart, we discover another hemiacetal as intermediate. Taking that apart we find we have split off a molecule of biacetyl and the compound we are left with is the dimer with some stereochemistry drawn in.

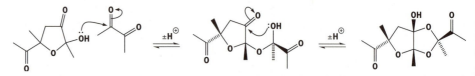

Evidently, the dimer can react with another molecule of biacetyl and the product can cyclize again to form a new hemiacetal containing none of the unstable 1,2-diketone functional groups. We can start the mechanism with the dimer as we have already made that.

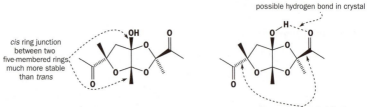

We have not previously considered stereochemistry. Hemiacetal (and acetal) formation is under thermodynamic control as all the reactions are reversible. The dimer and trimer crystallize from the liquid so the stereochemistry may be governed by the formation of the most stable possible compound or by the fact that the crystallization of the least soluble diastereoisomer removes it from the equilibrium and so more is formed. We can see some reasons why the diastereoisomer shown might be the most stable. The *cis* ring junction between the two five-membered rings is much more stable than the *trans*, the two acetyl groups may prefer to be *trans* to each other, and there may be an H bond in the crystal. We cannot be sure of these reasons but they are explored more in Chapter 33.

possible hydrogen bond in crystal

cis ring junction
between two
five-membered rings
much more stable
than *trans*

trans arrangement of the two
acetyl groups may be more stable

28

Problem 1

Attempted acylation at carbon often fails. What would be the true products of these attempted acylations, and how would you actually make the target molecules?

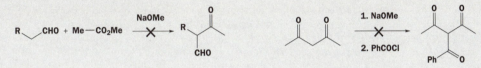

Purpose of the problem

Revision of simple enolate reactions (Chapter 21) and aldol reactions (Chapter 27). Clear thinking about what happens when you put carbonyl compounds in basic solution is essential.

Suggested solution

In the first case we want the aldehyde to form an enolate and we want the enolate to attack the ester. The first part is all right; the aldehyde will form an enolate more readily than the ester. But under these equilibrating conditions, only a small amount of enolate will be formed and it will react faster with the aldehyde than with the less electrophilic ester. The aldehyde will self-condense in an aldol reaction.

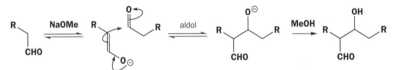

To make the product required we shall need to convert the aldehyde into a specific enol equivalent. There are various alternatives, of which the best are either an enamine or a silyl enol ether. Esters do not acylate either of these and an acyl chloride should be used. Don't forget the Lewis acid if you use a silyl enol ether.

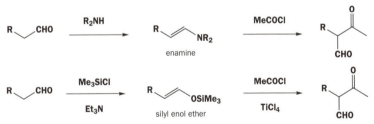

The enolate formation in the second example is a separate step and will work well because the two carbonyl groups combine to form a stable enolate and NaOMe is quite strong enough to convert the diketone entirely into the enolate. The problem is the acylation step. With a sodium enolate and a reactive acylating agent, a charge-controlled (hard/hard interaction) reaction will occur at the oxygen atom to give an enol ether.

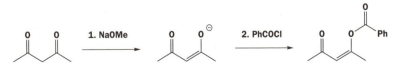

■ Several problems in this chapter ask you to suggest ways to make molecules by Claisen ester condensations and related reactions. We give one possible answer and sometimes comment on alternatives, but you should realize that there are many possible right answers to this sort of question. Make sure you have understood the principles behind each question and, if your answer is different from ours, check it with someone who knows.

The escape route from this difficulty suggested in the chapter was to use a magnesium enolate. Magnesium is chelated by the two oxygen atoms of the stable enolate and blocks reaction at either. *C*-Acylation occurs even with acyl chlorides.

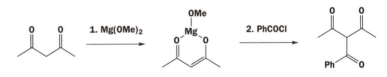

Problem 2

The synthesis of six-membered heterocyclic ketones by intramolecular Claisen condensation was described in the chapter and we pointed out that it doesn't matter which way round the cyclization happens as the product is the same. For example:

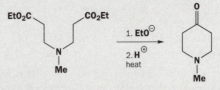

Strangely enough, five-membered heterocyclic ketones can be made by a similar process. The starting material is not symmetrical and two possible cyclized products can be formed. Draw structures for these two products and explain why it is unimportant which is formed.

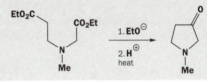

↳ other reactant exist v hat would happen ?

Purpose of the problem

To make sure you understand how extra ester groups can solve apparently complex acylation problems.

Suggested solution

The cyclization can occur in two ways to give two different products as either ester can form an enolate that attacks the other ester in an intramolecular acylation. We should draw the two products.

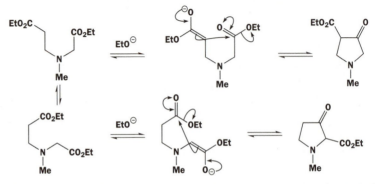

Though these compounds are different, each gives the same ketone after hydrolysis and decarboxylation of the ester group as the ketone carbonyl is on the same position in the ring in both compounds.

Problem 3

The synthesis of corylone was outlined in the chapter but no mechanistic details were given. Suggest mechanisms for the first two steps. The last step is a very unusual type of reaction and you have not met anything quite like it before. However, organic chemists should be able to draw mechanisms for new reactions and you might like to try your hand at this one. There are several steps.

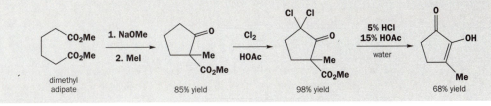

Purpose of the problem

General revision of mechanisms and a check that you realize what is the exact product of the Claisen ester condensation under the reaction conditions.

Suggested solution

The first step is an intramolecular condensation between two identical esters (sometimes called the Dieckmann condensation). Under the reaction conditions the product is the stable enolate of the keto-ester. The base used is methoxide and this is strong enough (pK_{aH} about 16) to form the enolate of a β-keto-ester (pK_{aH} about 12). This enolate is alkylated in this sequence *before* the usual aqueous acid work-up.

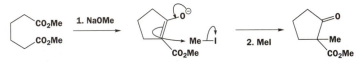

The chlorination is a straightforward chlorination of an enol (note enol not enolate as the catalyst is an acid) as described in Chapter 21. The only unusual thing is that enough chlorine is used to get a second chlorination. As explained in Chapter 21, this is slower than the first but is no trouble here as all other enolizing positions are blocked.

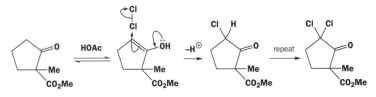

You were warned that the last step would be weird but at least the hydrolysis and decarboxylation are straightforward. We hope you remembered to use the cyclic mechanism for the decarboxylation step.

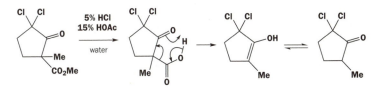

The product will mainly exist in the more stable keto-form but the unusual hydrolysis of the dichloride to the ketone might be easier in the enol form. Loss of a chloride in an S_N1 reaction (Chapter 17) gives a stable allylic cation. The second chloride is lost more easily because the OH group pushes it out. The final product is stable as the mono-enol with the most substituted double bond (thermodynamic control).

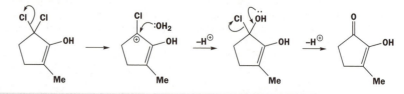

Problem 4

Acylation of the phenolic ketone gives a compound A, which is converted into an isomeric compound B in base. Cyclization of B in acid gives the product shown. Suggest mechanisms for the reactions and structures for A and B.

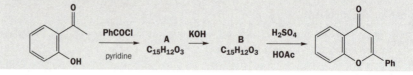

Purpose of the problem

Predicting products of acylation reactions. This is always more difficult than just drawing mechanisms but here you can work backwards from the final product as well as forwards.

Suggested solution

■ The true product of the acylation is the anion of the phenol: this helps the reaction since KOH is more basic than ArOH (Chapter 8).

The starting material is $C_8H_8O_2$ so A has an extra C_7H_4O. This looks like the addition of PhCOCl and the loss of HCl. The most obvious reaction is acylation at the phenolic oxygen rather than enolate formation as OH is more acidic than CH. This is a very acidic phenol as the carbonyl group helps to stabilize the oxyanion. Compound A is simply the benzoate ester of the phenol. Treatment with KOH isomerizes A to B and this is the heart of the problem. An intramolecular acylation of the only possible enolate can be catalysed by KOH even though it produces only a little enolate since cyclization to a six-membered ring is so easy.

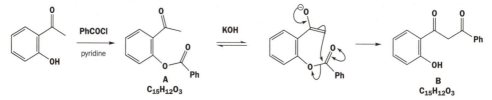

The final step is acid-catalysed and clearly involves the attack of the phenolic OH on one of the ketones. This intramolecular reaction prefers to form a six-membered ring rather than a strained four-membered one, and dehydration of the adduct gives an aromatic ring – two electrons each from the double bonds and two from a lone pair on the oxygen atom making six in all – the delocalization shown should help if you don't see this.

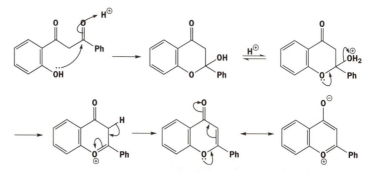

Problem 5

How could these compounds be made using the acylation of an enol or enolate as a key step?

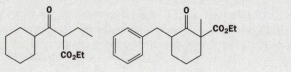

Purpose of the problem

Practice in using acylation at carbon to make compounds.

Suggested solution

The first problem has two possible solutions by direct acylation, labelled A and B in the diagrams. A would have to be controlled as the straight-chain ester could self-condense, but B needs no control as only the ketone can enolize, diethyl carbonate, $CO(OEt)_2$, is more electrophilic than a ketone, and only the wanted product can enolize again and form a stable enolate under the reaction conditions. However, route B adds only one carbon atom in the key step.

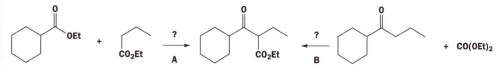

Route A can be realized with either a lithium enolate or a silyl enol ether, as explained in the chapter, using an acyl chloride as the electrophile.

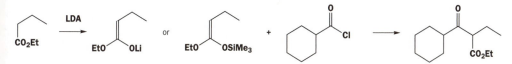

Route A requires the synthesis of the ketone starting material and this could be done by Grignard methods (see Chapter 9) or by acylation of a copper compound with an acyl chloride. The final step, acylation of the enolate with diethyl carbonate, requires no special control.

Problem 6

In a synthesis of cubane, a key step was the intramolecular acylation of this symmetrical diester. Explain why a strong base (the anion of DMSO, $MeSOCH_2^-$, was actually used) is necessary for this cyclization.

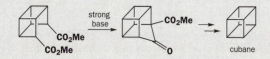

The starting material has both the ester groups on the outside of the molecule so that cyclization is impossible. What preliminary step must first occur for it to become possible?

Purpose of the problem

More serious thinking about the details of acylation reactions.

Suggested solution

The strong base is necessary in the cyclization because no stable enolate can be formed from the product. In other acylations of esters by esters the product has at least one hydrogen atom on the carbon atom between the two carbonyl groups and forms a stable enolate under the reaction conditions. There are several examples in the chapter and the answer to Problem 3 makes a special point of this. The strong base is needed to convert one of the esters completely into its enolate.

The stereochemical point is that one of the esters becomes an enolate and so loses its stereochemistry but the other must be pointing inwards for cyclization to occur. This can happen by reversible formation of the enolate anion.

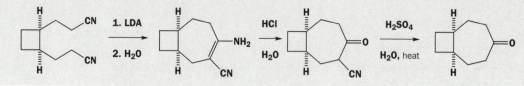

Problem 7

Suggest mechanisms for this sequence leading to a bicyclic compound with four- and seven-membered rings *cis*-fused to each other.

Purpose of the problem

Practice at acylation reactions with groups other than carbonyl.

Suggested solution

The acylation of RLi or RMgBr by nitriles (cyanides) is an effective way to make ketones (Chapter 9) and here we see the cyanide version of an intramolecular Claisen ester condensation. One cyanide makes an 'enolate', which attacks the other. The resulting imine tautomerizes to the conjugated enamino nitrile.

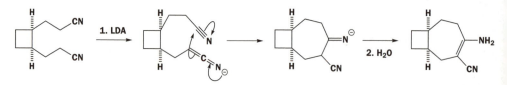

The enamine is hydrolysed to the ketone and then the nitrile is hydrolysed under strongly acidic conditions to the β-keto-acid, which decarboxylates to give the final product.

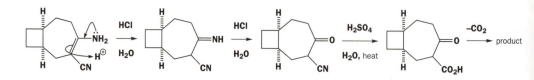

Problem 8

Give mechanisms for the steps used in this synthesis of the natural product bullatenone. Comment on the reagents used for the acylation step, on the existence of the first intermediate as 100% enol, on the mechanism of the cyclization, and on how the decarboxylation is possible.

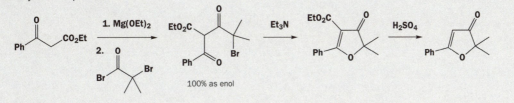

Purpose of the problem

Acylation and related reactions in the synthesis of a natural product.

Suggested solution

The first reaction is the acylation of a magnesium enolate of the kind we have seen before. The magnesium enolate is used to block the oxygen atoms of the delocalized enolate by chelation and to ensure *C*-acylation. The unusual acyl *bromide* is used as it is easy to prepare by direct bromination of the acid. The product has an exceptionally stable enol with conjugation covering three carbonyl groups (two ketones and the ester).

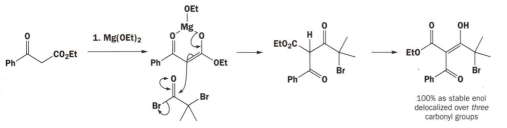

Cyclization occurs by intramolecular S_N2 reaction, probably using the enolate as nucleophile as even an amine can make this very stable enolate. Normally, tertiary halides do not react by the S_N2 reaction but this is intramolecular and next to a carbonyl group, which helps the S_N2 reaction but hinders the S_N1 reaction (Chapter 17).

■ It is possible that this cyclization is better described as a pericyclic reaction (Chapters 35 and 36).

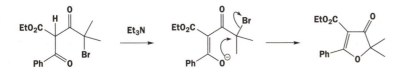

The hydrolysis of the ester happens by the usual mechanism (Chapter 12), but the decarboxylation is tricky. We need the ketone carbonyl group for our cyclic mechanism but we can't use it while the double bond is there. The solution is to protonate the enol ether first.

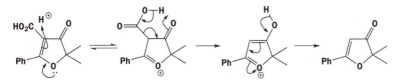

Problem 9

Suggest how the following reactions might be made to work. You will probably have to select a specific enol equivalent.

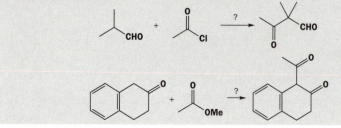

Purpose of the problem

Making reactions work is an important part of organic chemistry.

Suggested solution

The first reaction is a standard acylation of an aldehyde creating a quaternary centre. You might have used a silyl enol ether but an enamine, such as the one made from a cyclic secondary amine, is probably better.

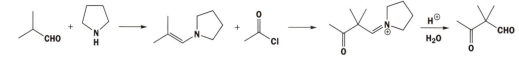

The second example might just go with simple base (MeO^-) catalysis as the conjugated ketone enolate is much more stable than that of the ester. However, it's probably safer to use a lithium enolate or a silyl enol ether (though then you'd have to use an acyl chloride as the electrophile).

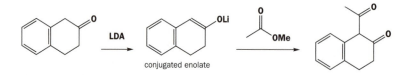

conjugated enolate

Problem 10

Suggest mechanisms for these reactions, explaining why these particular products are formed.

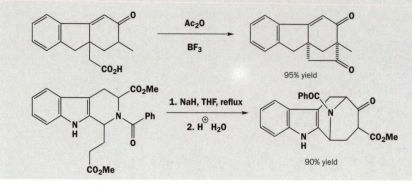

Purpose of the problem

Practice at drawing mechanisms for slightly unusual acylations on much more interesting molecules. The mechanisms are the same, however complicated the setting.

Suggested solution

The acid is converted into the mixed anhydride with acetic anhydride and the Lewis acid. The anhydride could enolize but no good intramolecular reaction can then happen. If the ketone enolizes on the only side possible, a stable five-membered ring can be formed. The stereochemistry is inevitable as the two-atom bridge has to be *cis* on the six-membered ring (and diaxial; see Chapter 18).

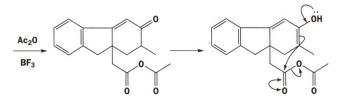

The second example features an alkaloid – a type of natural product to be discussed in Chapter 51. Sodium hydride is a strong base and could form an enolate from either ester. The product clearly results from enolate formation from the ester on the side chain and an intramolecular acylation with the other ester. As usual, this is because a stable enolate can be formed from the product under the reaction conditions.

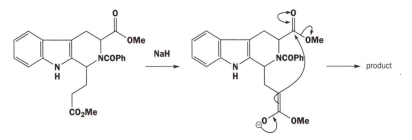

Problem 11

Sodium enolates generally react with acid chlorides to give enol esters. Give a mechanism for this reaction and explain the selectivity.

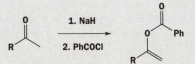

If the enol ester is treated with an excess of the sodium enolate, *C*-acylation occurs. Give a mechanism for this reaction. Why does the *C*-acylated product predominate?

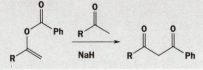

Purpose of the problem

Analysis of a mechanistic problem concerning acylation. Revision of kinetic and thermodynamic control.

Suggested solution

The first part is what you should expect. A simple enolate anion has the largest charge density on the oxygen atom and acyl chlorides are attracted by negative charge as the carbon atom of the carbonyl group has a substantial charge itself. The reaction is a charge-controlled combination of a hard nucleophile and a hard electrophile.

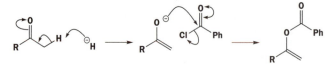

The second part is the interesting bit. Now the enol ester itself acts as the acylating agent and an equilibrium is set up between the *O*- and *C*-acylated products. The *C*-acylated product forms a stable enolate between the two carbonyl groups and this tips the equilibrium in its favour.

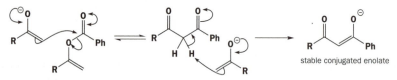

Problem 12

This is a *C*-acylation route to a simple ketone. Why was NaH chosen as the base? Why did *O*-acylation not occur? Why were *t*-butyl esters used? What would have probably happened if the more obvious Friedel–Crafts (Chapter 22) route had been tried instead?

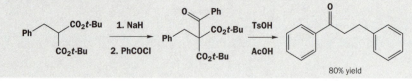

Purpose of the problem

Analysis of a mechanistic problem concerning acylation. Revision of kinetic and thermodynamic control.

Suggested solution

The expected result from acylation of a stable sodium enolate would be *O*-acylation as in the previous problem. However, the two *t*-butyl esters are very large and in the plane of the enolate so they will exert considerable steric hindrance at oxygen. The *t*-butyl groups are much further from the carbon atom. We show one possible arrangement of the enolate below.

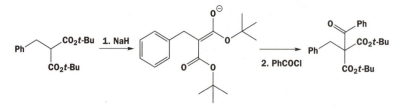

The *t*-butyl esters were also used to make ester 'hydrolysis' easier: the reaction with acid involves protonation of the carbonyl oxygen and loss of the *t*-butyl cation in an S_N1 reaction (Chapter 17) or, more likely under these conditions with no good nucleophile, an E1 reaction (Chapter 19).

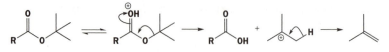

The Friedel–Crafts route would have required benzene to react with the acyl chloride below. But an intramolecular reaction to give a stable five-membered ring is much more likely.

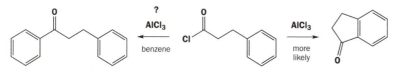

Problem 13

Base-catalysed reaction between these two esters allows the isolation of one product in 82% yield. Predict its structure.

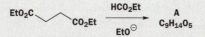

The NMR spectrum of the product shows that two species are present. Both show two 3H triplets at about $\delta_H = 1$ and two 2H quartets at about $\delta_H = 3$ p.p.m. One has a very low field proton and an ABX system at 2.1–2.9 with J_{AB} 16 Hz, J_{AX} 8 Hz, and J_{BX} 4 Hz. The other has a 2H singlet at 2.28 and two protons at 5.44 and 8.86 coupled with J 13 Hz. One of these protons exchanges with D_2O. Any attempt to separate the mixture (for example, by distillation or chromatography) gives the same mixture. Both compounds, or the mixture, on treatment with ethanol in acid solution give the same product. What are these compounds?

$$A \xrightarrow[\text{EtOH}]{H^{\oplus}} B$$
$$C_9H_{14}O_5 \qquad C_{13}H_{24}O_6$$

Compound B has IR 1740 cm^{-1}, δ_H (p.p.m.) 1.15–1.25 (four t, each 3H), 3.45 (2H, q), 3.62 (2H, q), 4.1 (two q, each 2H), 2.52 (2H, ABX system, J_{AB} 16 Hz), 3.04 (1H, X of ABX split into a further doublet by J 5 Hz), and 4.6 (1H, d, J 5 Hz). The couplings between A and X and between B and X are not quoted in the paper. Nevertheless, you should be able to work out a structure for compound B.

Purpose of the problem

Revision of enol structure by NMR and a further exploration of what happens to acylation products.

Suggested solution

Only the diester can form an enolate and ethyl formate (HCO_2Et – it is half an aldehyde) is much more electrophilic than the diester. We should expect the diester to be acylated by ethyl formate.

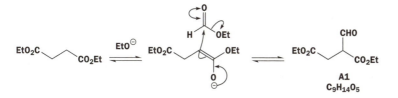

This compound (A1) fits the formula for A ($C_9H_{14}O_5$) and the NMR spectrum of the compound with the low field signal (CHO) and an ABX system also fits this structure. But what is the other compound? It obviously equilibrates easily with A2, which lacks both the aldehyde proton and the

ABX system, and it sounds like an enol. Compound A1 is chiral so the CH_2 appears as an ABX system but the enol is not chiral so the CH_2 is a singlet. Here are the structures with the NMR assignments (in each case the OEt groups provide the 3H triplets and 2H quartets).

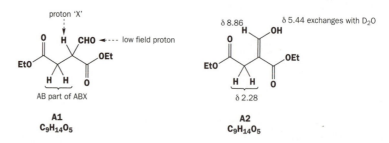

Treatment with acidic ethanol simply makes the diethyl acetal from the aldehyde group of A1. Since A1 and A2 are in equilibrium, all A2 is eventually converted into B via A1. Compound B is again chiral so the ABX system reappears with further coupling of H^x to the acetal proton. There are now four triplets and quartets for the four OEt groups.

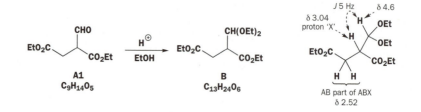

Suggested solutions for Chapter 29

Problem 1

Write full mechanisms for these reactions mentioned earlier in the chapter.

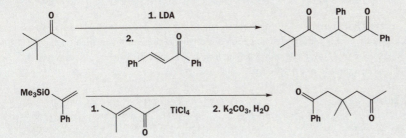

■ Several problems in this chapter ask you to suggest ways to make molecules by conjugate additions of enols or enolates. We give one possible answer and sometimes comment on alternatives, but you should realize that there are many possible right answers to this sort of question. Make sure you have understood the principles behind each question and, if your answer is different from ours, check it with someone who knows.

Purpose of the problem

An exercise in writing out the full mechanisms of two standard conjugate addition reactions.

Suggested solution

In the first example LDA makes the lithium enolate on the only side possible. Evidently the lithium enolate does a conjugate addition rather than the expected direct addition, possibly because of steric hindrance.

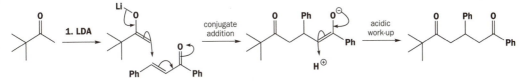

The second example is a typical Lewis acid-catalysed conjugate addition of a silyl enol ether. The exact details of the silyl transfer to the product need not bother you as long as you put the Lewis acid on the carbonyl group of the enone and used the silyl enol ether as the nucleophile. The silyl enol ether produced by the conjugate addition is hydrolysed to the product during the aqueous work-up.

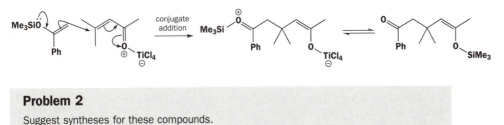

Problem 2

Suggest syntheses for these compounds.

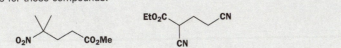

Purpose of the problem

An exercise in using standard conjugate addition reactions in making molecules.

Suggested solution

The nitro group and the tertiary centre next to it leave only one option for this synthesis. The nitro group is so anion-stabilizing that a tertiary amine is a strong enough base for this conjugate addition.

There are two alternatives for the next compound but conjugate addition is so good with acrylonitrile that we should prefer that route especially as the anion can then be stabilized by CN and CO_2Et. The reaction needs only catalytic base as the initially formed adduct removes a proton from the next molecule of cyanoacetate.

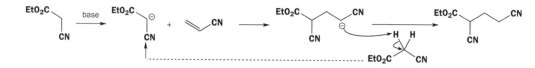

Problem 3

Suggest two different approaches to these compounds by conjugate addition of an enol(ate). Which do you prefer?

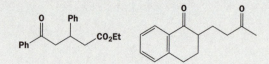

Purpose of the problem

Exercising judgement in using standard conjugate addition reactions in making molecules.

Suggested solution

There is little to choose between the two routes to the first compound. Both unsaturated compounds are easy to make and enol(ate) formation is easy in both cases. Perhaps the unsaturated ester will be more reliable at conjugate addition.

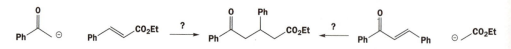

The second compound could also be done either way, but both starting materials for route A are available while the unsaturated ketone needed for route B will have to be made. In both cases, as an enolizable ketone is being added to an enolizable unsaturated ketone, specific enol(ate)s might be wise.

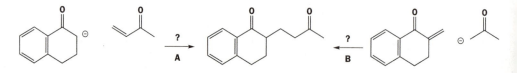

Problem 4

How could you use the Robinson annelation to make these compounds?

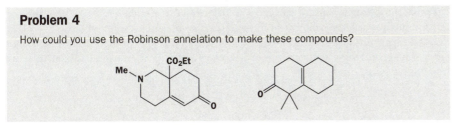

Purpose of the problem

Using a very important reaction to make cyclic compounds.

Suggested solution

In each case we need first to unravel the aldol cyclization reaction and then the conjugate addition. If we are lucky we shall find no problems in putting the molecule back together again. In the first case the starting material was discussed in Chapter 28 (p. 727) and the ester group makes the conjugate addition easy. The cyclization is unambiguous as no other stable cyclic enone can be formed.

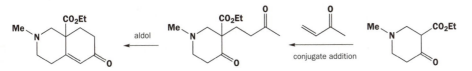

The second case looks very difficult because it does not look like the product of an aldol reaction. But we can imagine making it by dehydration of an aldol. The double bond may not be conjugated with the ketone but it is in a very stable place common to both rings. Again, only one aldol cyclization product is stable and the conjugate addition can easily be managed with an enamine, a silyl enol ether, or an extra CO_2Et group added to cyclohexanone.

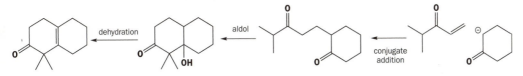

Problem 5

Predict the product that would be formed in these conjugate additions.

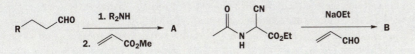

Purpose of the problem

Predicting products: difficult but necessary!

Suggested solution

The first one is relatively straightforward. The amine will make an enamine from the aldehyde, which should give an efficient conjugate addition to a reliable Michael acceptor. A full mechanism would complete a better answer.

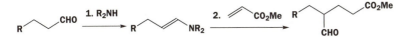

The second demands a bit of thought. The NH proton or else the proton between the two carbonyl groups might be removed. In the latter case, such a stable carbanion (actually enolate) should be all right at conjugate addition even with an aldehyde. If you said the carbanion would do a direct addition or that the amide anion would either add direct or in a conjugate fashion, those are reasonable answers.

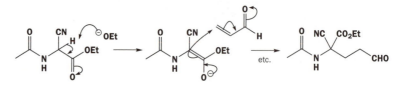

Problem 6

Suggest mechanisms for this reaction, commenting on any selectivity.

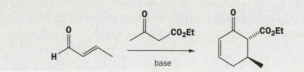

Purpose of the problem

This is easier because you are given the product but, as the reactions get more complicated, it is easier to overlook selectivity. Here is a reminder of the three main types (see p. 615 of the textbook): chemoselectivity (between separate functional groups); regioselectivity (between different aspects of the same functional group); and stereoselectivity (stereochemistry of the product).

Suggested solution

The mechanism is a conjugate addition of the central enolate of the keto-ester to the enal followed by an aldol cyclization of the ketone enolate on to the aldehyde and dehydration.

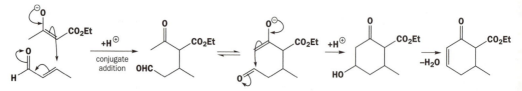

In the first step there is chemoselectivity in that the enolate is formed at the centre of the keto-ester and not on the aldehyde. You may not at first think that the aldehyde can form an enolate, but think again. There is regioselectivity in that the keto-ester could have formed another enolate. The reason for all these selectivities is the same: the conjugated enolate is much more stable than either of the alternatives.

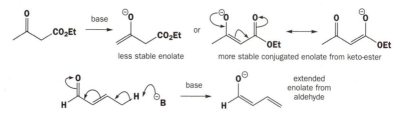

In the cyclization step (the aldol reaction) there is chemoselectivity in the formation of an enolate from the ketone and not from the aldehyde and regioselectivity in the formation from the enolate at the methyl group rather than between the ester and the ketone. This last enolate would be most stable for the same reasons as above.

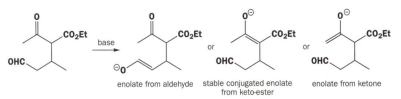

enolate from aldehyde   stable conjugated enolate   enolate from ketone
from keto-ester

We cannot argue this time that the most stable enolate is formed because it is the less stable enolate at the methyl group of the ketone that leads to the observed reaction. This time we must argue that all the enolates are formed in equilibrium with each other and that the cyclization determines the selectivity. The other cyclizations lead to less stable four-membered rings.

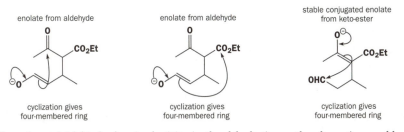

cyclization gives   cyclization gives   cyclization gives
four-membered ring   four-membered ring   four-membered ring

There is a trivial kind of regioselectivity in the dehydration – the alternative would give a nonconjugated enone and could not take place by the E1cB mechanism (Chapter 19).

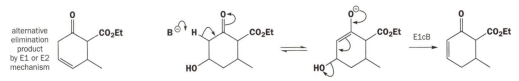

Finally, stereoselectivity. One diastereoisomer is formed, the one with the two substituents (Me and CO$_2$Et) *trans*. This cannot be controlled in the conjugate addition, the reaction that forms both centres, because there would be no reason for such control in an open-chain compound. Far more likely is that the final product is formed as a mixture of *cis* and *trans* compounds that equilibrates to the more stable *trans* compound by reversible enolate ion formation.

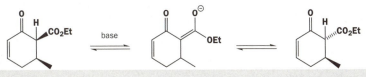

Problem 7

This example of the use of the Mannich reaction was given in the chapter. Draw detailed mechanisms for the two key steps shown here.

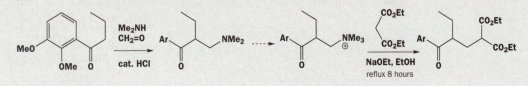

Purpose of the problem

Revision of the Mannich reaction.

Suggested solution

The first step is the Mannich reaction itself. It requires an iminium salt as the electrophile for a reaction with an enol.

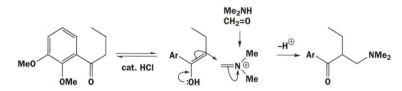

The second step involves an elimination of the tertiary amine (E1cB mechanism) and a conjugate addition of the enolate anion of diethyl malonate to the resulting enone. This device prevents the reactive enone from combining with itself by releasing it only in the presence of an excess of the nucleophile.

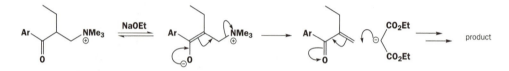

Problem 8

This symmetrical bicyclic ketone can easily be synthesized in two steps from simple precursors. What is the structure of the intermediate and what are the mechanisms of the reactions?

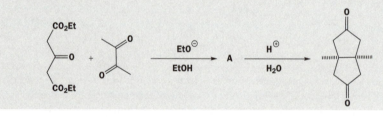

Purpose of the problem

Looking at this complicated product, you might be forgiven for assuming that it is difficult to make. This problem should convince you otherwise and also show you how powerful is the combination of aldol and conjugate addition.

Suggested solution

The diketone can only furnish the central rib of the bicyclic compound so the two 'wings' must come from the keto-diester. One thing immediately obvious is that four ester groups must be lost by hydrolysis and decarboxylation and that must be what the second step (aqueous acid) is doing. The product A of the first reaction must be the compound in the square box.

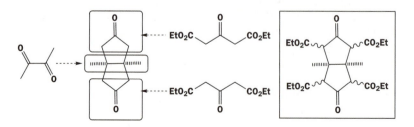

The first step must be an aldol reaction between the keto-diester (as enolate) and the very electrophilic 1,2-diketone. This reaction must give an enone as we are going to have to do a conjugate addition to add the second molecule of keto-diester.

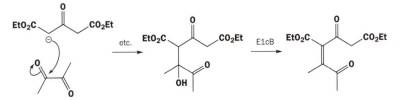

If this is a good reaction, and it is, a second intramolecular aldol reaction to form a five-membered ring ought to be a better one. This gives a compound that can do a double conjugate addition with the second molecule of keto-diester.

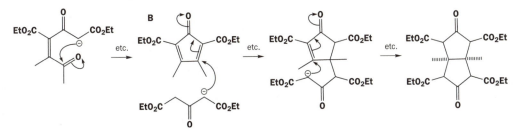

We do not, of course, mean that all these reactions happen at once and there is in fact a good reason (margin) why they do not happen exactly in the order given. The final stage is a normal hydrolysis and decarboxylation. The stereochemistry simply comes from the preference for two five-membered rings to be *cis*-fused (Chapter 33).

■ The cyclopentadienone B is anti-aromatic (four delocalized electrons) and is probably not formed. The first conjugate addition probably starts before the dehydration of the second aldol reaction.

Problem 9

Suggest ways to make these compounds using conjugate addition of enol(ate)s.

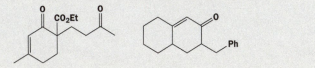

Purpose of the problem

These innocent-looking compounds need a bit of thought as several stages are needed for each compound.

Suggested solution

The last conjugate addition in the synthesis of the first compound is fairly obvious as the combination of ester and ketone provides a good enolate. Then we have a cyclohexenone rather like the compound in Problem 6. The compound needed for the intramolecular aldol reaction can be made by a second conjugate addition using a stable enolate.

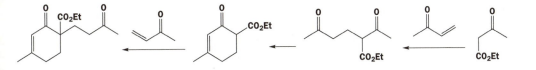

The final stage in the synthesis of the second compound must be an aldol reaction as the product is an enone. That leaves us with a diketone (the aldol cyclization is shown with a dotted arrow) and this could be made by two different conjugate additions.

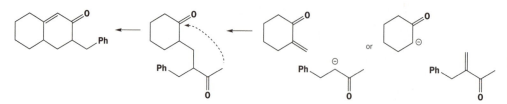

Since the enones will be made from the parent ketones, it doesn't really matter which way we do it. Cyclohexanone is symmetrical but the other ketone is not and, whichever way we do the reaction, we shall have to get selective enolization of 4-phenylbutan-2-one. We could carry out a Mannich reaction on 4-phenylbutan-2-one and then use the enamine, the silyl enol ether, or the keto-ester from cyclohexanone in the conjugate addition. The final cyclization could be carried out in acid or base.

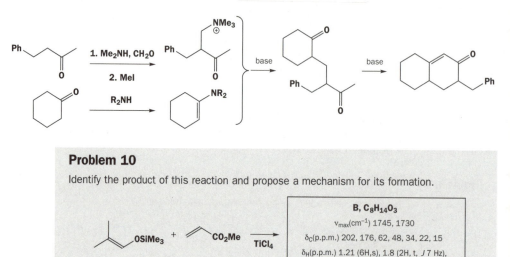

Problem 10

Identify the product of this reaction and propose a mechanism for its formation.

B, $C_8H_{14}O_3$
$v_{max}(cm^{-1})$ 1745, 1730
δ_C(p.p.m.) 202, 176, 62, 48, 34, 22, 15
δ_H(p.p.m.) 1.21 (6H,s), 1.8 (2H, t, J 7 Hz),
2.24 (2H, t, J 7 Hz), 4.3 (3H, s), 10.01 (1H, s)

Purpose of the problem

NMR revision in the context of conjugate addition.

Suggested solution

The IR suggests a dicarbonyl compound and the ^{13}C NMR suggests one aldehyde or ketone (202) and one acid derivative (178). The proton NMR confirms an aldehyde (10.1) and suggests that the acid derivative is a methyl ester (3H singlet at δ 4.3). Both NMRs show that there is no double bond any more. The proton NMR reveals a CMe$_2$ group on saturated carbon with no adjacent hydrogens (6H singlet at δ 1.21). There are two CH$_2$ groups coupled to each other but to nothing else (the two triplets). All this adds up to this aldehyde-ester.

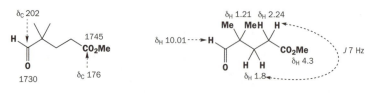

The mechanism is the usual Lewis acid-catalysed conjugate addition of a silyl enol ether on to an aldehyde, rather like the one we saw in Problem 1.

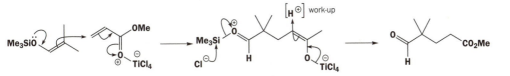

Problem 11

Suggest a synthesis for the starting material for this reaction, a mechanism for the reaction, and an explanation for the selectivity.

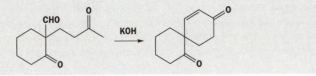

Purpose of the problem

Combination of synthesis, mechanisms, and selectivity – the whole works.

Suggested solution

The β-keto-aldehyde combination suggests a stable enolate adding to an enone. The enone is available and the keto-aldehyde can be made by an unambiguous acylation of cyclohexanone with unenolizable but electrophilic ethyl formate.

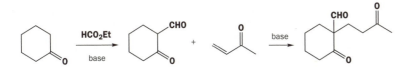

The mechanism for the cyclization is straightforward: the ketone forms an enolate at the methyl group and attacks the aldehyde.

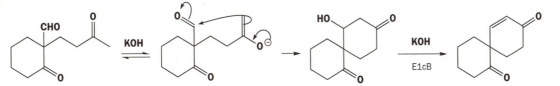

The selectivity is tricky. There is an alternative stable enone that might be the product after enolate formation at the same site but attack on the ketone in the ring. Presumably, the cyclization is kinetically controlled and the enolate prefers to attack the more electrophilic aldehyde.

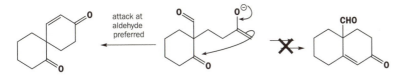

Problem 12

Suggest a mechanism for this reaction.

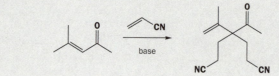

Purpose of the problem

Rehearsal of the mechanism of a conjugate addition with an extended enolate.

Suggested solution

The enone could form an enolate on the methyl group (kinetic) or a more stable extended enolate by removal of a proton from the far end (γ) of the molecule. Evidently, this is what happens here and the extended enolate attacks at the first carbon along the chain (α) to add one cyanoethyl group. Repetition adds the second cyanoethyl group and blocks the α position against any further enolate formation.

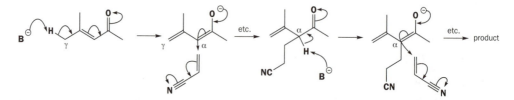

Problem 13

Suggest a mechanism for this reaction. How would you convert the product into the antibiotic anticapsin?

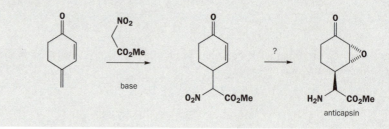

Purpose of the problem

Extension of conjugate addition to a new functional group (nitro) and a remote electrophilic site.

Suggested solution

The nitro-ester will form a very stable anion that could add to the β or δ positions on the dienone. Addition to the δ position removes the very reactive 'exo-methylene' group from the ring and leaves only the much more stable conjugated alkene inside the ring. Addition to the β position would destroy the conjugation.

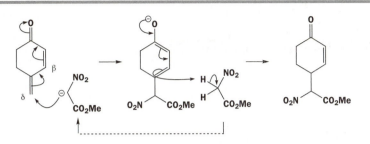

Completion of the synthesis requires epoxidation of the enone. This electrophilic alkene (Chapter 23) is best epoxidized by alkaline hydrogen peroxide (p. 588). Nucleophilic attack on the enone occurs from the opposite face to the large substituent already present. The nitro group must also be reduced to an amino group. It is best to do this after the epoxidation to avoid oxidation of the amino group and because of the danger of hydrogenation of the alkene.

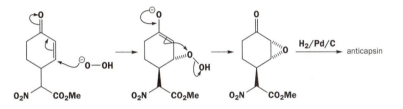

Suggested solutions for Chapter 30 30

Problem 1

Suggest ways to make these two compounds. Show your disconnections and don't forget to number the relationships.

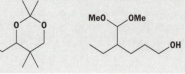

Purpose of the problem

First steps in designing a synthesis.

Suggested solution

Both compounds have a 1,1-diX relationship between the two oxygen atoms attached to the same carbon atom. You might have seen this in another way by recognizing the *acetal* functional group. In any case, both compounds can be made from a diol (or two molecules of an alcohol) and an aldehyde or ketone.

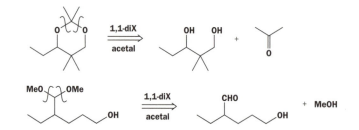

Continuing with the first compound we must next count the relationship between the two functional groups. We have a 1,3-diol so we need to think of various aldol or Claisen ester styles of reactions. These require a preliminary FGI and a few suggestions appear below.

■ FGI stands for functional group interconversion.

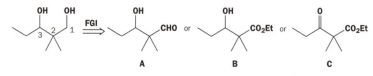

Since we shall reduce all the carbonyl groups to alcohols, it doesn't really matter which route we follow. Compound A would require a crossed aldol reaction between two aldehydes so we'll avoid that. Compound B is easier as we can use a specific enol equivalent (we have chosen the zinc enolate as it is less basic than most; see p. 706) from an ester to add to the aldehyde.

■ TM stands for target molecule, the molecule you are trying to make.

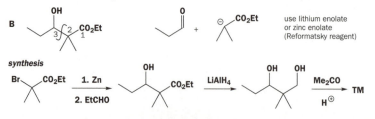

Direct disconnection of compound C suggests a crossed Claisen ester condensation but, if we remove one methyl group first, a great simplification results as we can use a self-condensation.

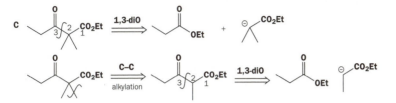

The synthesis is straightforward but there is an extra step required – the alkylation of the malonate-like diester. There are, of course, many other possibilities and, if you have a different answer, your synthesis may also work.

synthesis

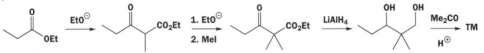

The other compound has a 1,5-relationship between the two functionalized carbon atoms and will require some sort of conjugate addition (Chapter 29). This time we want to reduce only one of the two carbonyl groups so we must make sure they are different. We choose an ester to help with the conjugate addition.

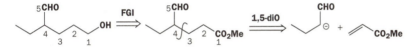

The aldehyde must be used as a specific enol equivalent in the conjugate addition. An enamine or a silyl enol ether would be fine. Now we must reduce the ester in the presence of the aldehyde, but we are going to convert the aldehyde into an acetal anyway so, if we do that first, there will be no chemoselectivity problem.

■ We chose the *methyl* ester to avoid ester exchange in this step.

synthesis

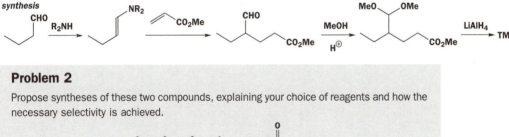

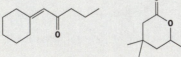

Problem 2

Propose syntheses of these two compounds, explaining your choice of reagents and how the necessary selectivity is achieved.

Purpose of the problem

First steps in designing a synthesis with selectivity required.

Suggested solution

The first compound is an α,β-unsaturated carbonyl compound and this is one of the most important functional group combinations for you to recognize in planning syntheses. It is the product of an aldol reaction – simply disconnect the alkene and write a new carbonyl group at the far end from the old. Don't lose any carbon atoms!

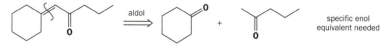

We find that we need a crossed aldol condensation between two ketones so we need chemoselectivity. We also need to make one enol(ate) from an unsymmetrical ketone so we need regioselectivity too. The obvious solutions are a lithium enolate, a silyl enol ether, or a compound with an extra ester group.

synthesis

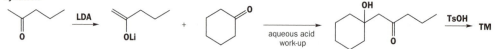

The second compound contains another functional group you will often meet – a lactone (cyclic ester). Simply disconnect the C–O bond and number the relationship between the two functional groups. When we discover a 1,5-relationship we know at once that we must use conjugate addition and so we must have two carbonyl groups. There are alternative disconnections (a and b) of the key intermediate, the keto-acid in the box.

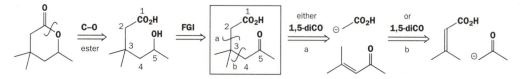

The synthesis will be easier if we use esters instead of acids. Syntheses based on either disconnection a or b will be fine. We prefer a because the enone starting material is a dimer of acetone and very easily made (it can be bought). The best specific enolate is probably malonate and there is no problem with the chemoselective reduction as NaBH$_4$ reduces ketones but not esters. The hydroxy-ester will cyclize on work-up.

synthesis

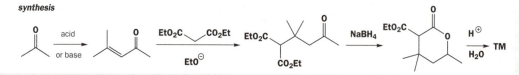

Problem 3

The reactions to be discussed in this problem were planned to give syntheses of these three target molecules.

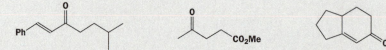

In the event, each reaction gave a different product shown below. What went wrong? Suggest syntheses that would give the target molecules.

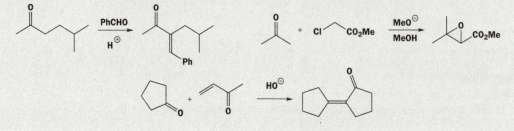

Purpose of the problem

Finding out what goes wrong is an important part of planning a synthesis.

Suggested solution

The aldol reaction planned for the first synthesis looks reasonable but the ketone is unsymmetrical and condensation has taken place on the wrong side. This is not surprising in acid solution as the more substituted enol is more stable. Use alkali instead.

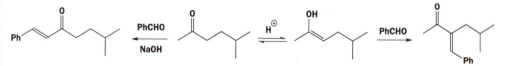

In the second case, alkylation of the enolate of the ketone was planned but evidently it is easier to form the enolate of the chloroester. The reaction that actually took place is the Darzens condensation. To avoid the problem, use a specific enol equivalent of the ketone such as an enamine or a keto-ester.

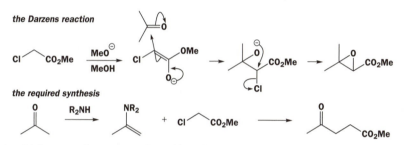

In the third case, cyclopentanone has self-condensed and ignored the enone to which it was supposed to add in a conjugate fashion and continue with a Robinson annelation (p. 761). The answer again is to use a specific enolate such as an enamine though the simplest here is a keto-ester as that can be easily prepared by intramolecular Claisen ester condensation (p. 727).

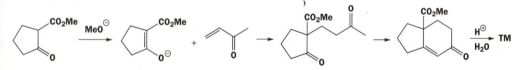

Problem 4

The natural product nuciferal was synthesized by the route summarized here.

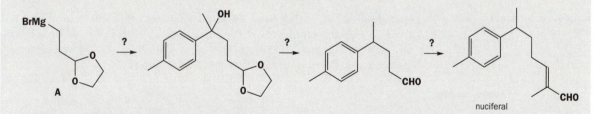

(a) Suggest a synthesis of the starting material A.
(b) Suggest reagents for each step.
(c) Draw out the retrosynthetic analysis giving the disconnections that you consider the planners had in mind and label them suitably.
(d) What synthon does the starting material A represent?

Purpose of the problem

Practice at an important skill – learning from published syntheses – as well as a popular style of exam question.

Suggested solution

(a) Grignard reagents are made from the corresponding halides and the further analysis is routine.

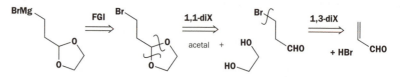

It turns out that the conjugate addition of bromide ion and acetal formation can be carried out as a single step since both are acid-catalysed.

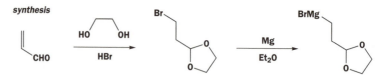

(b) The first step is the addition of the Grignard reagent to a ketone; the second involves the deprotection of the acetal but the OH group has also been replaced by hydrogen. Probably the easiest way to do this is via an acid-catalysed elimination and hydrogenation of the alkene.

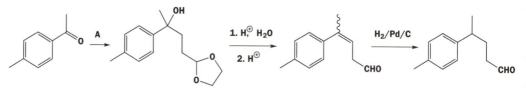

The last step is an aldol reaction between two aldehydes. The easiest way to do this would be by a Wittig reaction (p. 700) but a silyl enol ether of the aldehyde required to behave as an enol would also be fine.

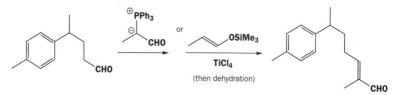

■ FGA stands for functional group addition.

(c) The retrosynthetic analysis involves a simple aldol disconnection and then an unusual FGA because the scientists wanted to use the simple aromatic ketone B, easily made by a Friedel–Crafts acylation (Chapter 22). The addition of an OH group allows a one-group C–C disconnection but then a reagent has to be found for synthon C.

nuciferal,analysis:

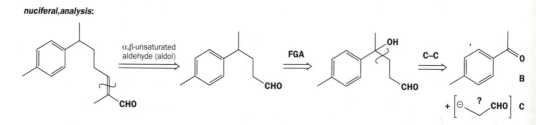

(d) And we know what the reagent was (A). This is a d³ synthon. Aldehydes are naturally electrophilic at C3 (by conjugate addition) so to make a reagent with unnatural polarity ('umpolung', p. 798), the aldehyde must be protected.

Problem 5

A synthesis of the enantiomerically pure ant pheromone is required. One suitable starting material might be the enantiomerically pure alkyl bromide shown. Suggest a synthesis of the pheromone based on this or another starting material.

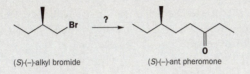

(S)-(–)-alkyl bromide   (S)-(–)-ant pheromone

Purpose of the problem

Synthesis with stereochemistry has an extra challenge. Here is an easy example.

Suggested solution

We know what the disconnection is, because we have been given one starting material, so this looks like an enolate alkylation. We must use a specific enolate to prevent the ketone from self-condensing. The only problem is that we must use a well-behaved specific enolate – in particular one that is nonbasic – so that there is no danger of elimination. The simplest is probably a keto-ester, easily made by Claisen ester condensation with diethyl carbonate (p. 730). The alkylation does not affect the chiral centre so the reaction will go with retention.

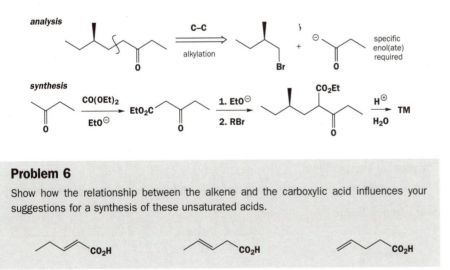

Problem 6

Show how the relationship between the alkene and the carboxylic acid influences your suggestions for a synthesis of these unsaturated acids.

Purpose of the problem

An exploration of the importance of functional group relationships.

Suggested solution

The first is an α,β-unsaturated carbonyl compound and can be made by a simple aldol reaction using a variety of specific enol(ate) equivalents for the acid part (Wittig, malonate, or silyl enol ether look the best).

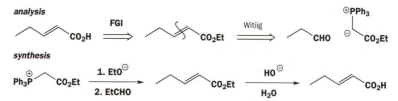

The second synthesis is the most difficult because the double bond can easily slip into conjugation with the carbonyl group. Perhaps the easiest synthesis is a cyanide displacement on an allylic halide, though there are other routes including alkyne reduction.

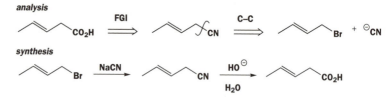

The third synthesis is best approached by the alkylation of a malonate using an allylic halide rather like the last example.

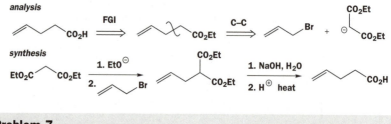

Problem 7

How would you make these compounds?

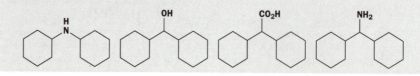

Purpose of the problem

Simple syntheses of apparently related compounds but the functional groups are very different.

Suggested solution

The secondary amine is best made by reductive amination via the imine (not usually isolated).

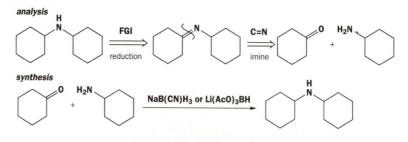

The secondary alcohol can be made by some sort of Grignard chemistry (Chapter 9). Cyclohexyl Grignard reagent can be added twice to ethyl formate or once to the cyclohexane aldehyde.

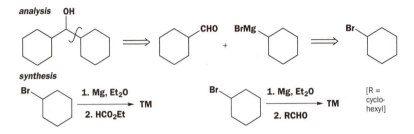

The carboxylic acid can best be made by a double alkylation of some specific enol(ate) equivalent. The stable enolate of diethyl malonate looks like a good bet.

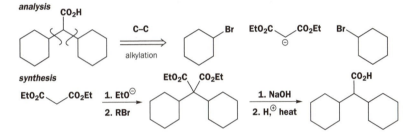

Finally, the primary amine could be made by reductive amination from the ketone with NH_4OAc and $NaB(CN)H_3$ or $NaBH(OAc)_3$. The ketone could be made by oxidation of the secondary alcohol we have just made. One of many alternatives would be the displacement of the toluene-p-sulfonate of the alcohol with azide ion and reduction of the azide.

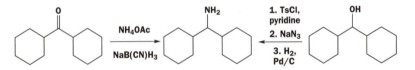

Problem 8

Show how the relationship between the two functional groups influences your suggestions for a synthesis of these diketones.

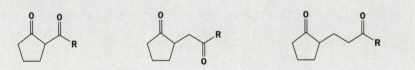

Purpose of the problem

A simple exercise in counting to show you that odd and even relationships really are different.

Suggested solution

The three diketones have 1,3-, 1,4-, and 1,5-dicarbonyl relationships. In each case the obvious disconnection is the bond joining the ring to the chain. For the 1,3-diketone, this means acylation of a ketone.

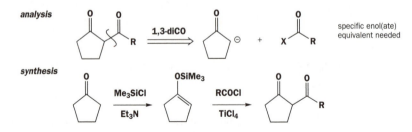

The same disconnection on the 1,4-diketone leads to totally different chemistry and requires a different specific enol(ate) equivalent for cyclopentanone.

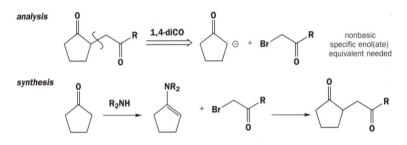

The 1,5-diketone requires a conjugate addition approach and we use a different specific enol(ate) equivalent more for variety than necessity.

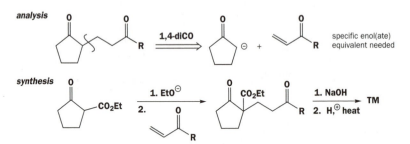

Problem 9

Suggest syntheses for these compounds. (*Hint.* Look out for a 1,4-dicarbonyl intermediate.)

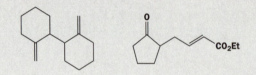

Purpose of the problem

Further exploration of the 1,4-diCO relationship but disguised by further chemistry.

Suggested solution

The diene looks like a Wittig product – the only easy way to get those *exo*-methylene groups. Wittig disconnection reveals a symmetrical 1,4-diketone and we really would prefer to disconnect the bond between the rings.

analysis

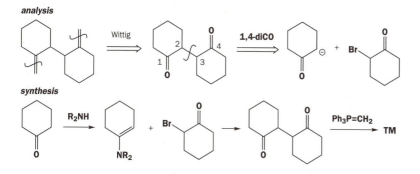

synthesis

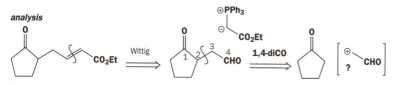

The second compound has an α,β-unsaturated ester so we are into aldol-style chemistry and here again the Wittig looks good, mainly because of the chemoselectivity needed in the aldol step. The enol(ate) equivalent must react with an aldehyde but not with a ketone and the phosphonium ylid will do just that.

analysis

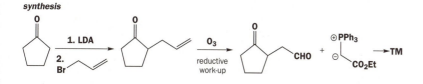

There is a problem with the aldehyde a² synthon. We don't want to use an unstable 2-haloaldehyde so an allyl halide is best. The alkene must be cleaved by ozonolysis with reductive work-up (Me₂S) to stop oxidation of the aldehyde.

synthesis

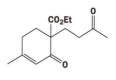

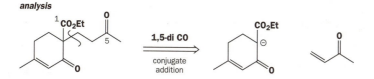

Problem 10

Suggest a synthesis of this diketo-ester from simple starting materials.

Purpose of the problem

The phrase 'from simple starting materials' in an exam warns you to go further back than you might otherwise.

Suggested solution

The first disconnection of the 1,5-diCO relationship is fairly obvious as it trims off the side chain from the ring. The enolate is also a good one.

analysis

Now we have a choice between dealing with the 1,3-diCO relationship or disconnecting the α,β-unsaturated ketone.

analysis 2

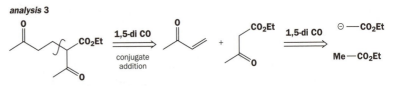

Disconnection b solves no problems – we still have to make the α,β-unsaturated carbonyl system and now we have to make one particular geometrical isomer (*cis*). Disconnection a reveals a new 1,5-diCO relationship. Disconnecting this at the branchpoint we get another simple available enone and a keto-ester (ethyl acetoacetate) that is the product of a self-condensation of ethyl acetate.

analysis 3

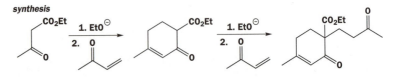

This synthesis is indeed from very simple starting materials: two molecules of ethyl acetate and two of butenone (methyl vinyl ketone, MVK).

synthesis

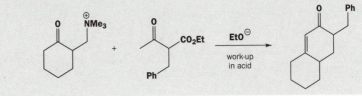

Problem 11

Explain what is happening in this reaction. Draw a scheme of retrosynthetic analysis corresponding to the synthesis. How would you make the starting materials?

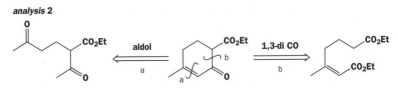

Purpose of the problem

Revision of the Mannich reaction cooperating with conjugate addition and an aldol reaction to complete the Robinson annelation.

Suggested solution

The base does two things: it makes the stable enolate from the β-keto-ester and it eliminates Me₃N from the tetraalkyl ammonium salt by the E1cB mechanism. Each molecule of enone is thus released only in the presence of a large excess of the enolate anion with which it is to react.

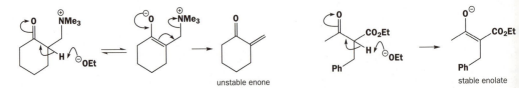

These two now react together in a conjugate addition and the rest of the Robinson annelation follows.

Stage 1: The conjugate addition

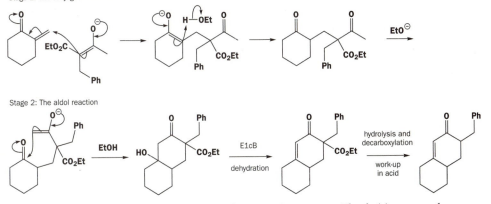

Stage 2: The aldol reaction

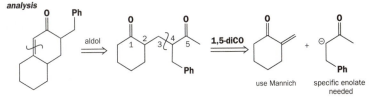

The scheme of retrosynthetic analysis retraces these steps in summary. The decision to use the Mannich reaction for the enone and to add a CO_2Et group to stabilize the enolate would be taken at the end of the analysis.

analysis

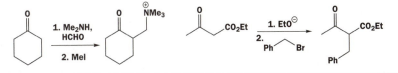

One starting material would be made by a Mannich reaction and the other by alkylation of ethyl acetoacetate. This is an added bonus of the extra CO_2Et group.

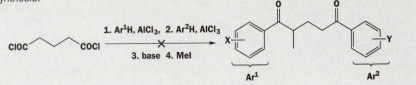

Problem 12

These diketones with different aryl groups at the ends were needed for a photochemical experiment. The compounds could be prepared by successive Friedel–Crafts acylations with a diacid dichloride but the yields were poor. Why is this a bad method? Suggest a better synthesis.

Purpose of the problem

You need to see what happens when syntheses are not carefully planned.

Suggested solution

The planning takes no account of chemoselectivity. The first Friedel–Crafts reaction will give a mixture of three products, two of which can react again in the second Friedel–Crafts reaction. The diketone fraction of this product will be a mixture of three compounds: two dimers and the mixed product that is wanted. The alkylation could go at either end of the molecule as there will be little difference in the stability of the enolates. Even if clean monoalkylation occurs (unlikely) there will be four final products.

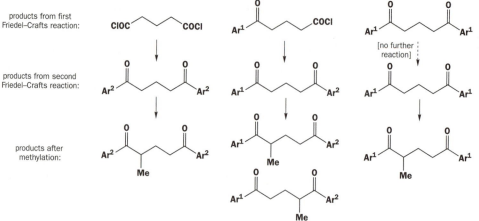

products from first Friedel–Crafts reaction:

products from second Friedel–Crafts reaction:

products after methylation:

■ The original authors used the morpholine enamines of the ketone Ar²COMe (J. C. Scaiano and group, *J. Am. Chem. Soc.*, 1980, **102**, 727).

The mixed diketone required is a 1,5-dicarbonyl compound and can be specifically made by conjugate addition. We can use the Mannich reaction for the enone as in Problem 11 and we leave you to choose the specific enolate.

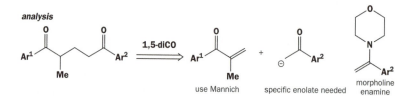

analysis

1,5-diCO

use Mannich   specific enolate needed   morpholine enamine

Problem 13

This is a synthesis for the ladybird defence compound coccinelline.

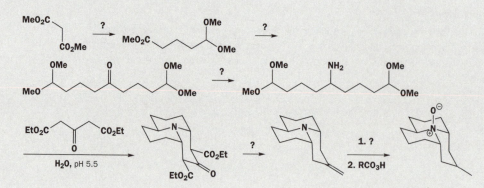

Suggest reagents for the reactions marked '?' (several steps may be needed) and give mechanisms for those that are not.

Purpose of the problem

Now a complete synthesis that was carefully planned. There is a variety of reactions typical in a complete synthesis.

Suggested solution

The first step gives a partly protected 1,5-dicarbonyl compound and so must be a conjugate addition to an aldehyde followed by protection of the aldehyde as an acetal. The second step is a Claisen ester condensation followed by hydrolysis and decarboxylation of the remaining ester group (strategy examined on pp. 727–8 of the textbook), and the third step is a reductive amination.

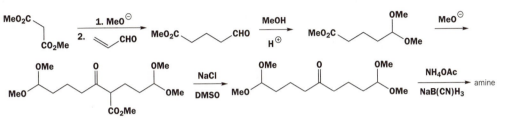

The last two stages require hydrolysis and decarboxylation of both ester groups and a Wittig reaction and then a reduction of the alkene by catalytic hydrogenation. The mechanism of the step that forms the tricyclic amine looks complicated but isn't. The conditions (pH 5.5 in water) will hydrolyse both acetals to aldehydes, which will immediately start to form imines with the primary amine. The stable enol of the keto-diester adds to these imines. You must just take it carefully stage by stage.

■ The details are in R. V. Stevens and group, *J. Am. Chem. Soc.*, 1979, **101**, 7032.

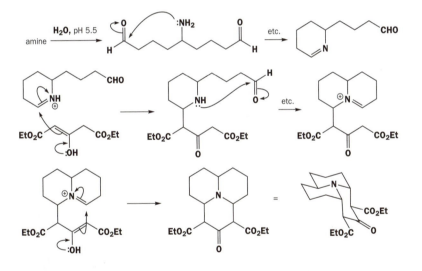

The stereochemistry comes from the intramolecular stage (last mechanism shown). The enol must attack the iminium ion from above or below, and it can reach only if the chain is joined axially to the ring already in existence (Chapter 18). The two ester groups are equatorial on the new ring because they equilibrate by enolization to the most stable arrangement.

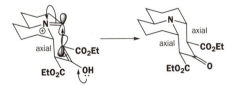

Problem 14

Suggest syntheses for these compounds.

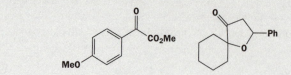

Purpose of the problem

A brief exploration of the 1,2-diCO relationship.

Suggested solution

The first compound is probably most simply derived from an alkene by dihydroxylation and then oxidation.

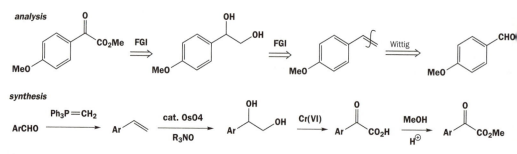

■ See D. W. Knight and
G. Pattenden, *J. Chem. Soc., Perkin Trans.* **1**, 1979, 84.

In fact, it was made by a more interesting strategy using a dithiane as an acyl anion equivalent (d^1 reagent).

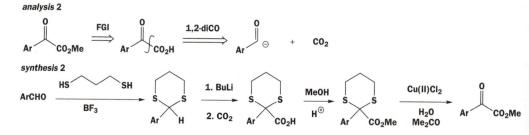

■ See J. E. Baldwin and group,
J. Chem. Soc., Chem. Commun.,
1976, 736.

The second compound is not a lactone as it might first appear, but a ketone. Disconnection of a C–O bond reveals an α,β-unsaturated ketone made by an aldol reaction from a hydroxy-ketone with a 1,2-diO relationship. Again, an acyl anion equivalent looks best. The published synthesis offers the two alternatives shown. Reaction of the hydroxy-ketone with PhCHO and base (not shown), followed by acid-catalysed cyclization, gives the target molecule.

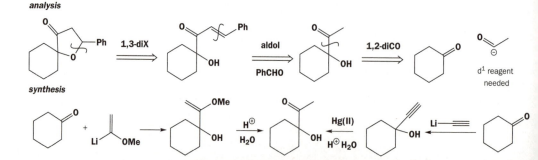

Problem 1

Deduce the structure of the product of this reaction from the spectra and explain the stereochemistry. Compound A has δ_H 0.95 (6H, d, J 7 Hz), 1.60 (3H, d, J 5 Hz), 2.65 (1H, double septuplet, J 4 and 7 Hz), 5.10 (1H, dd, J 10 and 4 Hz), and 5.35 (1H, dq, J 10 and 5 Hz).

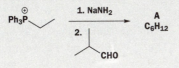

■ Several problems in this chapter ask you to suggest ways to make molecules. We give one possible answer and sometimes comment on alternatives but you should realize that there are many possible right answers to this sort of question. Make sure you have understood the principles behind each question and, if your answer is different from ours, check it with someone who knows.

Purpose of the problem

Just checking on a simple way to make Z-alkenes with a bit of NMR revision.

Suggested solution

This is obviously a Wittig reaction and we should expect a Z-alkene product as the ylid is not stabilized by further conjugation. The evidence is plain: the signals at 5.10 and 5.35 p.p.m. must be the alkene hydrogens and the coupling between them is 10 Hz. Definitely a Z-alkene.

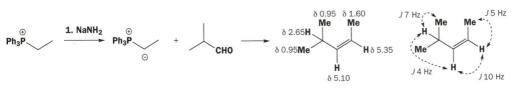

Problem 2

A single diastereoisomer of an insect pheromone was prepared in the following way. Which isomer is formed and why? Outline a synthesis of one other isomer.

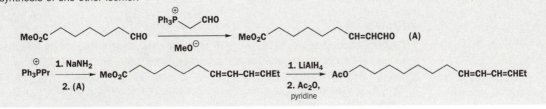

Purpose of the problem

Testing your knowledge of the stereochemistry of the Wittig reaction.

Suggested solution

The first Wittig, with a stabilized ylid, gives the E-enal (A). The second, with an unstabilized ylid, gives a Z-alkene, and the final structure is an E,Z-diene.

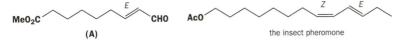

(A)   the insect pheromone

There are many ways to make the other isomers (*Z,Z*-; *E,E*-; *Z,E*-dienes) and you might have used alternative styles of Wittig reactions or reduction of alkynes (pp. 815–19) or other methods described in Chapter 31.

Problem 3

How would you prepare samples of both geometrical isomers of this compound?

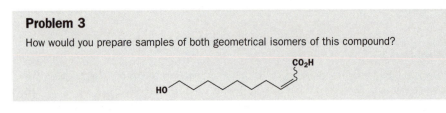

Purpose of the problem

A simple stereocontrolled alkene synthesis but both isomers are wanted.

Suggested solution

■ TM stands for target molecule.

There are many methods that can be used to tackle this question. The only snags are protecting the OH group if necessary and care in releasing the *Z*-compound as it will isomerize easily to the *E*-compound by reversible conjugate addition. Here is one way to the *Z*-alkene using reduction of an alkyne to control the stereochemistry. The OH group is protected as a benzyl ether removable by hydrogenation, perhaps under the same conditions as the reduction of the alkyne.

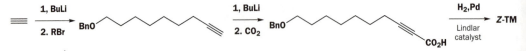

The *E*-alkyne might be produced by reduction of the alkyne with alkali metal in ammonia but a Wittig reaction is probably easier. Either an ylid or a phosphonate could be used. Protection of the OH as an acetate allows hydrolysis of both esters in the same step.

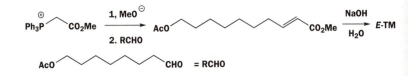

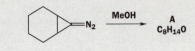

Problem 4

Decomposition of this diazocompound in methanol gives an unstable alkene A ($C_8H_{14}O$) whose NMR spectrum contains these signals: δ_H (p.p.m.) 3.50 (3H, s), 5.50 (1H, dd, *J* 17.9 and 7.9 Hz), 5.80 (1H, ddd, *J* 17.9, 9.2, and 4.3 Hz), 4.20 (1H, m), and 1.3–2.7 (8H, m). What is its structure and geometry? You are not expected to work out a mechanism for the reaction.

Purpose of the problem

NMR revision and a surprising structure – a test of your belief in physical methods, especially NMR.

Suggested solution

The starting material is $C_7H_{10}N_2$ so it has lost nitrogen (N_2) and gained CH_4O – one molecule of methanol. We can see the MeO group at 3.5 and the four CH_2 groups in the ring are still there (8H m at 1.3–2.7). All that is left is a multiplet at 4.2, obviously next to MeO, and a pair of alkene signals at 5.5 and 5.8, coupled with $J = 17.9$ Hz – obviously an E-alkene. One end of the alkene (5.5) is coupled to one proton, the other (5.8) to 2. We now have these fragments.

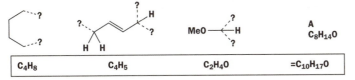

| C_4H_8 | C_4H_5 | C_2H_4O | $=C_{10}H_{17}O$ |

These add up to C_2H_3 too much! Clearly, the CH attached to OMe and the CH attached to the alkene must be the same and the CH_2 at the other end of the alkene must be one end of the chain of four. We now have a structure, but it doesn't join up.

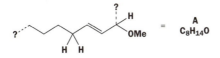

This is the test of your belief – the dotted ends must be joined up to give the structure of A. Yes, this does put an E-alkene in a seven-membered ring, and it is difficult to draw, but you were warned that the compound is unstable. The CH_2 group next to the CHOMe group is diastereotopic and so the coupling constants are different.

■ This was the discovery of H. Jendralla, *Angew. Chem., Int. Ed. Engl.*, 1980, **19**, 1032. If you were really on the ball, you might have noticed that a *trans*-cycloheptene is chiral so this compound must be a single diastereoisomer too, though we don't know which.

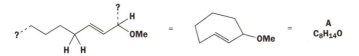

Problem 5

Why do these reactions give different alkene geometries?

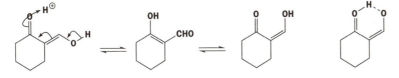

Purpose of the problem

Revision of enol chemistry from Chapters 21 and 27 with the question of enol geometry added.

Suggested solution

The geometry of the enol product of the first reaction changes easily because of delocalization and tautomerism. It settles down in the more favourable geometry under thermodynamic control and the Z-alkene is favoured because of hydrogen bonding.

The silyl enol ether does not enjoy the advantage of the hydrogen-bonded structure so it prefers the *E*-alkene for steric reasons. It might form only in this shape from the *E*-enolate but it might also equilibrate by silyl transfer.

Problem 6

Here is a synthesis of a prostaglandin analogue. Suggest reagents for the steps marked '?', give mechanisms for those not so marked, and explain any control of alkene geometry.

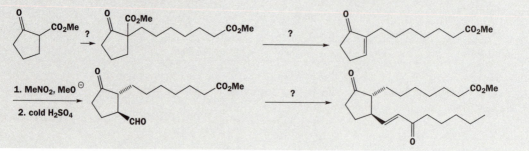

Purpose of the problem

Making double bonds both inside rings and in open chain.

Suggested solution

There is no difficulty in alkylating the β-keto-ester and removing the ester group but that leaves the question of how to introduce the double bond into the five-membered ring.

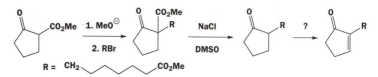

The most obvious way is a halogenation in acid solution (Chapter 21) on the more substituted side of the ketone and elimination of HX in base. The double bond will prefer to go inside the ring and there is no argument about its stereochemistry. Typical bases would be tertiary amines or hindered alkoxides (e.g. *t*-BuO⁻).

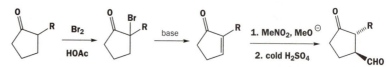

The mechanism for this last step must involve conjugate addition of the nitromethane anion and conversion of the nitro group to an aldehyde. The stereochemistry is set during the conjugate addition when protonation of the enolate from the less hindered side creates the *anti* relationship. Notice that the nitromethane anion is regenerated in this step.

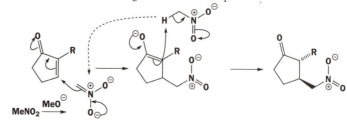

Conversion into the aldehyde involves hydrolysis of the 'enol' form of the nitro compound. This is rather like the hydrolysis of any imine and particularly like that of an oxime (Chapter 14).

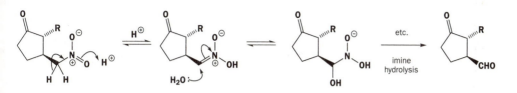

Finally, the second alkene must be inserted stereospecifically *trans*. Here a Wittig reagent using a stabilized ylid is best as these react with aldehydes and not with ketones.

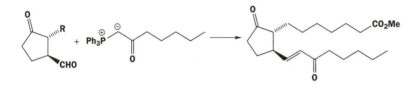

Problem 7

Isoeugenol, the flavouring principle of cloves, occurs in the plant in both the *E* (solid) and *Z* (liquid) forms. How would you prepare a pure sample of each and how would you purify each from any of the other isomer?

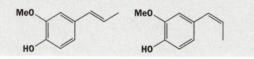

Purpose of the problem

Making simple double bonds with full stereochemical control.

Suggested solution

There are many possibilities. A Wittig approach is particularly attractive as the required aldehyde is vanillin, the common flavouring. Since the ylid would not be stabilized, this approach would give mostly *Z*-isoeugenol. Two molecules of base would be needed as the first would remove the proton from the phenol.

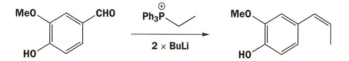

E-Isoeugenol could be made from the same aldehyde by a Julia reaction (p. 810) or by isomerizing *Z*-isoeugenol with light or iodine, or by a stereochemically controlled Peterson reaction (p. 812). Pure *Z*-isoeugenol could be separated from a small amount of *E*-isoeugenol by distillation or chromatography and pure *E*-isoeugenol could be separated from a small amount of *Z*-isoeugenol by crystallization or chromatography.

Problem 8

Thermal decomposition of this lactone gives mainly the *Z*-alkene shown with minor amounts of the *E*-alkene and an unsaturated acid. Suggest a mechanism for the reaction that explains these results.

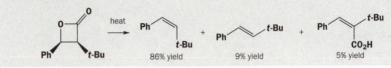

86% yield   9% yield   5% yield

Purpose of the problem

Practice in analysing a mechanism where stereochemistry gives a clue.

Suggested solution

The reaction is highly stereoselective but the minor products suggest that a simple dropping out of CO_2 from the lactone is not the answer. An E1 reaction with a zwitterionic intermediate formed by opening the lactone gives a way to all the products. The clue is that, at the moment CO_2 is lost, the bond linking it to the rest of the molecule must be parallel to the empty p orbital of the cation.

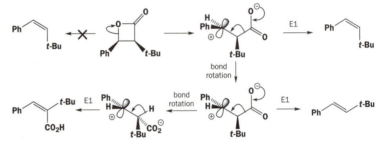

■ This was a mechanistic investigation and it uncovered more evidence that the zwitterion was indeed an intermediate in the decarboxylation (J. Mulzer and M. Zippel, *Tetrahedron Lett.*, 1980, **21**, 751).

If loss of CO_2 is faster than rotation about the central bond, the *Z*-acid is formed. If rotation occurs, then the *t*-Bu and Ph groups can adopt a favourable *anti* alignment and either CO_2 or H can be lost depending on which happens to be in the right place.

Problem 9

What controls the double bond geometry in these examples? In the second example, one alkene is not defined by the drawing.

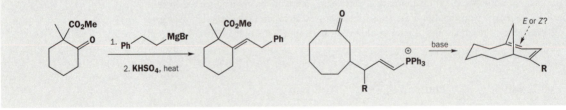

Purpose of the problem

Practice in explaining or even predicting the stereochemistry of an alkene.

Suggested solution

■ Note the chemoselectivity: the ketone is more electrophilic than the ester.

The first is simple. Addition of the Grignard reagent to the ketone produces a tertiary alcohol, which dehydrates in acid. The dehydration is reversible so the more stable *E*-alkene is formed selectively.

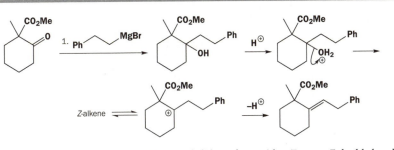

The second case is more interesting. The initial ylid can form with a *Z*- or an *E*-double bond. The *E*-ylid cannot cyclize but the *Z*-ylid can. The geometry of the second alkene is again determined simply by what is possible and what is not. This trisubstituted bicyclic alkene will inevitably be *cis* in one ring and *trans* in the other. We can have either a *cis*- or a *trans*-alkene in a ten-membered ring but we are forced to have a *cis*-alkene in the six-membered ring.

■ This work showed how easy it was to make strained alkenes by the Wittig reaction (W. G. Dauben and J. Ipaktschi, *J. Am. Chem. Soc.*, 1973, **95**, 5088).

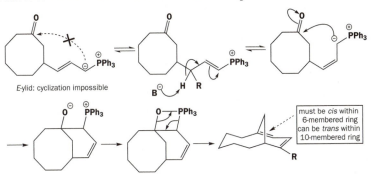

Problem 10

Treatment of this epoxide with base gives the same *E*-alkene regardless of the stereochemistry of the epoxide. Comment.

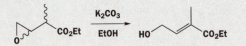

Purpose of the problem

To show that stereoselective reactions are sometimes better than stereospecific reactions in producing a single stereoisomer of a product.

Suggested solution

This is an E1cB elimination. The intermediate is the enolate of the ester so one of the stereogenic centres in the starting material is destroyed. No matter from which diastereoisomer it comes, the enolate can adopt a shape (C–O bond in the epoxide more-or-less parallel to the π orbitals of the enolate) that allows elimination to the *E*-alkene (see margin).

■ This ester was used in the synthesis of steroids (Chapter 51) by the Diels-Alder reaction (Chapter 35); Z. Valenta and group, *Can. J. Chem.*, 1979, **57**, 3354.

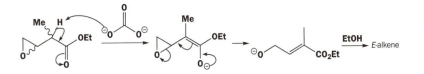

Problem 11

Which alkene would be formed in each of the following reactions? Explain your answer mechanistically.

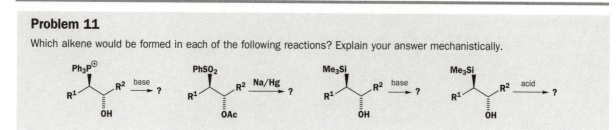

Purpose of the problem

Revision of three main methods for stereoselective (or specific) alkene bond formation.

Suggested solution

The first is a Wittig reaction with an unstabilized ylid, the second a Julia reaction, and the last two are Peterson reactions under different conditions. Each reaction is described in detail in the chapter. The Wittig reaction is under kinetic control and is a stereospecifically *cis* elimination. In this case the product is the *Z*-alkene.

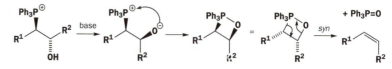

The Julia reaction is under thermodynamic control as the equilibration occurs under the reaction conditions. The product, formed stereoselectively, is the *E*-alkene.

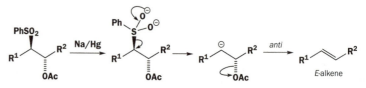

The Peterson reaction is a *syn* elimination under basic conditions, giving the *Z*-alkene, and an *anti* elimination under acidic conditions, giving the *E*-alkene.

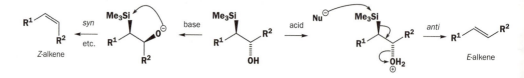

Problem 12

Comment on the difference between these two reactions:

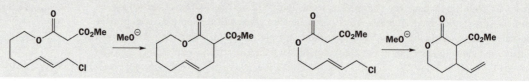

Purpose of the problem

To show that ring size combined with double bond geometry can affect the regioselectivity of cyclization reactions.

Suggested solution

First we should draw the mechanisms of the reactions and see what is different.

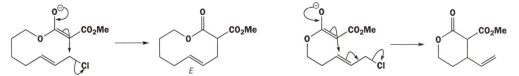

■ This question is explored in detail on pp. 604–11.

■ This chemistry made the synthesis of medium - and large-ring lactones much easier (J. Tsuji and group, *J. Am. Chem. Soc.*, 1978, **100**, 7424).

The first example is conjugate alkylation and the second direct alkylation. We might expect stable conjugated enolate anions to do one or the other and not chop and change. The first product has an *E*-alkene in a ten-membered ring. This is much more favourable than the *E*-alkene in a eight-membered ring that would be the product from the same reaction on the second compound. Direct addition gives an alkene with no geometry.

Problem 13

The elimination of alcohols 13A to give cinnamic acids *trans*-13B in aqueous sulfuric acid has been studied. If optically active 13A is used and the reaction stopped at 10% completion, the starting material is found to be completely racemized. What can you deduce about the mechanism of the elimination step?

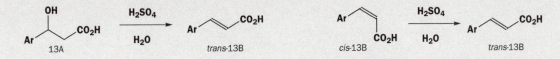

The *cis* cinnamic acids *cis*-13B also isomerize to the *trans* acids under the same conditions but more slowly. What is the mechanism of this reaction?

Purpose of the problem

Revision of Chapter 13 and a challenge to your ability to think through a mechanistic problem in connection with elimination reactions and the geometry of alkenes.

Suggested solution

The elimination of the secondary benzylic alcohol in acid solution must be an E1 reaction since the stable intermediate cation is formed easily. Racemization of the starting material suggests that the formation of the cation is rapidly reversible and the loss of the proton is the slow step in the E1 mechanism.

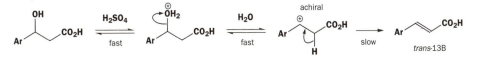

As the cation is achiral, any alcohol formed from it must necessarily be racemic. The isomerization of the *Z*-alkene must occur by protonation of the alkene to give the same cation. Whereas the protonation of the alcohol had to be fast, the protonation of the alkene can be slow (it is the reverse of the slow step in the elimination of the alcohol). The isomerization of *Z*- to *E*-alkene is slower than the dehydration of the alcohol because the protonation of the *Z*-alkene is even slower than the deprotonation of the cation.

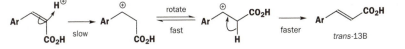

Problem 14

Give mechanisms for these stereospecific reactions on single geometrical isomers of alkenes.

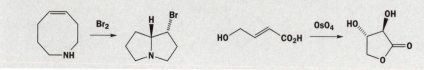

Purpose of the problem

To show that double bond geometry can affect the stereochemistry of tetrahedral centres by stereospecific reactions.

Suggested solution

■ S. R. Wilson and R. A. Sawicki, J. Org. Chem., 1979, **44**, 287.

First, we must draw mechanisms for the reactions. The first reaction clearly starts by bromination of the Z-alkene to form a bromonium ion (Chapter 20) that is attacked by the amine. The amine can choose which end it attacks (and it prefers to form two five-membered rings rather than even one strained four-membered ring) but it has no choice about stereochemistry. We must have *trans* addition to the alkene.

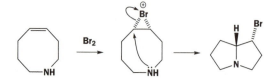

The second reaction starts with *cis* dihydroxylation by OsO$_4$. This is rather like epoxidation and is discussed in full in Chapter 35 (pp. 936–7). The starting material cannot form a lactone as the OH group is held away from the CO$_2$H group by the *E*-alkene. Once the molecule is dihydroxylated there is no longer any barrier to lactone formation and a five-membered ring is again preferred to the alternatives.

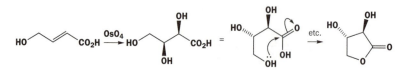

Suggested solutions for Chapter 32

Problem 1

A revision problem to start you off easily. A Pacific sponge contains 2.8% dry weight of a sweet-smelling oil with the following spectroscopic details. What is its structure and stereochemistry?

■ All NMR shifts are in p.p.m. and coupling constants are quoted in Hertz. The usual abbreviations are used: d = doublet; t = triplet; and q = quartet.

Mass spectrum gives formula: $C_9H_{16}O$

IR 1680, 1635 cm^{-1}

δ_H 0.90 (6H, d, J 7), 1.00 (3H, t, J 7), 1.77 (1H, m), 2.09 (2H, t, J 7), 2.49 (2H, q, J 7), 5.99 (1H, d, J 16), and 6.71 (1H, dt, J 16, 7)

δ_C 8.15 (q), 22.5 (two qs), 28.3 (d), 33.1 (t), 42.0 (t), 131.8 (d), 144.9 (d), and 191.6 (s)

Purpose of the problem

A gentle introduction to the determination of stereochemistry with two dimensions only.

Suggested solution

The IR suggests a conjugated carbonyl compound, confirmed by the two alkene and one carbonyl signals in the carbon NMR with the additional information that it is an aldehyde or ketone (δ_C about 200). The proton NMR shows that this is a ketone (no CHO proton) and that the alkene has two protons (5.99 and 6.71) and that they are *trans* (J 16). We also see an ethyl group (2H q and 3H t) attached to something with no Hs (looks like the carbonyl group). This suggests the unit in the margin, which leaves only C_4H_8. We know we have Me$_2$CH- from the 6H d and that leaves only CH$_2$. We have a structure.

■ Details are in D. J. Faulkner and B. N. Ravi, *Tetrahedron Lett.*, 1980, **21**, 23.

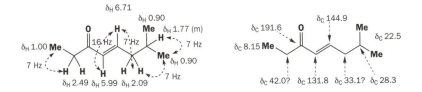

Problem 2

Reaction between this aldehyde and ketone in base gives a compound A with the ^{1}H NMR spectrum: δ 1.10 (9H, s), 1.17 (9H, s), 6.4 (1H, d, J 15), and 7.0 (1H, d, J 15). What is its structure? (Don't forget stereochemistry!) When this compound reacts with HBr it gives compound B with this NMR spectrum: δ 1.08 (9H, s), 1.13 (9H, s), 2.71 (1H, dd, J 1.9, 17.7), 3.25 (dd, J 10.0, 17.7), and 4.38 (1H, dd, J 1.9, 10.0). Suggest a structure, assign the spectrum, and give a mechanism for the formation of B.

Purpose of the problem

Slightly more difficult determination of stereochemistry moving from two dimensions to three. Revision of the Karplus relationship and of conjugate addition.

Suggested solution

The structure of A is easy. It has a *trans* (*E*-) double bond with two Hs (*J* 15) and two tertiary butyl groups. There isn't much else it can be and we suspect an aldol reaction between the enolizable ketone and the unenolizable aldehyde.

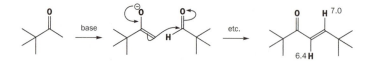

- J_{HH} values of 15+ are usually *trans*-alkenes or geminal.

B is more difficult. The alkene has obviously gone (no signals beyond 4.48) and there is one extra H. It looks as though HBr has added. The coupling of 17.7 Hz cannot be a *trans*-alkene as there isn't any kind of alkene so it must be geminal (^2J) coupling. This means the molecule is chiral so that a CH_2 group is diastereotopic. In fact, the expected conjugate addition of HBr (Chapter 23) has occurred.

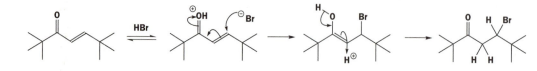

- See E. R. Kennedy and R. S. Macomber, *J. Org. Chem.*, 1974, **39**, 1952.

The three marked hydrogens form an ABX system: AB are the diastereotopic CH_2 group (J_{AB} = 17.7) and H is the CHBr proton (J_{AX} = 10; J_{BX} = 1.9). It is not normally possible to say which proton is A and which B but here with such large groups we may guess that there is one favoured conformation with one dihedral angle about 180° and one about 60°.

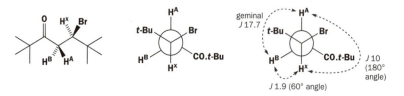

Problem 3

One of the sugar components in the antibiotic kijanimycin has the gross structure (margin) and the NMR spectrum shown below. What is its stereochemistry? All couplings in Hz; signals marked * exchange with D₂0.

δ_H 1.33 (3H, d, *J* 6), 1.61* (1H, broad s), 1.87 (1H, ddd, *J* 14, 3, 3.5), 2.21 (1H, ddd, *J* 14, 3, 1.5), 2.87 (1H, dd, *J* 10, 3), 3.40 (3H, s), 3.47 (3H, s), 3.99 (1H, dq, *J* 10, 6), 4.24 (1H, ddd, *J* 3, 3, 3.5), and 4.79 (1H, dd, *J* 3.5, 1.5).

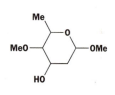

Purpose of the problem

Your first attempt at a serious assignment of three-dimensional stereochemistry from NMR.

Suggested solution

We can make some preliminary assignments from a combination of shift and coupling.

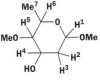

Signal	Integral and splitting	Comments	Assignment
1.33	3H, d, J 6	3H d must be CH*Me*	Me[7]
1.61*	1H, broad s	exchanges so must be OH	OH
1.87	1H, ddd, J 14, 3, 3.5	14 Hz looks like geminal (^2J) coupling	H[2] or H[3]
2.21	1H, ddd, J 14, 3, 1.5	2.21 and 1.87 are CH$_2$ group	H[2] or H[3]
2.87	1H, dd, J 10, 3	must be axial proton (10 Hz)	H[4] or H[5]
3.40	3H, s	one OMe group	OMe
3.47	3H, s	the other OMe group	OMe
3.99	1H, dq, J 10, 6	q means H[7], must be axial (10 Hz)	H[6]
4.24	1H, ddd, J 3, 3, 3.5	all small J, must be equatorial	H[4] or H[5]
4.79	1H, dd, J 3.5, 1.5	all small J, must be equatorial	H[1]

We don't mind which is H[2] or H[3] as they don't affect the stereochemistry, but we do mind which is H[4] or H[5]. Since H[6] is a 10 Hz doublet with H[5], we know that H[5] is at 2.87 and is axial. This gives us the entire assignment and the stereochemistry: H[5] and H[6] are axial; H[1] and H[4] are equatorial. That is why there are no large vicinal (^3J) couplings to the diastereotopic CH$_2$ group (H[2] and H[3]).

■ A. K. Mallams and group, *J. Am. Chem. Soc.*, 1981, **103**, 3938.

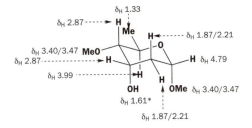

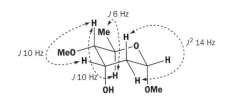

all couplings not shown are <4 Hz and are
axial/equatorial or equatorial/equatorial

Problem 4

Two diastereoisomers of this cyclic keto-lactam have been prepared. The NMR spectra have many overlapping signals but the proton marked in green in the textbook and here ringed can clearly be seen. In isomer A it is δ_H 4.12 (1H, q, J 3.5) and isomer B has δ_H 3.30 (1H, dt, J 4, 11, 11). Which isomer has which stereochemistry?

Purpose of the problem

Assignment of three-dimensional stereochemistry from NMR when only one signal can be made out clearly.

Suggested solution

The two isomers have *cis* and *trans* ring junctions so we should first make conformational drawings. The *trans* compound is easy as it has a fixed conformation like a *trans*-decalin (p. 463). The *cis* compound can have two conformations as both rings can flip.

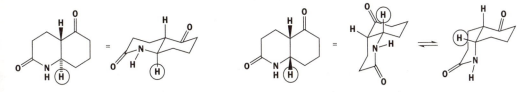

■ This compound was used in a total synthesis of luciduline by J. Szychowski and D. B. MacLean, *Can. J. Chem.*, 1979, **57**, 1631.

The vital proton is clearly axial in isomer B as it has two couplings to other axial hydrogens (10 Hz) and this must be the *trans*-isomer. Isomer A has three equal small couplings (3.5 Hz) and fits one of the conformers of the *cis*-isomer.

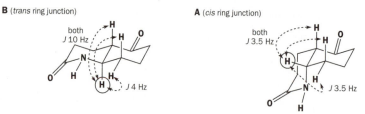

B (*trans* ring junction)　　　　　　　　　　　　　**A** (*cis* ring junction)

Problem 5

How would you determine the stereochemistry of these two compounds?

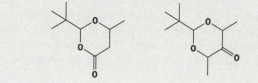

Purpose of the problem

An approach from the other end: how would you do the job? Also to remind you that we can determine relative stereochemistry (i.e. which diastereoisomer do we have?) but not absolute stereochemistry (i.e. which enantiomer do we have?) by NMR.

Suggested solution

■ It is easier, if we can, to put carbonyl groups at the ends of the rings. The structures in Problem 4 show how much better this looks. Here we can easily put the carbonyl groups at the ends.

By NMR, of course. Both compounds are six-membered rings so we should first make conformational drawings of all the possibilities. The first compound can have the *t*-butyl and methyl groups *cis* or *trans* and the *t*-butyl group will go equatorial. The second compound can have both methyl groups on the same side as the *t*-butyl group, both on the other side, or one on each side. Two of these are *meso* compounds, though this doesn't affect the assignment.

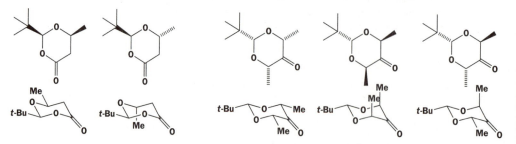

The key H atoms in the NMR are those we can see! In the first compound they are marked Hᵃ. Hᵈ tells us nothing as it is coupled to nothing. Hᵇ and Hᶜ are useful in that they tell us about Hᵃ. Hᵃ is easily identified by its quartet coupling to the methyl group. If it has a large axial-axial coupling (about 10 Hz) to Hᵇ then we have the *cis* compound, but if all its couplings are small (perhaps < 4 Hz, as in Problem 4) then it is the *trans* compound.

In the second compound a difficulty emerges: there is no coupling! We can tell by symmetry whether we have, on the one hand, the *cis, cis-* or the *trans, trans*-compound or, on the other hand, the *cis, trans*-compound. But how can we tell which of the symmetrical compounds we have? The chemical shifts will be different but we won't know which is which. However, if we irradiate the signal for the methyl groups we should get a strong NOE effect (p. 844) at H^a in the all-*cis* compound and nothing in the *trans* compound.

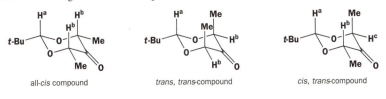

all-*cis* compound   *trans, trans*-compound   *cis, trans*-compound

Problem 6

The structure and stereochemistry of the antifungal antibiotic ambruticin was in part deduced from the NMR spectrum of this simple cyclopropane. Interpret the NMR spectrum and show how it gives definite evidence on the stereochemistry.

δ_H 1.21 (3H, d, J 7 Hz), 1.29 (3H, t, J 9), 1.47 (9H, s), 1.60 (1H, t, J 6), 1.77 (1H, ddq, J 6, 13, 7), 2.16 (1H, dt, J 6, 13), 4.18 (2H, q, J 9), 6.05 (1H, d, J 20), and 6.62 (1H, dd, J 13, 20).

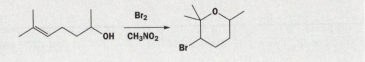

Purpose of the problem

Assigning a more complex NMR and making deductions about stereochemistry in small rings.

Suggested solution

In cyclopropanes the *cis* coupling is usually larger than the *trans* coupling because the dihedral angle for *cis* Hs is 0° but that of *trans* Hs is not 180°. Assigning the three ring hydrogens depends on (a) the quartet coupling to the methyl group and (b) the 13 Hz coupling to the proton on the alkene. This means that the third (6 Hz triplet) must be next to the carbonyl group. The two *trans* couplings round the ring are the same (6 Hz) and smaller than the *cis* coupling (7 Hz). The geometry of the double bond is on more certain grounds (20 Hz can only be a *trans* coupling).

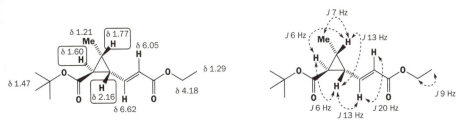

Problem 7

In Chapter 20 we set a problem asking you what the stereochemistry of a product was. Now we can give you the NMR spectrum of the product and ask: how do we *know* the stereochemistry of the product? You need only the partial NMR spectrum: δ_H 3.9 (1H, ddq, J 12, 4, 7) and 4.3 (1H, dd, J 11, 3).

Purpose of the problem

You can prove for yourself that this deduction made in earlier chapters is true.

Suggested solution

A saturated six-membered ring means conformational drawings again. The proton next to O must be the ddq because it is next to the methyl group and the one at 4.3 must therefore be the proton next to Br. Both have one large coupling (12 or 11 Hz) and this must be axial-axial so both protons must be axial.

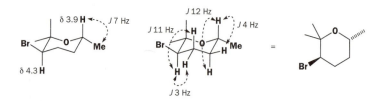

Problem 8

The structure of a Wittig product intended as a prostaglandin model was established by the usual methods – except for the geometry of the double bond. Irradiation of a signal at 3.54 (2H, t, J 7.5) led to an enhancement of another signal at δ_H 5.72 (1H, t, J 7.1) but not to a signal at δ_H 3.93 (2H, d, J 7.1). What is the stereochemistry of the alkene? How is the product formed?

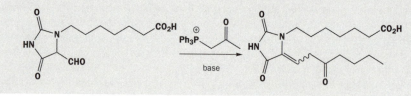

Purpose of the problem

Exploration of the usefulness of the NOE when coupling constants are of little use.

Suggested solution

The signal at 5.72 is obviously the proton on the alkene. The 2H triplet must be next to the N atom and the 2H doublet must be between the alkene and the ketone. The NOE falls off as the sixth power of the distance so the first (*E*-) alkene is correct. The second (*Z*-) alkene would give the opposite result. The NOE also suggests that the predominant conformation of the top chain must be as shown with the chain bent away from the alkene.

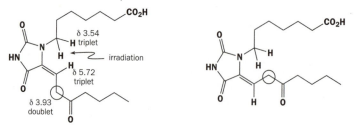

The product must be formed by a Wittig reaction of the usual sort (probably *E*-selective as this is a stabilized ylid) followed by base-catalysed isomerization of the alkene into conjugation with the ring carbonyl group. This must be under thermodynamic control as too must be the geometry of the final alkene.

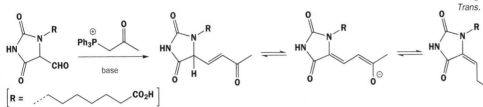

■ This work was done by chemists at the Wellcome Research Centre and is described by A. G. Caldwell and group, *J. Chem. Soc., Perkin Trans. 1*, 1980, 495.

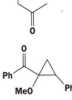

Problem 9

How would you determine the stereochemistry of this cyclopropane? The NMR spectra of the three protons on the ring are given: {δ_H 1.64 (1H, dd, J 6, 8), 2.07 (1H, dd, J 6, 10), and 2.89 (1H, dd, J 10, 8).

■ This work was done by chemists at University College, Dublin and is described by J. A. Donnelly and group, *J. Chem. Soc., Perkin Trans. 1*, 1979, 2629.

Purpose of the problem

Finding out how to determine stereochemistry when there is a quaternary chiral centre.

Suggested solution

Irradiation of the MeO signal (2.07) should give an NOE and reveal which of H^1 and H^2 is on the same side of the ring and may show that H^3 is also on the same side. Irradiation of H^1 and H^2 in turn (you were given the NMR shifts of this part of the molecule so you could see that, fortunately, H^1, H^2, and H^3 are well separated in the NMR spectrum) should give definite information on the relative stereochemistry. The original work shows clearly that the compound is one diastereoisomer but which one was not determined.

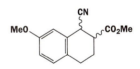

Problem 10

A chemical reaction produces two diastereoisomers of the product. Isomer A has δ_H 3.08 (1H, dt, J 4, 9, 9) and 4.32 (1H, d, J 9) while isomer B has δ_H 4.27 (1H, d, J 4). The other protons overlap. Isomer B is converted into isomer A on treatment with base. What is the stereochemistry of A and B?

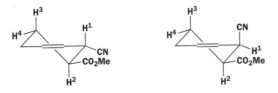

Purpose of the problem

Determining stereochemistry on the minimum of information.

Suggested solution

There are two diastereoisomers and the difference in coupling constants is striking. These are not true cyclohexanes but partly flattened by the benzene ring and best drawn as cyclohexenes (p. 469). You should imagine the benzene ring coming towards you where the double bond is drawn.

The two protons we can see in isomer A must be H^1 and H^2 as they have the largest shifts. The proton with only one coupling must be H^1 as it has only one neighbour (H^2). Isomer A has a 9 Hz coupling between H^1 and H^2 so both these protons must be axial. Isomer A is therefore the *trans* isomer. The signal for H^2 is a dt because it has two axial neighbours (H^1 and H^3) and one equatorial neighbour (H^4). Isomer B shows H^1 only and it is clearly equatorial ($J_{12} = 4$). Though we can't see H^2 it must be axial as B must be the *cis*-isomer since it is different from A.

$$A =$$ $$B =$$

Problem 11

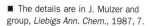

Muscarine, the poisonous principle of the death cap mushroom, has the following structure and proton NMR spectrum. Assign the spectrum. Can you see definite evidence for the stereochemistry? All couplings in Hz; signals marked * exchange with D$_2$O.

δ_H 1.16 (3H, d, J 6.5), 1.86 (1H, ddd, J 12.5, 9.5, 5.5), 2.02 (1H, ddd, J 12.5, 2.0, 6.0), 3.36 (9H, s), 3.54 (1H, dd, J 13, 9.0), 3.74 (1H, dd, J 13, 1.0), 3.92 (1H, dq, J 2.5, 6.5), 4.03 (1H, m), 4.30* (1H, d, J 3.5), and 4.68 (1H, m).

Purpose of the problem

Demonstrating that it can be very difficult to determine stereochemistry even with all the information.

Suggested solution

■ The details are in J. Mulzer and group, *Liebigs Ann. Chem.*, 1987, 7.

Coupling constants around five-membered rings tend to be much the same whether they are ^2J (geminal), $^3J_{cis}$, or $^3J_{trans}$ (vicinal). Even so, the two diastereotopic CH$_2$ groups are easy to find with their large ^2J couplings of 13 and 12.5. The one with one extra coupling must be in the side chain and the others in the ring. Here is the full analysis. You will see that it is, in fact, very difficult to get conclusive evidence on stereochemistry though you should see that, in general, *cis* couplings tend to be larger than *trans*.

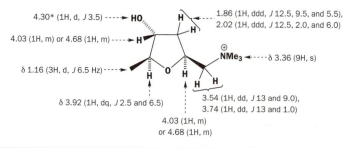

4.30* (1H, d, J 3.5) ----► HO,,,,,,   1.86 (1H, ddd, J 12.5, 9.5, and 5.5),
2.02 (1H, ddd, J 12.5, 2.0, and 6.0),

4.03 (1H, m) or 4.68 (1H, m) ----► H

NMe$_3$ ◄----δ 3.36 (9H, s)

δ 1.16 (3H, d, J 6.5 Hz) ----►

δ 3.92 (1H, dq, J 2.5 and 6.5)

3.54 (1H, dd, J 13 and 9.0),
3.74 (1H, dd, J 13 and 1.0)

4.03 (1H, m)
or 4.68 (1H, m)

Problem 12

An antifeedant compound that deters insects from eating food crops has the gross structure shown below. Some of the NMR signals that can clearly be made out are also given. Since NMR coupling constants are clearly useless in assigning the stereochemistry, how would you set about it?

δ_H 2.22 (1H, d, J 4), 2.99 (1H, dd, J 4, 2.4), 4.36 (1H, d, J 12.3), 4.70 (1H, dd, J 4.7, 11.7), 4.88 (1H, d, J 12.3)

Purpose of the problem

Assigning stereochemistry when there are isolated spin systems.

■ S. V. Ley and group, *J. Chem. Soc., Chem. Commun.*, 1981, 1001.

Suggested solution

The signals given are all of protons next to electronegative atoms. There are two AB systems for the diastereotopic CH_2 groups in the epoxide and the CH_2OAc group. The CH_2OAc group is 'normal' with ^2J 12.3 and a large shift (δ_H 4.36 and 4.88). The epoxide is unusual with a much smaller ^2J of only 4 Hz (actually normal for epoxides) and smaller shifts (δ_H 2.22 and 2.99). Both appear isolated but one of the epoxide protons shows long-range coupling to some other proton. These are no help with the stereochemistry but the proton next to the other OAc group (δ_H 4.70) is clearly axial with typical axial/axial (11.7) and axial/equatorial (4.7) couplings.

The answer came from NOE experiments There were large NOEs between the δ_H 4.70 axial proton and one of the epoxide protons and between the axial methyl group and one of the protons on the CH_2OAc group. The stereochemistry is like this.

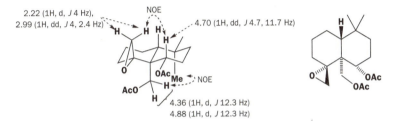

2.22 (1H, d, *J* 4 Hz),
2.99 (1H, dd, *J* 4, 2.4 Hz)
4.70 (1H, dd, *J* 4.7, 11.7 Hz)
4.36 (1H, d, *J* 12.3 Hz)
4.88 (1H, d, *J* 12.3 Hz)

Problem 13

The seeds of the Costa Rican plant *Ateleia herbert smithii* are avoided by all seed eaters (except a weevil that adapts them for its defence) because they contain two toxic amino acids (IR spectra like other amino acids). Neither compound is chiral. What is the structure of these compounds? They can easily be separated because one (A) is soluble in aqueous base but the other (B) is not.

A is $C_6H_9NO_4$ (mass spectrum) and has δ_C 34.0 (d), 40.0 (t), 56.2 (s), 184.8 (s), and 186.0 (s). Its proton NMR has three exchanging protons on nitrogen and one on oxygen and two complex signals at δ_H 2.68 (4H, A_2B_2 part of A_2B_2X system) and 3.37 (X part of A_2B_2X system) with J_{AB} 9.5, J_{AX} 9.1, and J_{BX} small.

B is $C_6H_9NO_2$ (mass spectrum) and has δ_C 38.0 (d), 41.3 (t), 50.4 (t), 75.2 (s), and 173.0 (s). Its proton NMR spectrum contains two exchanging protons on nitrogen and δ_H 1.17 (2H, ddd, *J* 2.3, 6.2, 9.5), 2.31 (2H, broad m), 2.90 (1H, broad t, *J* 3.2), and 3.40 (2H, broad s).

Because the coupling pattern did not show up clearly as many of the coupling constants are small, decoupling experiments were used. Irradiation at δ_H 3.4 simplifies the δ_H 2.3 signal to (2H, ddd, *J* 5.8, 3.2, 2.3), sharpens each line of the ddd at 1.17, and sharpens the triplet at 2.9.

Irradiation at 2.9 sharpens the signals at 1.17 and 2.9 and makes the signal at 2.31 into a broad doublet, *J* about 6. Irradiation at 2.31 sharpens the signal at 3.4 slightly and reduces the signals at 2.9 and 1.17 to broad singlets. Irradiation at 1.17 sharpens the signal at 3.4 slightly so that it is a broad doublet, *J* about 1.0, sharpens the signal at 2.9 to a triplet, and sharpens up the signal at 2.31 but irradiation here had the least effect.

This is quite a difficult problem but the compounds are so small (C_6 only), have no methyl groups, and have some symmetry so you should try drawing structures at an early stage.

Purpose of the problem

A difficult problem for those of you who like to try the real thing.

Suggested solution

■ Then read all about it in E. A. Bell
and group, *J. Am. Chem. Soc.*,
1980, **102**, 1409.

The compounds are unusual cyclobutane amino acids. The simplest thing is to give you the answers and let you see if your suggestions fit and if you can assign the spectra.

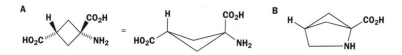

Problem 1

Comment on the control over stereochemistry achieved in this sequence.

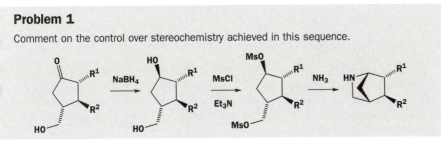

Purpose of the problem

A gentle introduction to stereochemical control in rings.

Suggested solution

The reducing agent could attack either side of the ring in the first step but it reacts with the free OH group to produce a new reducing agent and hydride delivery is intramolecular from the bottom face. The mesylation (mechanism on p. 484) does not affect the stereochemistry as no bonds are formed or broken to any of the chiral centres.

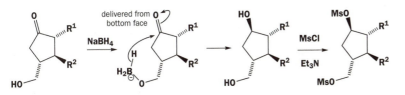

The reaction with ammonia probably starts with the primary alkyl mesylate and the second step is then intramolecular. It is also stereospecific – the S_N2 displacement must occur with inversion and, fortunately, the amine is already on the bottom face.

■ This chemistry is described by E. J. Corey and group in *Tetrahedron Lett.*, 1979, 671.

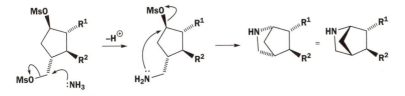

Problem 2

Explain the stereochemistry of this sequence of reactions, noting the second step in particular.

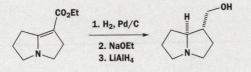

Purpose of the problem

To show how even 'non-reactions' can influence stereochemistry.

Suggested solution

There appears to be a mistake at first since the hydrogenation will add a molecule of hydrogen to one face of the alkene to give what appears to be the wrong product.

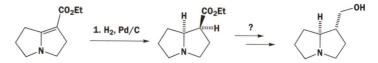

■ H. W. Pinnick and Y. Chang, *Tetrahedron Lett.*, 1979, 837.

The second step is now rather important. Ethoxide will form the enolate of the ester reversibly and turn it to the outside, the convex face, of the molecule. Though nitrogen is not normally a fixed chiral centre because it undergoes rapid pyramidal inversion, here it is fixed by the need of the five/five fused ring system to have a *cis* ring junction. The last step is just the reduction of the ester with no change in the stereochemistry.

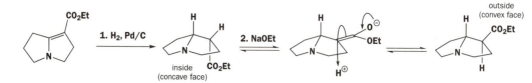

Problem 3

Explain how the stereo- and regiochemistry of these compounds are controlled. Why is the epoxidation only moderately stereoselective, and why does the amine attack where it does?

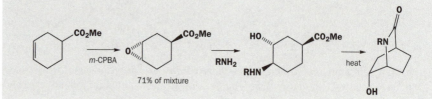

71% of mixture

Purpose of the problem

An exercise to remind you how important conformational analysis is in any stereoselective reaction of saturated six-membered rings.

Suggested solution

The conformation of the cyclohexene will have the CO_2Me group in an equatorial position, almost in the plane of the alkene so that it offers only slight steric hindrance. The opening of the epoxide is dominated by conformation. Approach a would give a twist-boat product but approach b gives the chair cyclohexane observed.

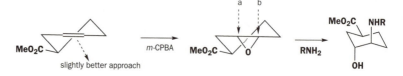

slightly better approach

Before cyclization, the compound must go into a boat form so that the amine and the ester can approach one another. This boat is fixed in the final cyclic amide. The cyclization does not affect the stereochemistry.

■ J. W. Huffman and group, *J. Org. Chem.*, 1967, **32**, 697.

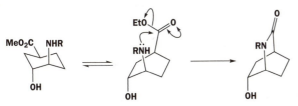

Problem 4

What controls the stereochemistry of this product? You are advised to draw a mechanism first and then consider the stereochemistry.

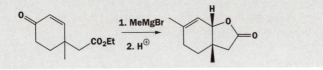

Purpose of the problem

To show how ring-closing reactions, particularly those on the side of an already existing ring, can give excellent stereochemical control. And again – the importance of conformation in six-membered rings.

Suggested solution

We'd better draw the mechanism first, as the question says. Grignard reagents tend to do direct addition to enones and the product shows that the methyl group has done just that. But the resulting OH is in the wrong position to cyclize to the ester and there doesn't seem to be much scope for stereochemical control so we probably get a mixture of diastereoisomers.

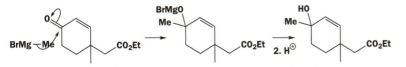

This first product is a tertiary allylic alcohol so it will lose water under the acidic work-up conditions and readdition of water to the other end of the allyl cation gives an alcohol that might cyclize to the final product.

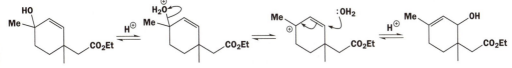

Cyclization of the alcohol with an OH group *cis* to the ester will occur much faster than if they are *trans*. As all the alcohols are in equilibrium via the allyl cation, only the *cis* lactone will be formed regardless of the stereochemistry of the tertiary alcohol. Another and probably better explanation is that the ester or the acid derived from it by hydrolysis adds to the allyl cation.

■ E. W. Colvin and R. A. Raphael, *J. Chem. Soc., Chem. Commun.*, 1971, 858.

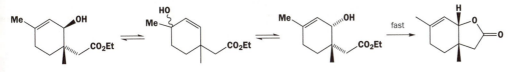

Problem 5

Why is one of these esters more reactive than the other?

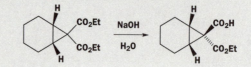

Purpose of the problem

Reminder of the power of folded molecules with concave and convex sides.

■ P. F. Hudrlik and group, *J. Am. Chem. Soc.*, 1973, **95**, 6848.

Suggested solution

The molecule is folded around the ring junction with one of the ester groups inside the fold (on the concave face) and the other out in the open on the convex face. Since the CO_2Et group gets *larger* in the rate-determining step in ester hydrolysis – attack of hydroxide ion on the carbonyl group – the outside CO_2Et group will be hydrolysed faster.

■ rds stands for 'rate-determining step'.

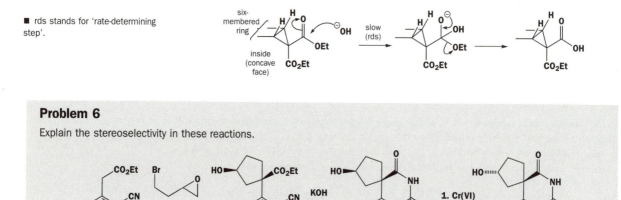

Problem 6

Explain the stereoselectivity in these reactions.

Purpose of the problem

Another cyclization reaction and an example of a controlled inversion of configuration.

Suggested solution

The first stereoselective reaction is surprising as it may appear that the initial alkylation decides the stereochemistry. But that is not the case as you will see if you draw the mechanism. The ester enolate is very easily formed as it is stabilized by the pyridine ring and the nitrile as well as by the ester. Even a weakish base like carbonate is good enough.

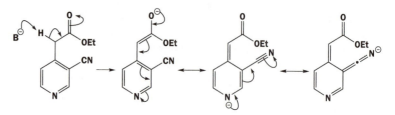

The first intermediate produced by alkylation with the primary alkyl bromide (or the epoxide) has two chiral centres and will no doubt be formed as a mixture of diastereoisomers. But this doesn't matter as the enolate has to be reformed for the next alkylation and that destroys one of the chiral centres. We are back to a single compound again.

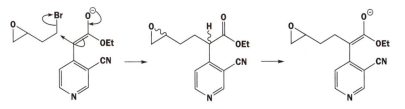

All depends on the arrangement of the molecule for the cyclization step. The mechanism is straightforward enough but drawing the transition state is tricky. We offer two drawings: an attempt at a conformational drawing and a Newman projection. The vital feature in both is that the enolate carbon, and the C–O bond of the epoxide must be collinear. The molecule folds so that the end of the five-membered ring bends upwards away from the large pyridine ring. You may have a better drawing. This is not obvious even when you know the answer.

■ It was not predicted by the chemists who did the work (A. S. Kende and T. P. Demuth, *Tetrahedron Lett.*, 1980, **21**, 715): 'The product was obtained as a single racemic diastereomer rather than the expected *cis/trans* mixture.'

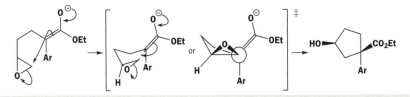

Problem 7

A problem from the chapter. Draw a mechanism for this reaction and explain why it goes so much better than the elimination on a β-lactone.

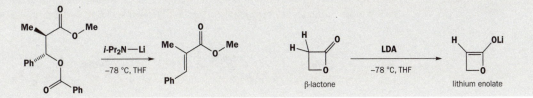

Purpose of the problem

A reminder that cyclic molecules may stop some reactions as well as helping to control others.

Suggested solution

Formation of the lithium enolate leads to elimination in the first example. If we rotate the central bond of the starting material we see that this would be a *cis* elimination so it must be E1cB and not E2. The stereochemistry of the product is determined stereoselectively by the best arrangement in the transition state for the elimination step.

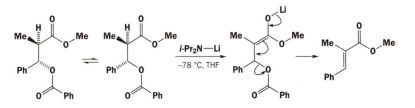

■ J. Mulzer and T. Kerkmann, α
Deprotonation of β-lactones – an
example of a 'forbidden' β
elimination, *J. Am. Chem. Soc.*,
1980, **102**, 3620.

In the elimination step the C–O bond joining the benzoate leaving group to the rest of the molecule must be parallel to the p orbitals of the lithium enolate. There is no problem about this. In the β-lactone, however, the C–O single bond is forced to be orthogonal to the p orbitals of the enolate and elimination is nearly impossible.

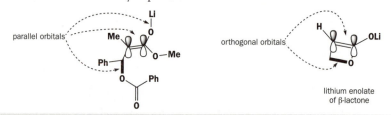

Problem 8

Another problem from the chapter. The synthesis of the starting material for this reaction is a good example of how cyclic compounds can be used in a simple way to control stereochemistry. Draw mechanisms for each reaction and explain the stereochemistry.

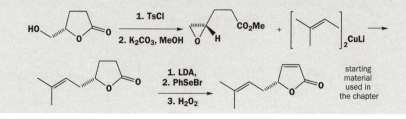

Purpose of the problem

Reinforcement of material from the chapter (p. 854) and some important reactions.

Suggested solution

Tosylation of the primary alcohol is followed by ester exchange with methanol to release the anion of a secondary alcohol, which promptly closes to the epoxide. There is no change at the chiral centre.

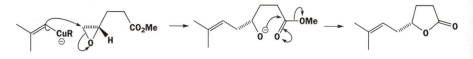

Now, the vinyl cuprate attacks the epoxide at its less substituted end releasing the same oxyanion, which promptly closes the lactone again. Once more there is no change at the chiral centre. This switching of an oxygen atom from lactone to epoxide and back again is a popular method for stereochemical control.

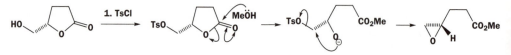

Finally, the double bond is introduced by selenium chemistry (p. 1271). The steps are straightforward and the geometry of the alkene is dictated by the ring.

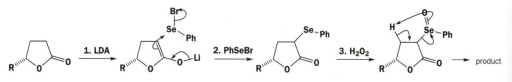

Problem 9

A revision problem. Suggest mechanisms for the reactions used to make this starting material used in the chapter.

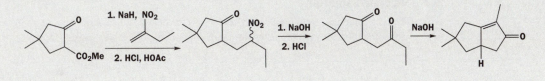

Purpose of the problem

Revision of mechanisms in the context of Chapter 33 (p. 864).

Suggested solution

The stable enolate of the keto-ester adds to the nitroalkene in a conjugate addition for which both partners are excellent (Chapter 29). Hydrolysis and decarboxylation gives the first intermediate. You can discover how this starting material was made in Problem 40.4, p. 1076.

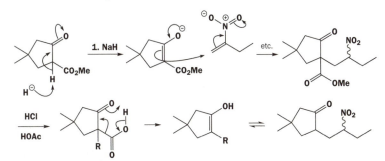

Conversion of the nitro compound to the ketone by hydrolysis of the 'enol' derivative is followed by an extremely good intramolecular aldol reaction giving the fused five-membered ring.

■ J. Froborg and G. Magnusson, *J. Am. Chem. Soc.*, 1978, **100**, 6728.

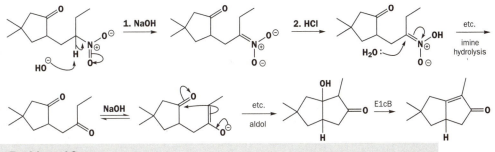

Problem 10

And another problem from the chapter. Here also draw a mechanism for the formation of the starting material. You have never seen the cyclopropane reagent, but think how it might react...

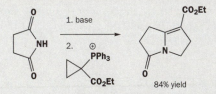

84% yield

Stuck? The first step opens the three-membered ring and the second step is a well-known alkene-forming reaction...

Purpose of the problem

Revision of mechanisms in the context of Chapter 33 (p. 865).

Suggested solution

The last reaction is, of course, a Wittig reaction so the first must be nucleophilic attack by the anion of the imide on the cyclopropane with the phosphonium ylid as the leaving group.

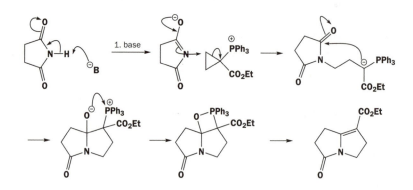

Problem 11

In the chapter we introduced the selective reduction of the Wieland–Miescher ketone (p. 869). The problem is: can you suggest a reason for this stereoselectivity?

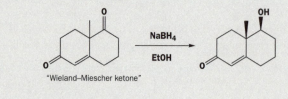

"Wieland–Miescher ketone"

Purpose of the problem

Back to the main subject of the chapter: explaining stereoselectivity in reactions of cyclic compounds.

Suggested solution

This is an example of a small nucleophile adding to a ketone in a six-membered ring. The bit added (H) is smaller than the bit already there (O) and the reagent (NaBH$_4$) is also small so axial attack is preferred so that the larger substituent OH) ends up equatorial. A larger reagent, say NaAlH(OBu-t)$_3$, would add equatorially (see p. 852).

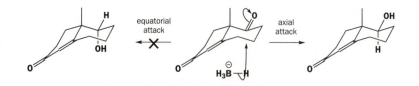

Problem 12

Suggest mechanisms for these reactions and explain the stereochemistry.

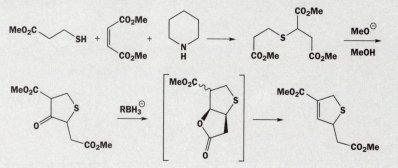

Purpose of the problem

Though ultimately it disappears, explaining the stereoselectivity of the reduction is interesting and applicable elsewhere.

Suggested solution

This looks complicated but it's just a question of working through each reaction. The stereochemistry needs some thought. The first reaction is a conjugate addition of a sulfur nucleophile to a very electrophilic alkene. The base used, piperidine, is an ordinary secondary amine but that is basic enough (pK_{aH} about 11) to produce reasonable amounts of anion from the thiol (pK_a also about 10). There is no stereochemistry in this step.

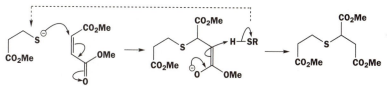

The next stage is an intramolecular Claisen ester condensation. We can easily discover which enolate reacts with which ester by drawing the starting material in the shape of the product. The alternatives are three- or six-membered rings: five-membered rings are more stable than three- and more rapidly formed than six-membered. Under the reaction conditions there is no stereochemistry as the product exists as a stable conjugated enolate ion (p. 724).

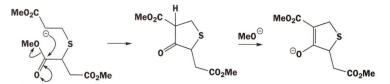

The reduction is the stereoselective step and the hydrogen atom is added from the face of the ketone opposite the larger substituent. The oxyanion so formed reacts with the ester to give a five-membered lactone. The other ester group equilibrates via the stable enol or enolate until the ketone is reduced. Then it is fixed in whatever configuration it happens to be.

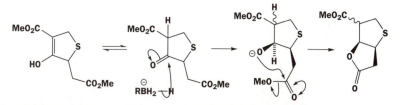

Finally, the second ring is opened again by an E1cB elimination via the enolate of the ester. This removes all the stereochemistry and gives the starting material for the chemistry described in the chapter (p. 876).

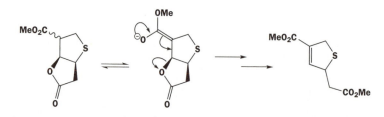

Problem 13

Hydrolysis of a bis-silylated ene-diol gives a hydroxy-ketone A whose stereochemistry is supposed to be as shown. Reduction of A gives a diol B. The ^{13}C NMR spectrum of B has five signals: one in the 100–150 p.p.m. range, one in the 50–100 p.p.m. range, and three below 50 p.p.m. The proton NMR of the three marked hydrogens in A is given below with some irradiation data. Does this information give you confidence in the stereochemistry assigned to A? You may wish to consider the likely stereochemical result of the reduction of A.

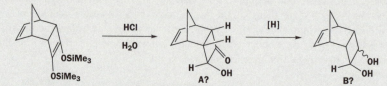

A has δ_H 4.46 (1H, dd, J 9.0, 3.8 Hz), 3.25 (1H, ddd, J 9.0, 7.5, 4.5 Hz), and 3.48 (1H, ddd, J 7.5, 5.5, 3.8 Hz). Irradiation at 3.48 collapses the signal at 4.46 to (d, J 9.0 Hz) and the signal at 3.25 to (dd, J 9.0, 4.5 Hz); irradiation at 4.46 collapses the signal at 3.48 to (dd, J 7.5, 5.5) and the signal at 3.25 to (dd, J 7.5, 4.5).

Purpose of the problem

NMR revision. In determining stereochemistry, as in other parts of chemistry, the evidence does matter.

Suggested solution

The structure of B is obviously symmetrical and the two OH groups must be *cis* (both in or both out). We should expect A to be reduced to the *trans* compound as the nucleophile would attack from the convex face of the folded molecule. This suggests that B is the *cis* (inside) diol and A has its one OH group inside too.

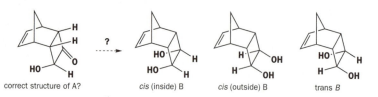

| correct structure of A? | *cis* (inside) B | *cis* (outside) B | *trans* B |

The NMR spectrum of A has two large couplings (7.5 and 9.0) between the marked Hs. These look more like *cis* than *trans* couplings (*cis* couplings are larger than *trans* couplings in four-membered rings). The signal at 4.46 must be the proton next to OH and the other two must have additional couplings to protons H^X and H^Y. All this data fits if protonation of the silyl enol ether occurs on the *exo* face to give A and reduction of A also occurs on the *exo* face.

■ Details in B. M. Trost and group, *J. Org. Chem.*, 1978, **43**, 4559.

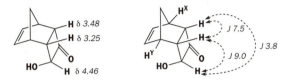

Suggested solutions for Chapter 34 **34**

Problem 1

How would you make each diastereoisomer of this product from the same alkene?

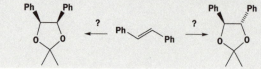

Purpose of the problem

A gentle introduction to stereochemical control in open-chain compounds.

Suggested solution

The compounds are acetals and can both be made from the corresponding diols with no change in stereochemistry. The question really is: how do you make *cis* and *trans* diols from the alkene?

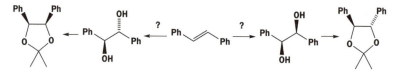

The *cis* diol is best made by dihydroxylation with OsO_4 as the reagent and a co-oxidant to regenerate it. The *trans* diol comes from the epoxide by nucleophilic attack with water.

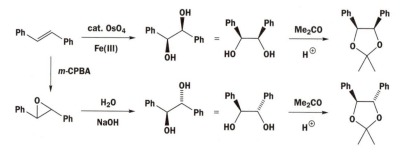

Problem 2

Explain the stereoselectivity shown in this sequence of reactions.

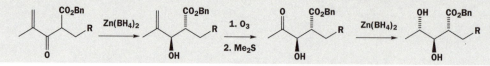

Purpose of the problem

Chelation-controlled reduction is an important method for stereochemical control in open-chain compounds.

Suggested solution

In both reductions the zinc atom is coordinated to the oxygen of the nearer functional group (CO_2Et in the first and OH in the second) and the oxygen atom of the ketone being reduced. This fixes the conformation of the molecule and the borohydride ion attacks from the less hindered side (p. 893). *Anti* stereochemistry results in both cases.

■ T. Nakata and group, *Tetrahedron Lett.*, 1983, **24**, 2657.

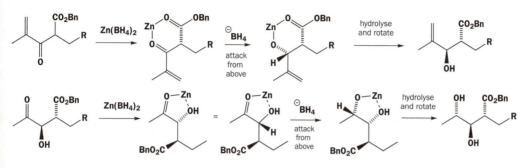

Problem 3

How is the relative stereochemistry of this product controlled? Why was this method chosen?

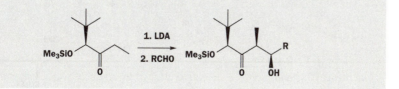

Purpose of the problem

This may seem trivial but the principle is important.

Suggested solution

The relationship between the two chiral centres in the product is 1,5 and this is too remote for any realistic hope of control. The only way is to disconnect between the two centres and add a removable anion-stabilizing group to the nucleophilic synthon and a leaving group to the electrophilic synthon. Sulfone and iodide are good choices. The starting materials must, of course, be single enantiomers – then only one diastereoisomer can be produced.

■ K. Mori and group, *Tetrahedron*, 1983, **39**, 2439.

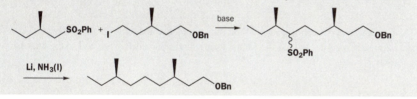

Problem 4

Explain the stereochemical control in this reaction, drawing all the intermediates.

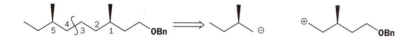

Purpose of the problem

The aldol is a versatile and important way of controlling open-chain stereochemistry by way of a cyclic transition state.

Suggested solution

The geometry of the enolate is all important (p. 898) and here the large *t*-butyl group will give just the *Z*-lithium enolate. Then the mechanism has a saturated six-membered cyclic (Zimmerman–Traxler) transition state with the R group of the aldehyde RCHO taking up an equatorial position. This gives the *syn* aldol.

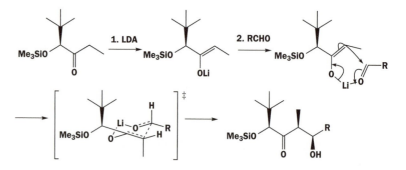

Notice that in the transition state: (1) R chooses to go equatorial; (2) the methyl group is forced axial because it is *cis* to OLi; (3) the *t*-butyl group forces the aldehyde to add to the back face of the enolate as drawn.

Problem 5

When this hydroxy-ester is treated with a twofold excess of LDA and then alkylated, one diastereoisomer of the product predominates. Why?

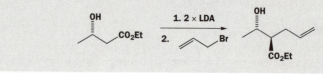

Purpose of the problem

Analysis of an apparently simple case where chelation has the last word.

Suggested solution

The first LDA molecule removes the OH proton and only the second gives the enolate. The enolate is held in a ring by chelation to the first lithium atom so that the allyl group adds to the less hindered face – opposite the methyl group.

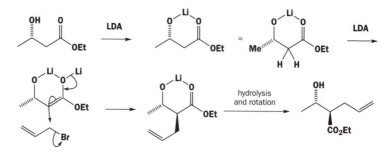

Problem 6

Explain how the stereochemistry of this epoxide is controlled.

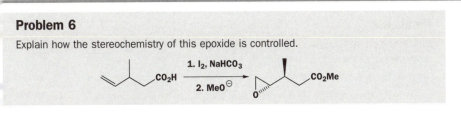

Purpose of the problem

An example of the important iodolactonization reaction.

Suggested solution

The bicarbonate ($NaHCO_3$) is a strong enough base to produce the anion of the carboxylic acid. Iodine attacks the alkene reversibly to give a mixture of diastereoisomers of the iodonium ion. If the I^+ and Me groups are on the same side of the chain, the CO_2^- group can attack the iodonium ion from the back and set up a *trans* iodolactone. The iodolactone is cleaved by methoxide and the oxyanion displaces iodide to give the epoxide.

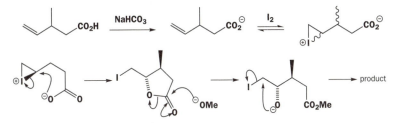

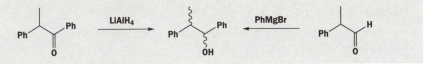

Problem 7

Explain how these two reactions give different diastereoisomers of the product.

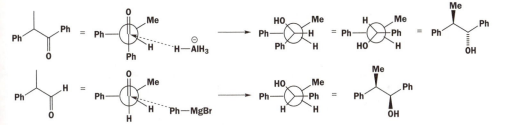

Purpose of the problem

Practice at the Felkin-Anh style of stereochemical control.

Suggested solution

In each case we have nucleophilic attack on a carbonyl group with a neighbouring chiral centre. The Felkin analysis tells us first to put the largest group perpendicular to the carbonyl group and then to bring the nucleophile in alongside the smallest substituent. This is best shown with a Newman projection (p. 888). In the first case it is better to rotate the front atom so that the two Ph groups are at 180° and we can draw the structures in the same way.

Problem 8

Explain the stereoselectivity in this reaction. What isomer of an epoxide would be produced on treatment of the product with base?

Purpose of the problem

Practice at an electronically controlled Felkin–Anh style of stereochemical control.

Suggested solution

In this case the chlorine dominates because it has an electronic interaction with the carbonyl group. The two alkyl chains come out opposite one another so it is easy to draw the product in a sensible fashion.

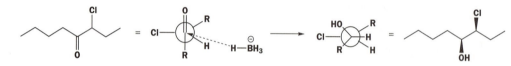

To draw the stereochemistry of the epoxide formation it is sensible to put the reacting groups in the plane of the paper and arranged so that the oxyanion can do an S_N2 displacement.

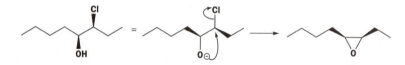

Problem 9

How could this cyclic compound be used to produce the open-chain compound with correct relative stereochemistry?

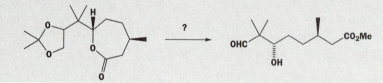

Purpose of the problem

Practice at relating the stereochemistry of cyclic and open-chain compounds.

Suggested solution

We should first discover which atoms in the cyclic compound provide which atoms in the product. Numbering the atoms is the easiest way and it shows little change except that C9 has gone and C8 has become an aldehyde.

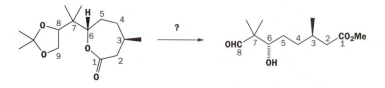

We need to hydrolyse the ester and the acetal and oxidize the 1,2-diol. The stereochemistry at C3 and C7 is unchanged and neither is threatened by any of the reaction conditions.

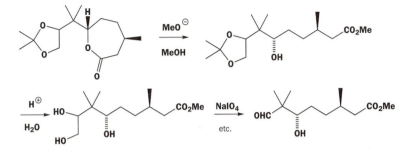

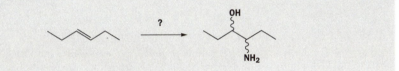

Problem 10

How would you transform this alkene stereoselectively into either of the diastereoisomers of the amino-alcohol?

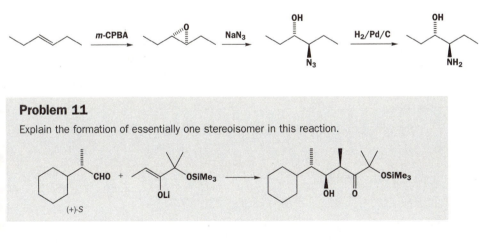

Purpose of the problem

A more difficult extension of Problem 1.

Suggested solution

Opening the epoxide with a nitrogen nucleophile makes one isomer. At least the alkene is symmetrical so it doesn't matter which end of the epoxide is attacked by the nucleophile. We have chosen azide ion (N_3^-) as the nucleophile. Ammonia will also do as, although it can react repeatedly, it usually behaves itself with epoxides. You were not asked to make both diastereoisomers, so we can stop there.

Problem 11

Explain the formation of essentially one stereoisomer in this reaction.

Purpose of the problem

A more difficult extension of Problem 4 with added Felkin complications.

Suggested solution

The *syn* selectivity of the aldol reaction comes from the chair conformation of the cyclic (Zimmerman–Traxler) transition state. Ignoring the stereochemistry of the aldehyde we have this simplified explanation.

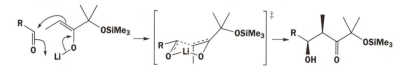

We have inevitably drawn the *syn* aldol product as one enantiomer but so far there is no control over absolute stereochemistry. The aldehyde is itself a single enantiomer and so the two faces of the carbonyl group are diastereotopic and which one the enolate will attack would normally be determined by the usual Felkin argument.

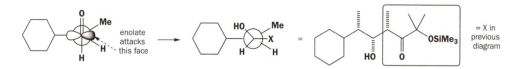

■ And to the surprise of Satoru Masamune and his co-workers, *Angew. Chem., Int. Ed. Engl.*, 1980, **19**, 557.

To our surprise this is not the preferred isomer. In fact the 'anti-Felkin' isomer predominates by about 3:1. The compound is entirely the *syn* aldol but attack has occurred on the aldehyde in the alternative conformation.

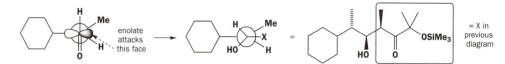

There is an important lesson to be learnt here. The principles we have been explaining are generally true but in any individual case the result may not follow the principle. This is particularly true of Felkin control with aldehydes as H and O are not that different in size. You should first apply the principle (here Felkin control) and then check the result. You should not be ashamed if you got this one wrong.

Problem 12

How would you attempt to transform this allylic alcohol into both diastereoisomers of the epoxide stereoselectively? You are not expected to estimate the degree of success.

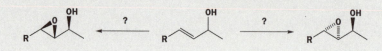

Purpose of the problem

An exercise in *true* control – getting whichever isomer you want.

Suggested solution

The OH group will direct a simple *m*-CPBA epoxidation by hydrogen-bonding to the reagent and delivering the oxygen atom to the same face in a Houk conformation (p. 877). To get the *anti* epoxide we must block the OH group with a large removable protecting group to get control by steric hindrance. You might have chosen a large silyl group, a benzyl group, or many others. In the book (p. 884) we suggested an acetate.

■ Following on from the last problem, though you do get these selectivities with the reagents given, they are not often very good. Depending on R, the *syn* selectivity with *m*-CPBA may be only 60:40. The principle works, but not well.

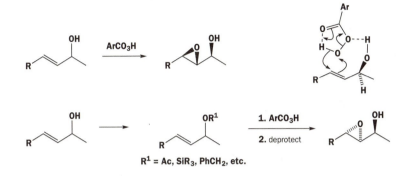

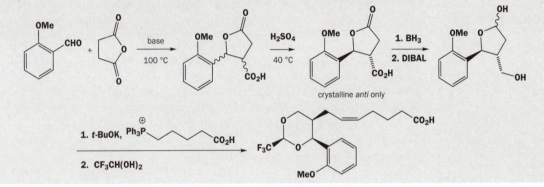

R¹ = Ac, SiR₃, PhCH₂, etc.

Problem 13

Revision. Here is an outline of the AstraZeneca synthesis of a thromboxane analogue. Explain the reactions, giving mechanisms for each step, and explain how the stereochemistry is controlled. In what way could this be considered an example of the control of open-chain stereochemistry when all of the molecules are cyclic?

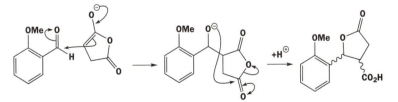

Purpose of the problem

Mainly revision but there is some stereochemical control both in two and three dimensions.

Suggested solution

The first reaction clearly involves the formation of an enolate from the only enolizable compound (the anhydride) and its attack on the aromatic aldehyde. The cyclization that follows may look a bit awkward but it forms a five-membered ring and the leaving group is a stable carboxylate anion. There is no stereochemical control in this step.

Treatment with acid allows the formation of the enol of the acid and the molecule adopts the more stable *anti* configuration under thermodynamic control. Borane reduces the acid (p. 618) and DIBAL reduces the lactone to the lactol (pp. 620–1).

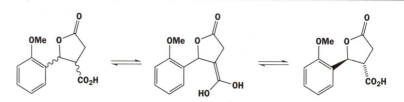

Now a Wittig reaction with an unstabilized ylid selectively gives the *Z*-alkene. Note that *three* molecules of base are needed: one to open the lactol, one to remove the proton from CO_2H, and one to make the ylid.

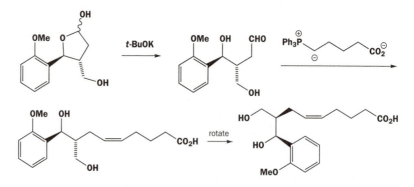

The last reagent is the hydrate of trifluoroacetaldehyde with which it is in equilibrium. The very reactive aldehyde forms an acetal with the diol we have just made and does so under thermodynamic control. The product has two equatorial groups and the one axial group has no 1,3-diaxial interactions as the relevant atoms in the ring are the oxygens of the acetal.

■ J. M. Lawlor and M. B. Macnamee, *Tetrahedron Lett.*, 1983, **24**, 2211. The product is a thromboxane antagonist developed by the then ICI and described in S. Lee and G. Robinson, *Process development*, Oxford University Press, Oxford, 1995, pp. 27–37.

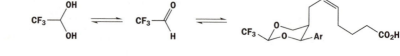

In what way could this be considered an example of the control of open-chain stereochemistry when all the molecules are cyclic? Well, they are all cyclic except the last intermediate and in every case the ring concerned is an ester (lactone) or an acetal that can be hydrolysed to an open-chain compound without disrupting the stereochemistry.

Suggested solutions for Chapter 35

35

Problem 1

Give mechanisms for these reactions, explaining the stereochemistry.

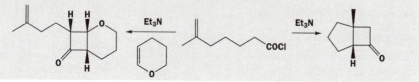

Purpose of the problem

Not the expected Diels–Alder reaction, but [2 + 2] cycloadditions of ketenes.

Suggested solution

Treatment of acid chlorides with tertiary amines produces ketenes. In this case an intramolecular [2 + 2] cycloaddition is possible. The stereochemistry is trivial: the *cis* ring junction is the only

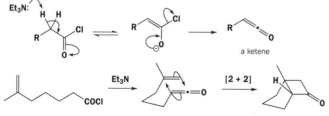

possible outcome.

If a more reactive alkene (electron-donating O makes the HOMO energy higher) is provided, the ketene adds to that instead. Notice that the alkene must be present when the ketene is generated. The mechanism and part of the stereochemistry is simple. Because the cyclic alkene has *cis* stereochemistry, the two hydrogens on the six-membered ring must be *cis* in the product. The regiochemistry arises because the alkene is an enol ether and the large coefficient in its HOMO interacts with the central atom of the ketene, that is, the largest coefficient in the LUMO.

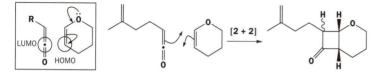

■ There is rather more in this than we can discuss here; see R. H. Bisceglia and C. J. Cheer, *J. Chem. Soc., Chem. Commun.,* 1973, 165.

The stereochemistry at the remaining centre comes from the way in which the two molecules approach each other. The two components are orthogonal and the dotted lines in the middle diagram below show how the new bonds are formed. The carbonyl group of the ketene will prefer to be in the middle of the ring and the side chain on the ketene will bend down away from the top ring. These [2 + 2] thermal cycloadditions normally give the all-*cis* product.

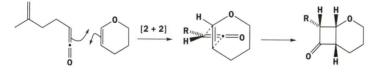

Problem 2

Predict the structure of the product of this Diels–Alder reaction.

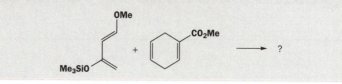

Purpose of the problem

Now the expected Diels–Alder reaction.

Suggested solution

The diene is electron-rich and will use its HOMO in the cycloaddition. It will therefore prefer the alkene with the lowest LUMO and that must be the unsaturated ester. Both substituents on the diene direct reaction to the same end. We can predict this from electron donation from the oxygen atoms.

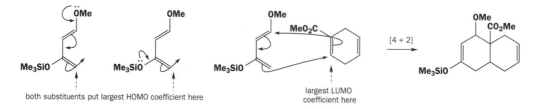

both substituents put largest HOMO coefficient here

largest LUMO coefficient here

■ This chemistry is part of a synthesis of the antitumour agent vernolepin by S. Danishefsky and group, *J. Am. Chem. Soc.*, 1976, **98**, 3028.

The stereochemistry of the alkene (H and CO₂Me *cis*) will be faithfully reproduced in the product. The stereochemistry of the OMe group comes from *endo* attack – we should tuck the ester group underneath the diene so that it can overlap with the orbitals of the middle two atoms. If you also said that this product will eliminate methanol on work-up and that only the stereochemistry at the ring junction really matters, you'd be quite right.

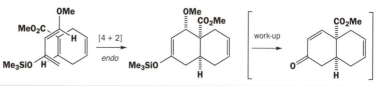

Problem 3

Comment on the difference in rate between these two reactions. It is estimated that the second goes about 10^6 times faster than the first.

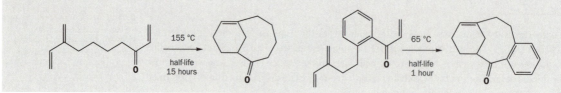

Purpose of the problem

More details of the intramolecular Diels–Alder reaction.

Suggested solution

The dienes are the same, the ring sizes are the same, and the only difference is the presence of the benzene ring in the faster reacting compound. We should draw a mechanism for one of the reactions just to see what is happening.

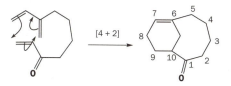

We are making two new rings. The six-membered ring containing a *cis*-alkene presents no problem. The eight-membered ring with a ketone in it might present a problem, and the ten-membered ring with the *trans*-alkene is definitely a problem. It is much easier to make medium (8- to 14-membered) rings when there is a *cis*-alkene elsewhere in the ring and the benzene ring helps there. It increases the population of conformations with the ends of their chains close together. It probably also lowers the energy of the LUMO.

■ This reaction is part of a synthesis of the taxane skeleton (K. J. Shea and P. D. Davis, *Angew. Chem., Int. Ed. Engl.*, 1983, **22**, 419).

Problem 4

Justify the stereoselectivity in this intramolecular Diels–Alder reaction.

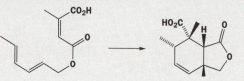

Purpose of the problem

Investigating the stereoselectivity of the intramolecular Diels–Alder reaction.

Suggested solution

Intramolecular Diels–Alder reactions can give *endo* or *exo* products. We should first discover which this is. Drawing the transition state for the *endo* product, we find that the *endo* product is indeed formed. So electronic factors dominate, perhaps because the dienophile has such a low-energy LUMO and has two carbonyl groups for secondary orbital overlap with the back of the diene.

■ J. D. White and B. G. Sheldon, *J. Org. Chem.*, 1981, **46**, 2273.

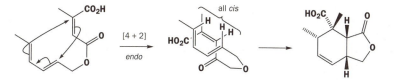

Problem 5

Explain the formation of single adducts in these reactions.

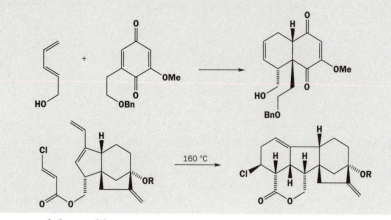

Purpose of the problem

Investigating the regio- and stereoselectivity of one inter- and one intramolecular Diels–Alder reaction.

Suggested solution

The stereoselectivity of the first reaction is straightforward. It gives the *endo* product.

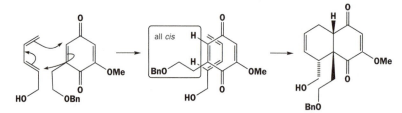

The regiochemistry is not quite so simple. The diene has the larger HOMO coefficient at the top end, as drawn, so we must deduce that the largest LUMO coefficient in the unsymmetrical quinone is at the top left as drawn. This is probably because conjugation with the MeO group makes the top carbonyl and the right-hand alkene less electrophilic and the bottom carbonyl activates the top end of the left-hand alkene. If you prefer, it is an '*ortho*' product.

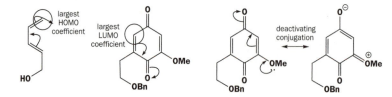

■ These are early steps in Corey's synthesis of the plant hormone gibberellic acid (E. J. Corey and group, *J. Am. Chem. Soc.*, 1978, **100**, 8031–6).

The second example is intramolecular so the regioselectivity is determined by that alone – the ester linkage between the diene and the dienophile is too short for them to join up the other way round. This same link ('tether') also forces the dienophile to approach the diene from the bottom face. All that remains is the *endo/exo* question and the diagram shows that the answer is *endo* with the carbonyl group tucked under the back of the diene.

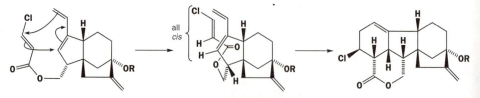

Problem 6

Revision elements. Suggest two syntheses of this spirocyclic ketone from the starting materials shown. Neither starting material is available.

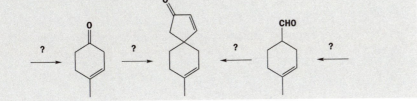

Purpose of the problem

Revision of synthesis (Chapters 25 and 30) with some cycloaddition. Helping you to realize that there are alternative ways to make six-membered rings.

Suggested solution

The most obvious disconnection is of the α,β-unsaturated ketone with an aldol reaction (Chapter 27) in mind. This reveals a 1,4-dicarbonyl compound. Direct disconnection to one of the starting materials is now possible and suggests a Diels–Alder reaction.

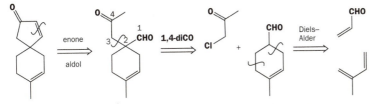

The Diels–Alder reaction has the right ('*para*') regioselectivity, especially if we use a Lewis acid such as $SnCl_4$, and we shall need a nonbasic specific enol equivalent for the alkylation – an enamine will do fine.

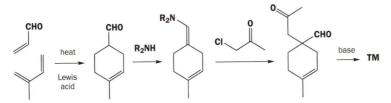

The other route demands a different disconnection of the keto-aldehyde plus one further aldol disconnection. The starting material is more easily made by Birch reduction than by a Diels–Alder reaction.

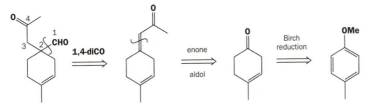

The Birch reduction gives the enol ether of the ketone and demands careful hydrolysis to avoid the double bond moving into conjugation with the carbonyl group. The aldol reaction requires some kind of control; perhaps the silyl enol ether of acetone will do. Then we need a reagent for

'⁻CHO' that will do conjugate addition. The most obvious choices are cyanide ion or nitromethane. The last step is the same as in the first synthesis.

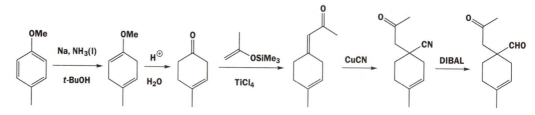

Problem 7

This reaction appeared in Chapter 33. Account for the selectivity.

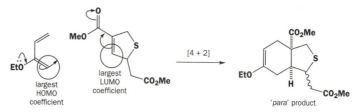

Purpose of the problem

Revision of synthesis (Chapters 25 and 30) with some cycloaddition. Helping you to realize that there are alternative ways to make six-membered rings.

Suggested solution

The regioselectivity comes from the contribution of the EtO group to the HOMO of the diene and the dominating influence of the conjugated CO_2Me group on the LUMO of the dienophile. Neither the other CO_2Me group nor the sulfur atom is conjugated to the alkene of the dienophile and neither has much influence on the reaction. Alternatively, you could say that the OEt and CO_2Me groups are '*para*' in the product.

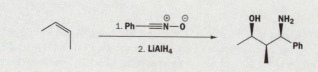

The only stereoselectivity is the *cis* relationship between H and CO_2Me in the dienophile, which is faithfully reproduced at the ring junction of the product.

Problem 8

Draw mechanisms for these reactions and explain the stereochemistry.

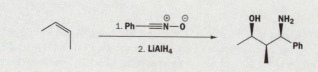

Purpose of the problem

Exploration of stereochemical control by 1,3-dipolar cycloadditions. Revision of the importance of cyclic compounds.

Suggested solution

The nitrile oxide adds in one step to the *cis*-alkene to give a single diastereoisomer of a 1,3-dipolar cycloadduct. This is also a [4 + 2] cycloaddition with the dipole supplying four electrons. The two methyl groups on the alkene start *cis* and remain so in the adduct.

■ V. Jäger and group, *Liebig's Ann. Chem.*, 1980, 122.

The first reduction must be of the imine as it also is stereoselective with hydride being transferred from the face of the five-membered ring opposite to the methyl groups. If reduction of the N–O bond occurred first, we should expect little control in the reduction of the imine.

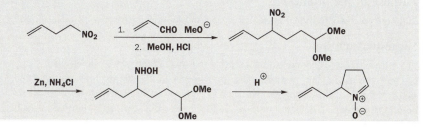

Problem 9

Revision. One of the nitrones used as an example in the chapter was prepared by this route. Explain what is happening and give details of the reactions.

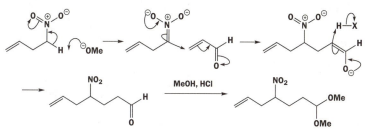

Purpose of the problem

Revision of mechanisms from previous chapters.

Suggested solution

The first reaction is a base-catalysed conjugate addition of the kind described in Chapter 29 (pp. 766–8) followed by the formation of an acetal by the mechanisms described in detail in Chapter 14 (pp. 342–5).

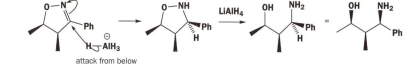

■ The product was used in a synthesis of tropane alkaloids (p. 1416) by J. J. Tufariello and group, *J. Am. Chem. Soc.*, 1979, **101**, 2435.

The next step is the reduction of the nitro group to the hydroxylamine by the very versatile zinc and the cyclization of the hydroxylamine on to the released aldehyde to give the nitrone.

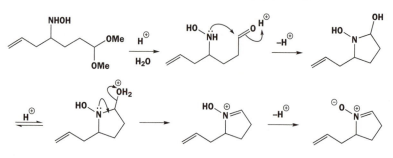

Problem 10

Explain why this Diels–Alder reaction gives total regioselectivity and stereospecificity but no stereoselectivity. What is the mechanism of the second step? What alternative route might you have considered if you wanted to make this final product and why would you reject it?

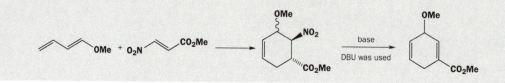

Purpose of the problem

More on selectivity in the Diels–Alder reaction and more comparison with Birch reduction.

Suggested solution

The regioselectivity is dominated by the nitro and MeO groups – we can tell that by the way they end up 'ortho' in the product. The stereospecificity is the faithful reproduction of the stereochemistry of the dienophile. The stereoselectivity that might have been observed would have been *exo/endo* stereochemistry but there are two groups (NO_2 and CO_2Me) that might have gone *endo* and apparently no preference.

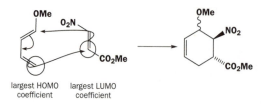

largest HOMO coefficient   largest LUMO coefficient

The base-catalysed elimination must be an E1cB loss of nitrite (NO_2^-) from the ester enolate.

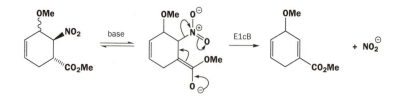

Alternative syntheses would include a direct Diels–Alder reaction with an alkyne, but that would give the wrong ('*para*') regioisomer, or the Birch reduction of a *m*-methoxybenzoate ester but that too would have given the wrong regioselectivity.

■ This is a general synthesis of *meta*-substituted aromatic compounds by S. Danishefsky and group, *J. Am. Chem. Soc.*, 1978, **100**, 2918.

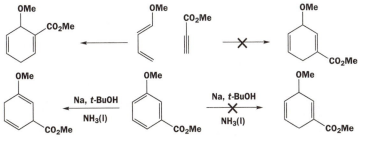

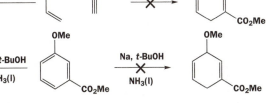

Problem 11

Give mechanisms for these reactions and explain the regio- and stereochemical control (or the lack of it!).

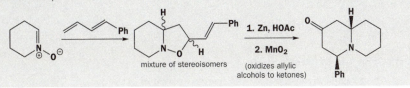

mixture of stereoisomers   (oxidizes allylic alcohols to ketones)

Purpose of the problem

Selectivity and application of 1,3-dipolar cycloadditions.

Suggested solution

The first thing to do is to sort out the mechanism for the cycloaddition. The nitrone uses its LUMO (the π^* of the C=N bond) to react with the HOMO of the diene whose largest coefficient is at the end away from the phenyl group. There is no stereoselectivity as there is no conjugating group and so no *exo/endo* selection.

■ The 1,3-dipolar cycloaddition was developed by J. J. Tufariello and R. G. Gatrone, *Tetrahedron Lett.*, 1978, 2753.

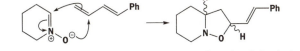

Reduction with zinc cleaves the N–O bond and MnO_2 oxidizes the allylic alcohol to the enone. At this point there is only one chiral centre so the mixture of diastereoisomers has become one compound. Conjugate addition of the amine gives the new ring.

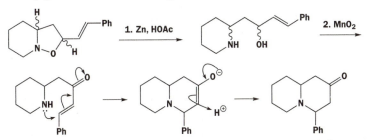

The stereochemistry is more difficult to explain. The product has a *trans* ring junction from choice but has the Ph group axial – obviously not from choice. This must be the kinetic product. It

■ This is part of a synthesis of various alkaloids by C. Kibayashi and group, *J. Chem. Soc., Chem. Commun.*, 1983, 1143.

looks as though the ring must close with the best overlap between the nitrogen lone pair and the π orbital of the enone to give a *cis* ring junction, which equilibrates by pyramidal inversion at nitrogen to the more stable *trans* ring junction.

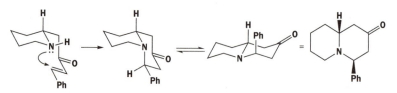

Problem 12

Suggest a mechanism for this reaction and explain the stereo- and regiochemistry. How would you prepare the unsaturated ketone starting material?

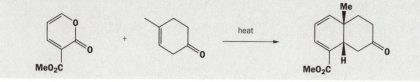

Purpose of the problem

Selectivity and application of forward and reverse Diels–Alder cycloadditions.

Suggested solution

■ This sequence was used by D. S. Watt and E. J. Corey in a synthesis of occidentalol (*Tetrahedron Lett.*, 1972, 4651).

The reaction is clearly a cycloaddition but at first sight the regioselectivity is all wrong. The answer comes from a realization that this is a reverse electron demand Diels–Alder reaction. The diene is very electron-deficient with two conjugated carbonyl groups so the dienophile needs to be electron-rich. The enone is not electron-rich enough but its enol is. The enone could be prepared by Birch reduction

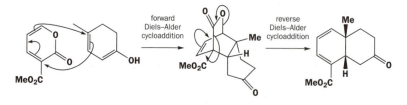

Problem 13

Photochemical cycloaddition of these two compounds is claimed to give the single diastereoisomer shown. The chemists who did this work claim that the stereochemistry of the adduct is simply proved by its conversion into a lactone on reduction. Comment on the validity of this deduction and explain the stereochemistry of the cycloaddition.

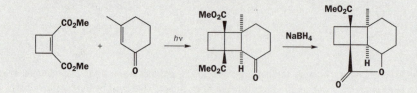

Purpose of the problem

Selectivity and application of photochemical [2 + 2] cycloadditions.

Suggested solution

Either of the two starting materials could absorb the light to provide the SOMO for the cycloaddition. This does not affect the stereochemistry of the reaction. There is no *endo* effect in [2 + 2] photocycloadditions so the molecules simply come together with the rings arranged in an *exo* fashion to give the least steric hindrance (diagram in frame in margin).

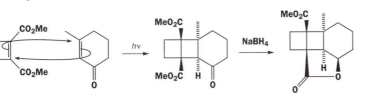

The stereochemistry of the reduction is easy to explain as the molecule is folded in such a way that only the bottom face of the carbonyl group is open to nucleophilic attack. The oxyanion produced can immediately cyclize to form the lactone. Clearly, this is possible only if the O⁻ group is up, but it is also possible only if the CO_2Me groups are on this side of the molecule. The formation of the lactone does prove the stereochemistry of the cycloaddition.

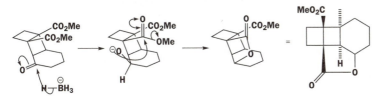

Problem 14

Thioketones, with a C=S bond, are not usually stable as we shall see in Chapter 46. However, this thioketone is quite stable and undergoes reaction with maleic anhydride to give the product shown. Comment on the stability of the starting material, the mechanism of the reaction, and the stereochemistry of the product.

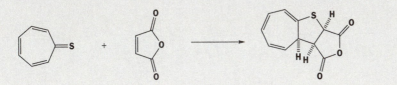

Purpose of the problem

Exploration of a new structure, revision of aromaticity, and encounter with [8 + 2] cycloadditions.

Suggested solution

This particular thioketone is stable because the C=S bond is very polarized by delocalization making the seven-membered ring an aromatic cation with six electrons in it. Though all the diagrams are correct, the one in the frame (margin) gives a better representation of the structure.

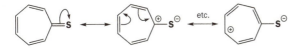

■ T. Machiguchi *et al.*, *J. Chem. Soc., Chem. Commun.*, 1973, 196.

But we can't use diagrams like that for mechanisms! The cycloaddition uses maleic anhydride as a two-electron component with a low LUMO. We could use one of the dienes in the ring to provide four electrons but their ends are far apart and the electron-deficient ring is a poor HOMO. If we include the sulfur atom we can provide eight electrons and an atom (S) with a large HOMO coefficient. The tricyclic product is clearly folded back on itself so that the triene in the seven-membered ring and the carbonyl groups in the anhydride are close. There must be an *endo* effect in [8 + 2] cycloadditions.

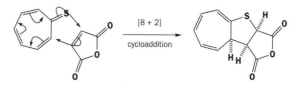

Problem 15

This unsaturated alcohol is perfectly stable until it is oxidized with Cr(VI): it then immediately cyclizes to the product shown. Explain.

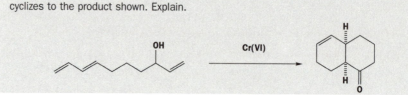

Purpose of the problem

Discovery of a common effect in intramolecular cycloadditions.

Suggested solution

■ J.-L. Gras and M. Bertrand, *Tetrahedron Lett.*, 1979, 4549.

If the starting material were heated it would no doubt undergo a Diels–Alder reaction but the diene and dienophile are poorly matched. Both have high-energy HOMOs but there isn't a low-energy LUMO in sight. Once the enone is formed, the energies match well and cycloaddition is fast. The stereochemistry comes from an *endo* arrangement (diagram in frame in margin).

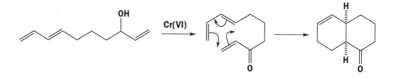

Problem 16

Suggest mechanisms for these reactions and comment on the stereochemistry of the first product.

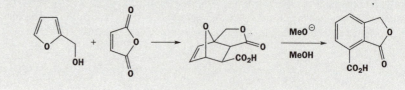

Purpose of the problem

Two unusual things: a stereoselectivity and a subsequent reaction.

Suggested solution

There has obviously been a cycloaddition between the furan as diene and maleic anhydride and an esterification of the free OH group by the anhydride. Whichever we do first, the stereochemistry is *exo*. This is because furans are aromatic and the Diels–Alder reaction is reversible and under thermodynamic control. We'll do that first.

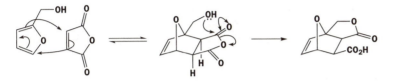

The second step is obviously some sort of elimination driven by the formation of a benzene ring. Both the lactone and the CO_2H group are still there so it looks as though MeO^- acts as a base and that the bridging oxygen atom is lost, presumably as water. The E1cB mechanism looks likely.

■ A. Pelter and B. Singaram, *J. Chem. Soc., Perkin Trans. 1*, 1983, 1383.

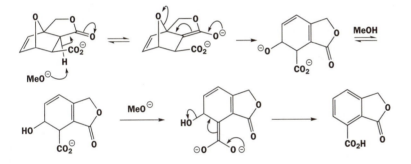

Problem 1

Give mechanisms for these steps, commenting on the regioselectivity of the pericyclic step and the different regioselectivity of the two metals.

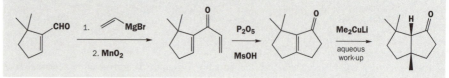

Purpose of the problem

A gentle introduction to an electrocyclic reaction without stereo- or regioselectivity.

Suggested solution

Grignard reagents generally prefer direct to conjugate addition especially with unsaturated aldehydes. MnO$_2$ specializes in oxidizing allylic alcohols and is the gentle oxidizing agent we need to produce the unstable dienone.

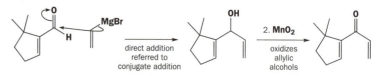

The pericyclic process comes next and it is a Nazarov reaction (p. 962), a conrotatory electrocyclic closure of a pentadienyl cation to give a cyclopentenyl cation. There is no stereochemistry and the only regiochemistry is the position of the double bond at the end of the reaction. Here it prefers the more substituted side of the ring.

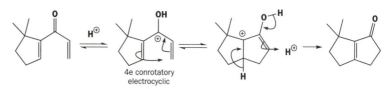

The final cuprate addition goes in a conjugate fashion as we should expect as this is the speciality of Cu(I) compounds. The *cis* 5/5 ring junction is much preferred to *trans* and can equilibrate on work-up by enolization.

Problem 2

Predict the product of this reaction.

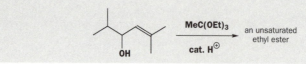

Purpose of the problem

A gentle introduction to an electrocyclic reaction without stereo- or regioselectivity.

Suggested solution

This is a classic Claisen [3,3]-sigmatropic rearrangement sequence starting with an allylic alcohol and forming a vinyl ether by acetal (or in this case, *orthoester*) exchange. The reaction is very *trans*-selective (p. 944).

■ This product was used in a synthesis of chrysanthemic acid by Jacqueline Ficini and Jean d'Angelo, *Tetrahedron Lett.*, 1976, 2441.

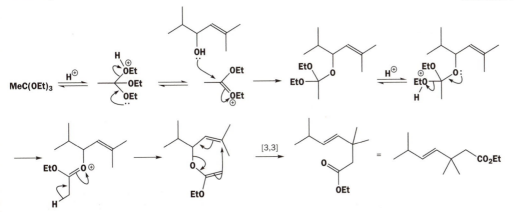

Problem 3

Give mechanisms for this alternative synthesis of two fused five-membered rings.

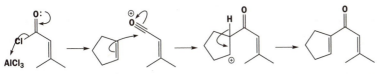

Purpose of the problem

A gentle introduction to an electrocyclic reaction without stereo- or regioselectivity.

Suggested solution

The first stage is an aliphatic Friedel–Crafts reaction with an acylium ion attacking the alkene.

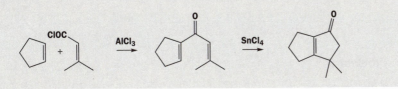

Next, a Nazarov reaction catalysed by a different Lewis acid closes the five-membered ring and puts the double bond in the only place it can go. The electrocyclic step is conrotatory but this has no meaning with this achiral product.

■ W. Oppolzer and K. Bättig, *Helv. Chim. Acta*, 1981, **64**, 2489.

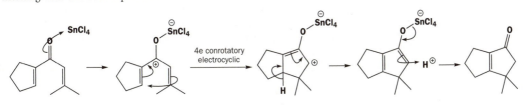

Problem 4

Explain what is going on here.

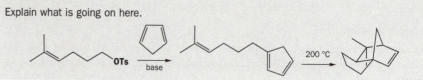

Purpose of the problem

Two pericyclic reactions: a sigmatropic shift and a cycloaddition in one reaction scheme.

Suggested solution

The aromatic anion of cyclopentadiene displaces tosylate from the alkyl group and then a [1,5] hydrogen shift gives the first product. Such a shift is allowed suprafacially on the ring (pp. 953–5).

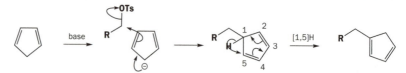

Now there is an intramolecular Diels–Alder reaction requiring a high temperature because the dienophile is not activated. The stereochemistry is not obvious but there is no *endo* effect so the molecule folds to give the new five-membered ring a *cis* ring junction with the old.

■ This sort of Diels–Alder reaction was used in a synthesis of cedrol by E. G. Breitholle and A. G. Fallis, *J. Org. Chem.*, 1978, **43**, 1964.

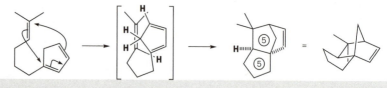

Problem 5

In Chapter 33, Problem 13, we used a tricyclic hydroxy-ketone whose stereochemistry had been wrongly assigned. Now we are going to show you how it was used and you are going to interpret the results. This is the correct result.

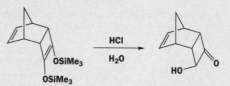

The hydroxy-ketone was first converted into a compound with PhS and OAc substituents. Explain the stereochemistry of this process.

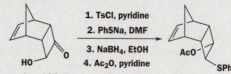

Pyrolysis of this compound at 460°C gave a diene whose NMR spectrum included δ_H (p.p.m.) 6.06 (1H, dd, *J* 10.3, 12.1 Hz), 6.23 (1H, dd, *J* 10.3, 14.7 Hz), 6.31 (1H, d, *J* 14.7 Hz), and 7.32 (1H, d, *J* 12.1 Hz). Does this agree with the structure given? How is this diene formed and why does it have that stereochemistry?

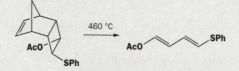

Purpose of the problem

Relating the material of this chapter to that of previous chapters and a bit of revision of basic mechanisms.

Suggested solution

The first sequence of reactions is simple. The S_N2 reaction goes with inversion and reduction occurs on the outside (convex, *exo*-face) of the folded ketone.

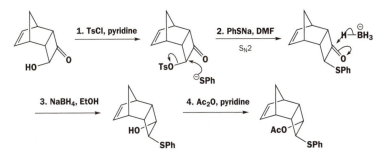

The second reaction involves a reverse Diels–Alder reaction and an electrocyclic opening of the cyclobutene product. This is a four-electron conrotatory process. The two substituents may both rotate out to give the *E,E*-diene or both in to give the *Z,Z*-diene.

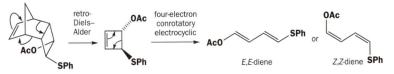

The NMR spectrum shows clearly that the *E,E*-diene is formed. The two coupling constants for the simple doublets must be for the terminal hydrogens and 14.7 Hz is definitely a *trans* coupling. We might wonder about 12.1 Hz as it is on the low side but the alkene has an electronegative oxygen (OAc) substituent and couplings will be low (pp. 269–70).

■ Control of diene stereochemistry by this method was very useful in cycloadditions (B. M. Trost and group, *J. Org. Chem.*, 1978, **43**, 4559).

Problem 6

Careless attempts to carry out a Claisen rearrangement on this allyl ether often give the compound shown instead of the expected product. What is the expected product? How is the unwanted product formed? Addition of a small amount of a weak base, such as $PhNMe_2$, helps prevent the unwanted reaction. How?

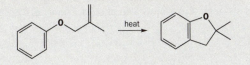

Purpose of the problem

Exploration of what may follow a simple [3,3]-sigmatropic rearrangement.

Suggested solution

The expected Claisen rearrangement is straightforward and evidently happens.

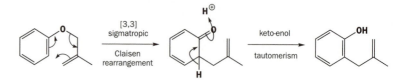

The ring is formed by addition of the phenol to the alkene. The alkene is an ordinary alkene that reacts with electrophiles, not nucleophiles, so this must be an acid-catalysed reaction and can be suppressed by a weak base.

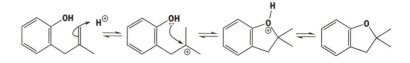

Problem 7

Treatment of this imine with base followed by an acidic work-up gives a cyclic product with two phenyl groups *cis* to one another. Why is this?

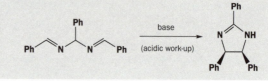

Purpose of the problem

An unusual example of an electrocyclic reaction on an anion.

Suggested solution

■ This was the first example of a stereochemical study in anionic electrocyclic reactions; see D. H. Hunter and S. K. Sim, *J. Am. Chem. Soc.*, 1969, **91**, 6202.

The proton from the middle of the molecule is removed to give an anion stabilized by two nitrogens and two phenyl groups. A six-electron electrocyclic reaction closes the five-membered ring and this must be disrotatory moving both phenyl groups up (or down).

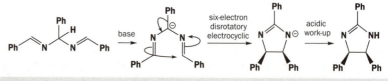

Problem 8

This question concerns the structure and chemistry of an unsaturated nine-membered ring. Comment upon its structure. Explain its different behaviour under thermal or photochemical conditions.

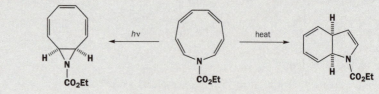

Purpose of the problem

Revision of aromaticity and two alternative electrocyclic reactions.

Suggested solution

The amine has eight electrons in double bonds and a lone pair on the nitrogen atom making ten in all. It could be aromatic with $4n + 2$ electrons ($n = 2$). The two reactions are clearly electrocyclic and must be disrotatory to get *cis* ring junctions (the only possible arrangement for joining two flat rings). Thermally this means a six-electron process, but photochemically an eight-electron process is all right. The nitrogen does not appear to be involved in either reaction.

■ This was an investigation into the aromaticity of the starting material by A. G. Anastassiou and J. H. Gebrian, *Tetrahedron Lett.,* 1969, 5239.

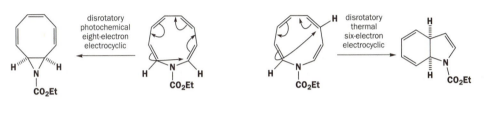

Problem 9

Propose a mechanism for this reaction that accounts for the stereochemistry of the product.

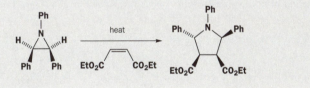

Purpose of the problem

Another electrocyclic/cycloaddition combination for you to work out.

Suggested solution

The three-membered ring opens using the lone pair on nitrogen in a four-electron conrotatory electrocyclic process. One phenyl group must rotate inwards and the other outwards. Then a cycloaddition of the four-electron dipole on to the two-electron dienophile goes without change of stereochemistry. The ester groups remain *cis* and the phenyls must be one up and one down.

■ This extensive study of the opening of three-membered heterocyclic rings came from Huisgen's group in Munich (*J. Chem. Soc., Chem. Commun.,* 1971, 1187, 1188, 1190, and 1192).

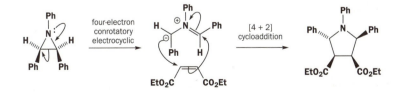

Problem 10

Treatment of cyclohexa-1,3-dione with this acetylenic amine gives a stable enamine in good yield. Refluxing this enamine in nitrobenzene gives a pyridine after a remarkable series of reactions. Fill in the details: give mechanisms for the reactions, structures for any intermediates, and suitable explanations for each pericyclic step. A mechanism is not required for the last step (nitrobenzene acts as an oxidant).

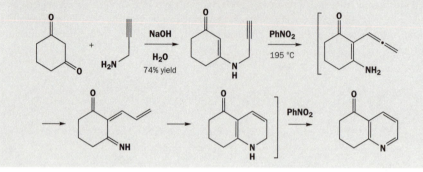

Purpose of the problem

Practice in unravelling complicated reaction sequences involving pericyclic steps.

Suggested solution

The formation of the enamine requires only a patient adding and subtracting of protons.

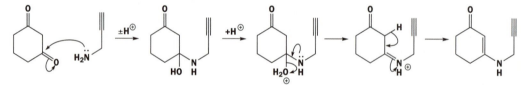

The cascade of reactions in hot nitrobenzene starts with a [3,3]-sigmatropic rearrangement that is unusual in that it forms an allene but is otherwise straightforward. To get to the next intermediate, the stable conjugated primary enamine, we must enolize and go back to the ketone again but move the double bond into conjugation.

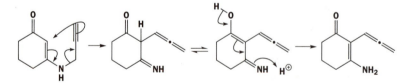

■ K. Berg-Nielsen and L. Skattebol, *Acta Chem. Scand.*, 1978, **B32**, 553.

Now we can transfer a proton from nitrogen to the middle of the allene. This is formally a [1,5]H shift and is, of course, allowed but it may be the movement of a proton as nitrogen is involved. This gives a diene that can twist round for a six-electron electrocyclic reaction. This would no doubt be disrotatory but we can't tell as there is no stereochemistry.

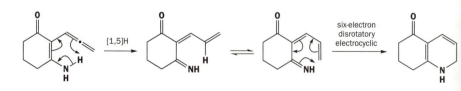

Problem 11

Problem 11 in Chapter 32 was concerned with two diastereoisomers of this compound that were formed in what we then called enigmatically 'a chemical reaction'.

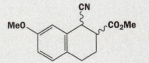

We can now let you into the secret of that 'chemical reaction'. A benzocyclobutene was heated with methyl acrylate to give a 1:1 mixture of the two isomers. What is the mechanism of the reaction and why is only one regioisomer but a mixture of stereoisomers formed? Isomer B is converted into isomer A on treatment with base. What is the stereochemistry of A and B?

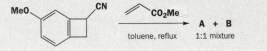

Purpose of the problem

Electrocyclic and cycloaddition reactions again with some stereochemical interest.

Suggested solution

The cyclobutene opens in a four-electron conrotatory electrocyclic reaction. The cyanide can go either way and it will prefer to go out. The diene so produced is unstable and is extremely reactive in cycloadditions as it regenerates the benzene ring that way. The regiochemistry is predictably 'ortho' and we should probably use the HOMO of the diene and the LUMO of the unsaturated ester.

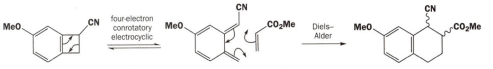

The stereochemistry shows that a 1:1 mixture of *exo* and *endo* products is formed. Presumably the diene is so reactive that it has little need of secondary orbital interactions and steric hindrance plays an equal part. The isomerization in base must be by enolization (of either group) so that both substituents can be equatorial. The diagrams omit the benzene ring for clarity. The *cis* compound is **B** and the *trans* **A**.

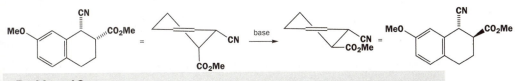

Problem 12

Treatment of this amine with base at low temperature gives an unstable anion that isomerizes to another anion above −35°C. Aqueous work-up gives a bicyclic amine. What are the two anions? Explain the stereochemistry of the product. Revision of NMR. In the NMR spectrum of the product the two green (in the book, ringed here) hydrogens appear as an ABX system with J_{AB} 15.4 Hz. Comment.

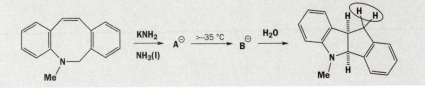

Purpose of the problem

An unusual electrocyclic reaction on an anion with stereochemistry and NMR revision.

Suggested solution

The first anion A is formed by removal of the only possible protons – those on the one CH_2 group in the molecule. This anion might be considered aromatic (six electrons from the three double bonds, two from the nitrogen, and two from the anion) but it is clearly unstable as it closes in an electrocyclic reaction at $> -35°C$. This is a six-electron process and must therefore be disrotatory. The rotating hydrogens are marked on the diagram of **A**. Both anions, **A** and **B**, are extensively delocalized and it is a matter of choice where you draw the negative charge.

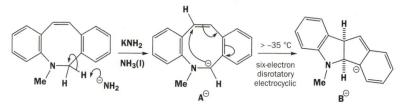

■ This study was originally aimed at the aromaticity of the starting material (A. G. Anastassiou and group, *J. Chem. Soc., Chem. Commun.*, 1981, 647).

Anion B is protonated by water with preservation of the right-hand aromatic ring. The final product is a chiral molecule having no plane of symmetry so the ringed CH_2 group is diastereotopic with J_{AB} 15.4 Hz. This is larger than usual because of the π contribution – a neighbouring π bond increases ^2J by about 2 Hz.

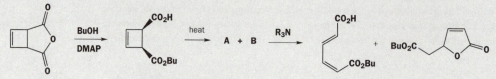

diastereotopic Hs
different shifts
couple to each other

Problem 13

How would you make the starting material for these reactions? Treatment of the anhydride with butanol gives an ester that gives two inseparable compounds on heating. On treatment with an amine, an easily separable mixture of an acidic and a neutral compound is formed. What are the components of the first mixture and how are they formed?

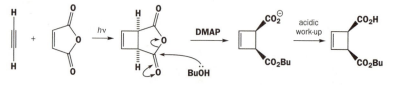

Purpose of the problem

Exploration of alternative conrotatory openings of a cyclobutene.

Suggested solution

DMAP

The starting material can be made by a photochemical $[2 + 2]$ cycloaddition of acetylene and maleic anhydride. Treatment with butanol and base leads to the monoester because, after the butanol has attacked once, the product is the anion of the carboxylic acid and cannot be attacked again by the nucleophile. DMAP is a base whose structure appears in the margin.

Heat opens the cyclobutene in a conrotatory four-electron electrocyclic reaction. As the two groups are *cis* on the cyclobutene, one must rotate outwards and one inwards. The two groups are similar but not the same so there is little selection in the rotation and both products are formed.

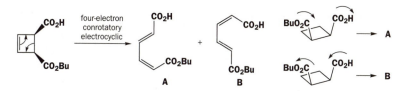

Treatment with the tertiary amine forms the anions of the carboxylic acids. That of B can do a conjugate addition to the unsaturated ester and form the lactone but the carboxylate of A cannot react.

■ This observation was vital in developing a synthesis of varrucarin A, a natural product with antitumour activity (B. M. Trost and P. G. McDougal, *J. Org. Chem.*, 1984, **49**, 458).

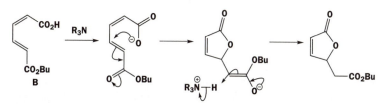

Problem 14

Treatment of this keto-aldehyde (which exists largely as an enol) with the oxidizing agent DDQ (a quinone – see p. 1192) gives an unstable compound that converts into the product shown. Explain the reactions and comment on the stereochemistry.

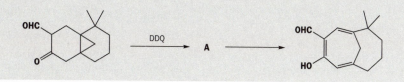

Purpose of the problem

Exploration of a less well defined pericyclic sequence.

Suggested solution

DDQ oxidizes the position between the two carbonyl groups to insert a double bond conjugated with both. We can now put in some stereochemistry as the three-membered ring has to be *cis* fused to both six-membered rings. This diene undergoes electrocyclic ring opening to form a seven-membered ring. This is a six-electron and therefore disrotatory reaction and the two bonds to the old three-membered ring are therefore both allowed to rotate inwards – the only rotation that can give the product.

■ This reaction is part of a synthesis of the skeleton of the natural product spiniferin by J. A. Marshall and R. E. Conrow, *J. Am. Chem. Soc.*, 1980, **102**, 4275.

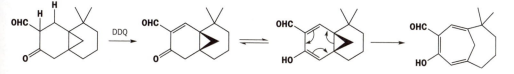

Problem 15

Explain the following observations. Heating this phenol brings it into rapid equilibrium with a bicyclic compound that does not spontaneously give the final aromatic product unless treated with acid.

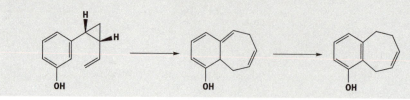

Purpose of the problem

Another electrocyclic reaction involving a three-membered ring. The σ bonds in three-membered rings are strained and more reactive than normal σ bonds and take part readily in pericyclic reactions.

Suggested solution

The first step is very like a Claisen rearrangement but, as there is no oxygen atom, it is strictly a Cope rearrangement, but in any case it is a favourable [3,3]-sigmatropic rearrangement because the σ bond involved is in a three-membered ring. This product cannot go directly to an aromatic compound as that would require a [1,3] (or a [1,7] depending on how you count) hydrogen shift. These would have to be antarafacial on the π system and cannot happen in this rigid structure.

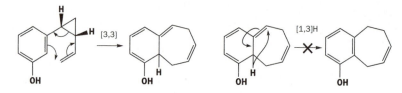

■ This reaction was carried out as part of mechanistic study by E. N. Marvell and S. W. Almond, *Tetrahedron Lett.*, 1979, 2777.

The aromatization can happen by an ionic mechanism. If the extended enol is protonated at its remote end it can then lose a proton from the ring junction and form the phenol.

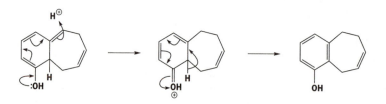

Problem 1

Rearrangements by numbers. This problem is just to help you acquire the skill of tracking down rearrangements by numbering. There are no complicated new reactions here. Just draw a mechanism.

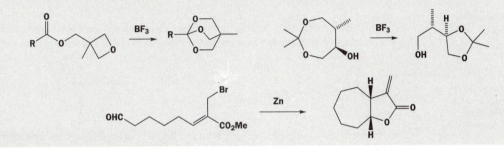

Purpose of the problem

As the problem says, to help you learn about simple rearrangements – those where the carbon skeleton doesn't change.

Suggested solution

The first reaction is the preparation of Corey's protecting group for carboxylic acids. The Lewis acid complexes one of the oxygens and all the atoms in the starting material end up in the product. Atoms 3 and 5 are easy to identify in the product and it doesn't much matter which of the CH$_2$ groups you label 1, 2, and 4. The dotted lines show the new bonds made or old bonds broken.

■ E. J. Corey and N. Raju,
Tetrahedron Lett., 1983, **24**, 5571.

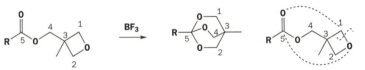

The mechanism can be drawn in various ways; here are two possibilities, the second being better.

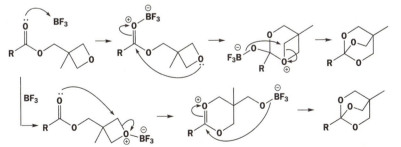

The second reaction is even easier to work out. Atoms 2 and 3 are easy to find and they identify 1 and 4. Just one C–O bond has been made, and one broken. As the compounds are acetals, we must use oxonium ion intermediates and *not* do S_N2 reactions (Chapter 14). Loss of BF_3 and rotation of the last intermediate though 180° gives the product.

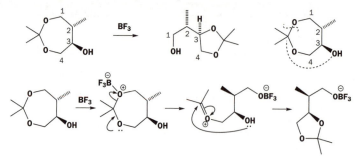

■ E. J. Corey and M. G. Bock, *Tetrahedron Lett.*, 1975, 2643.

The third reaction involves a cyclization. Atoms 1 and 7 clearly make the new bond and the rest of the atoms fit into place without change except that the Br is gone and the alkene has moved from 7/8 to 8/9.

■ This new reaction paved the way for the synthesis of the seven-membered ring antitumour compounds (M. F. Semmelhack and E. C. S. Wu, *J. Am. Chem. Soc.*, 1976, **98**, 3384).

Zinc will insert oxidatively into the C–Br bond and the mechanism follows from the nucleophilic nature of the organometallic compound.

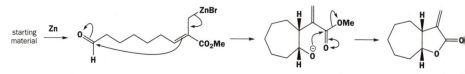

Problem 2

Explain this series of reactions.

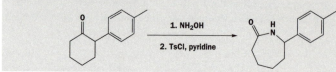

Purpose of the problem

Working out the stereochemistry and mechanism of the Beckmann rearrangement.

■ This example comes from a general investigation into the Beckmann and the related Schmidt rearrangements by R. H. Prager *et al.*, *Aust. J. Chem.*, 1978, **31**, 1989.

Suggested solution

The first reaction forms the oxime by the usual mechanism (Chapter 14). This reaction is under thermodynamic control so the OH group will bend away from the aryl substituent. Then we have the Beckmann rearrangement (pp. 997–9). The group *anti* to the OH group migrates from C to N and that gives the product after rehydration and adjustment of protons.

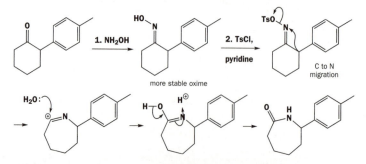

Problem 3

Draw mechanisms for the reactions and structures for the intermediates. Explain the stereochemistry, especially of the reactions involving boron. Why was 9-BBN chosen as the hydroborating agent?

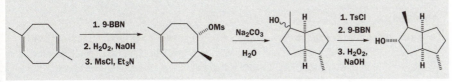

Purpose of the problem

Rearrangements involving boron and a ring-closing rearrangement of sorts plus stereochemistry.

Suggested solution

The starting material is symmetrical so it doesn't matter which face of which alkene you attack. The only important thing is that the boron binds to the more nucleophilic end of the alkene and that R_2B and H are added *cis*. Alkaline H_2O_2 makes the hydroperoxide anion (HO-O$^-$) and that attacks boron. The important migration follows: the whole ring moves from boron to oxygen with retention of configuration. Hydrolysis and mesylation (p. 484) give the first intermediate.

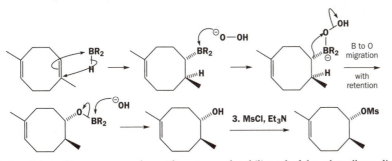

This mesylate cyclizes an aqueous base. The more nucleophilic end of the other alkene displaces the mesylate with inversion to make the *cis* ring junction much preferred by a 5/5 fused system. Water then attacks the tertiary cation to give the next intermediate.

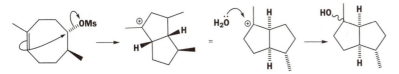

Elimination of the alcohol (E1, of course, as it is tertiary) gives the alkene and a repeat of the hydroboration from the outside (*exo*, convex) face of the folded molecule gives the final alcohol with five new stereogenic centres.

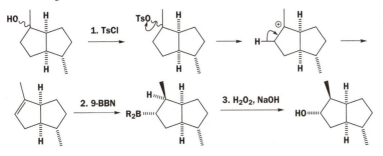

■ This product was used as a foundation for the synthesis of 'iridoid' terpenes by R. S. Matthews and J. K. Whitesell, *J. Org. Chem.*, 1975, **40**, 3313.

9-BBN

9-BBN was chosen because it is very large and reinforces the natural electronic preference of the borane to bind to the less substituted end of the alkene with a steric preference. It also has bridgehead atoms bound to the boron and they make poor migrating groups (p. 1283) preventing competitive migrations. The structure of 9-BBN is in the margin.

Problem 4

It is very difficult to prepare three-membered ring lactones. One attempted preparation, by the epoxidation of di-*t*-butyl ketene, gave an unstable compound with an IR stretch at 1900 cm^{-1} that decomposed rapidly to the four-membered lactone shown. Do you think they made the three-membered ring?

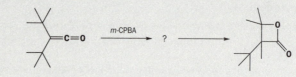

Purpose of the problem

Rearrangements as proof of structure? Surely not! Wait and see.

Suggested solution

The expected three-membered lactone would have a very high carbonyl stretching frequency because of ring strain (pp. 365–6). A three-membered cyclic ketone has a carbonyl stretch of about 1815 cm^{-1} and lactones have higher-frequency absorptions. So it might be the lactone. If it is, we must find a mechanism for the ring expansion to the four-membered ring and that will have to involve rearrangement because one of the *t*-butyl groups has gone. The general conclusion from this and other evidence is that R. Wheland and P. D. Bartlett did indeed make the first α-lactone.

■ *J. Am. Chem. Soc.*, 1970, **92**, 6057; see also J. K. Crandall and S. A. Sojka, *Tetrahedron Lett.*, 1972, 1641.

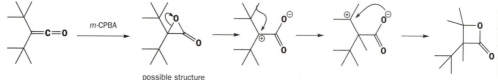

possible structure

Problem 5

Suggest a mechanism for this rearrangement.

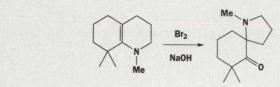

Purpose of the problem

Working out the mechanism of a new rearrangement.

Suggested solution

The starting material is an enamine and will react with bromine in the manner of an enol. Addition of hydroxide to the first product gives the starting material for the rearrangement. Notice that the nitrogen atom migrates rather than the carbon atom in the other ring and that suggests participation in the loss of the bromide.

■ This reaction was discovered by L. Duhamel and J.-M. Poirier, *J. Org. Chem.*, 1979, **44**, 3576.

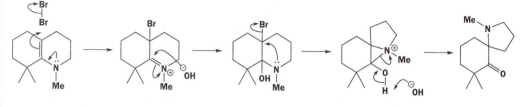

Problem 6

A single enantiomer of the epoxide below rearranges with Lewis acid catalysis to give a single enantiomer of product. Suggest a mechanism and comment on the stereochemistry.

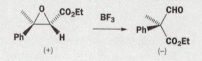

Purpose of the problem

An unusual group migrates and stereochemistry gives a clue to mechanism.

Suggested solution

The mechanism for the reaction must involve Lewis acid complexation of the epoxide oxygen atom, cation formation, and migration of CO_2Et. This last point may surprise you but inspection of the structure of the product shows that the CO_2Et group is indeed bonded to the other carbon atom.

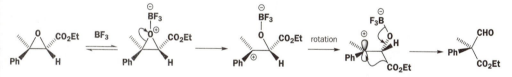

Although something like this must happen, our mechanism raises as many questions as it answers.

- Why does that bond of the epoxide open? *Answer.* Because the tertiary benzylic cation is much more stable than a secondary cation next to a CO_2Et group.
- Why does the CO_2Et group migrate rather than the H atom? *Answer.* For the same reason! If the H atom migrates, the product would be a cation (or at least a partial positive charge would appear in the transition state) next to the CO_2Et group.
- Surely the carbocation intermediate is planar and the product would be racemic? *Answer.* This was the purpose of the investigation. One chiral centre is lost in the reaction so only absolute stereochemistry is relevant. One explanation is that the cation is short-lived and that bond rotation is fast in the direction shown (the CO_2Et group is already down and has to rotate through only 30° to get in the right position for migration). The other is that the migration is concerted. This looks unlikely as the orbital alignments are poor.

■ This mechanism was investigated by R. D. Bach and co-workers; *J. Am. Chem. Soc.*, 1976, **98**, 1975; 1978, **100**, 1605.

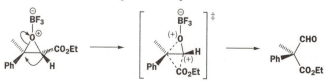

■ The pinacol dimerization appears on p. 1029.

Problem 7

The 'pinacol' dimer from cyclobutanone rearranges with the expansion of one of the rings to give a cyclopentanone fused *spiro* to the remaining four-membered ring. Draw a mechanism for this reaction. Reduction of the ketone then gives an alcohol that rearranges to the alkene in acid. Try working out a mechanism for this transformation. You might also like to think about why the rearrangement happens.

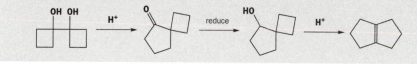

Purpose of the problem

For you to discover that four-membered rings rearrange easily to form five-membered rings.

Suggested solution

The first reaction is a simple pinacol rearrangement (p. 984). The molecule is symmetrical so protonation of either alcohol and migration of either C–C bond give the product.

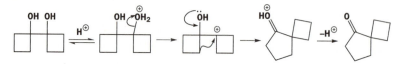

Reduction of the ketone to the alcohol is trivial and then acid treatment allows the loss of water and a ring expansion of the second four-membered ring. You may have drawn this as a concerted or stepwise process. The elimination gives the most highly substituted alkene. Both rearrangements happen very easily because of the relief of strain in going from four- to five-membered rings.

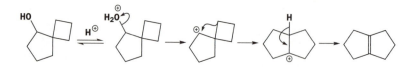

Problem 8

Give the products of Baeyer–Villiger rearrangement on these carbonyl compounds with your reasons.

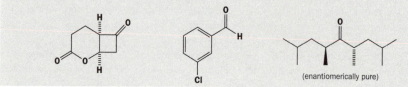

Purpose of the problem

Prediction in rearrangements is as important as elsewhere and the Baeyer–Villiger is one of the more predictable rearrangements.

Suggested solution

There are a few minor traps here that we're sure you avoided. The first compound has two carbonyl groups, but esters don't do Baeyer–Villiger rearrangements so only the ketone reacts. The more substituted carbon atom migrates with retention of configuration (pp. 992–7). The aldehyde rearranges with migration of the benzene ring in preference to the hydrogen atom. The last compound is C_2 symmetric so it doesn't matter which group migrates as long as you ensure retention of configuration (and take care when re-drawing as the migrating group is drawn the other way up). The products are:

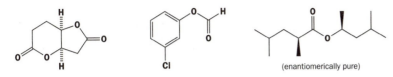

(enantiomerically pure)

Problem 9

Suggest mechanisms for these rearrangements explaining the stereochemistry in the second example.

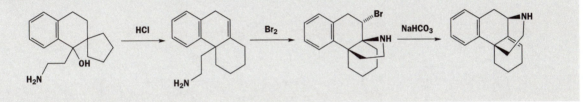

Purpose of the problem

Unravelling one rearrangement after another involving ring expansion and ring contraction with some stereochemistry.

Suggested solution

The first reaction is a simple ring expansion. The amine is not involved as it is completely protonated. The final loss of the proton may be concerted with the migration as this would explain the position of the alkene in the product.

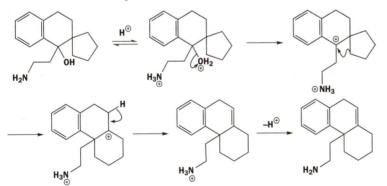

The second reaction starts with bromination of the alkene and interception of the bromonium ion by the amine. Only when the bromine adds to the opposite face of the alkene can the amine cyclize so this reaction resembles iodolactonization. Probably the initial bromination is reversible.

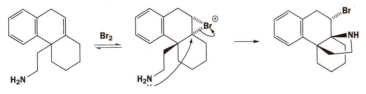

■ This work allowed the synthesis and trial of morphine analogues as painkillers in the Bristol Laboratories of Canada (I. Moncovic *et al.*, *J. Am. Chem. Soc.*, 1973, **95**, 7910).

Finally, the weak base bicarbonate (HCO_3^-) is enough to remove a proton from the cation formed by nitrogen participation in the displacement of bromide. The five-membered heterocycle expands to six-membered. This alkene is formed because the $C-N^+$ bond to the tertiary carbon is broken preferentially.

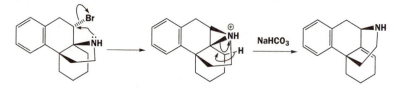

Problem 10

Give mechanisms for these reactions, commenting on any regio- and stereoselectivity. What controls the rearrangement?

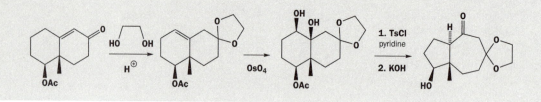

Purpose of the problem

Exploring the 'semipinacol' rearrangement with some revision of stereochemistry and acetal formation.

Suggested solution

The formation of the acetal is by the usual mechanism (p. 342) with one twist. The double bond is in conjugation with the ketone in the starting material but has moved in the acetal. There is no longer any conjugation possible but how and why does it move? How is easy. At several stages in the acetal mechanism, intermediates are formed that can enolize – we show one possibility. The conjugated and the unconjugated intermediate are in equilibrium and either can cyclize to the acetal. So the 'unconjugated' acetal must be more stable. This is partly because one of the C–O bonds must be axial and it is better for the acetal centre to be on a true chair than on a twist-chair conformation.

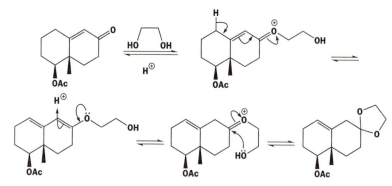

The next stage is a standard dihydroxylation with OsO_4. The two OH groups must be added in a *cis* fashion but why do they add to the top face of the alkene? This is easier to see in a conformational drawing. The top face of the alkene is virtually unhindered whereas the bottom face has most of the other ring – indeed all of the rest of the molecule except (just) the methyl group. The product is a *cis*-decalin (Chapter 18, p. 463).

■ This sequence was part of the synthesis of the antitumour lactone confertin by C. H. Heathcock and group, *J. Am. Chem. Soc.*, 1982, **104**, 1907.

The rearrangement starts with chemoselective tosylation of the secondary alcohol. Base then initiates the rearrangement itself by removing the proton from the tertiary alcohol. The stereochemistry follows the usual path – the migrating group goes with retention and the tosylate carbon is inverted. In an ordinary pinacol rearrangement either of the alcohols could be protonated but in this 'semi-pinacol' rearrangement only the secondary alcohol is tosylated and so it must leave. The migrating group is selected by whichever bond is *anti*-periplanar to the C–OTs bond. In this case it is the ring junction bond as the OTs group is equatorial (see diagram in the margin).

OAc omitted for clarity

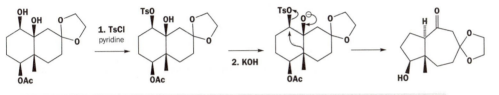

Problem 11

Suggest mechanisms for these reactions that explain any selectivity in the migration.

Purpose of the problem

To show that ring expansion from three- to four-membered rings and ring contraction the other way are about as easy.

Suggested solution

The first mechanism is a pinacol rearrangement and the compound is symmetrical so it doesn't matter which alcohol is protonated. Both three- and four-membered rings are strained and the σ-bonds are more reactive (higher HOMO energy) than normal. This makes ring contraction an easy reaction even though the strain is not relieved.

■ J. M. Conia and J. P. Barnier, *Tetrahedron Lett.*, 1971, 4981.

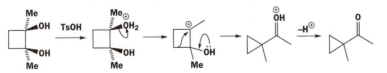

The second example looks at first to be a similar pinacol rearrangement. But the resulting ketone cannot easily be transformed into the product.

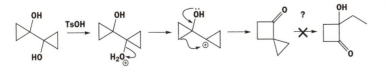

■ J. M. Denis and J. M. Conia,
Tetrahedron Lett., 1972, 4593.

Breaking one of the cyclopropane rings gets us off to a better start. This gives a hydroxy-ketone that can rearrange in a pinacol fashion to the product with ring expansion of the remaining cyclopropane.

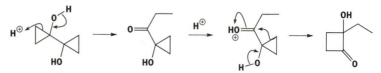

Problem 12

Attempts to produce the acid chloride from this unusual amino acid by treatment with $SOCl_2$ gave instead a β-lactam. What has happened?

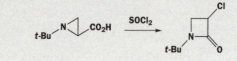

Purpose of the problem

To show that ring expansion from three- to four-membered rings is even easier in heterocycles because of participation.

Suggested solution

■ This surprising reaction is one way to make the important β-lactams present in penicillins and other antibiotics (J. A. Deyrup and S. C. Clough, *J. Am. Chem. Soc.*, 1969, **91**, 4590).

The formation of the acid chloride might go to completion or it might be some intermediate on the way to the acid chloride that rearranges. We shall use an intermediate. Whichever you use, it is participation by nitrogen that starts the ring expansion going, though the next intermediate is very unstable. When chloride attacks the bicyclic cation, it cleaves the most strained bond, the one in both three-membered rings.

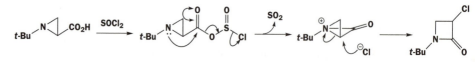

Problem 13

Revision content. Suggest mechanisms for these reactions, commenting in detail on the rearrangement step.

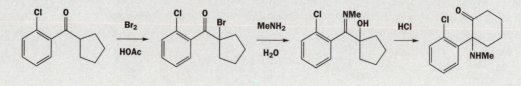

Purpose of the problem

To show that heteroatoms can migrate too given the chance and revision of Chapters 14 and 21.

Suggested solution

The first step is a standard acid-catalysed halogenation of a ketone via the enol (p. 533) and the second looks at first like a standard imine formation (p. 349) until you notice that the Br has been replaced by OH. Imine formation starts in the usual way but, before the carbonyl oxygen leaves, it is ideally placed to displace the bromine and form an epoxide. Opening the epoxide with the nitrogen atom completes the migration of oxygen from one atom to another.

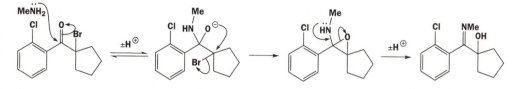

Protonation of the imine provides the perfect platform for a ring expansion as the OH group can push a ring bond across to the imine carbon. In this sequence we see participation by a heteroatom in the first rearrangement and then heteroatoms as electron sources and electron sinks in the second.

■ C. L. Stevens *et al.*, *J. Org. Chem.*, 1965, **30**, 2967.

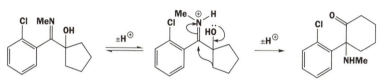

Problem 14

Suggest a mechanism for this rearrangement, comparing it with a reaction discussed in the chapter. What controls the stereochemistry?

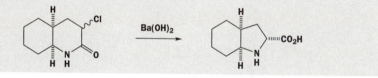

Purpose of the problem

Helping you to recognize the Favorskii reaction (p. 990).

Suggested solution

The most acidic proton in the molecule is the NH of the amide. Loss of the chloride ion from the resulting anion gives a sort of oxyallyl cation and this can close in a disrotatory electrocyclic two-electron process. The resulting cation is just a strange way of drawing a three-membered cyclic amide and attack of hydroxide on this strained molecule releases the best anion (nitrogen) and gives the product. We know that the CO_2H group in the product is down and so the conrotation must occur with the hydrogen (and the lone pair on nitrogen) rotating up. This puts the new three-membered ring on the outside (*exo*- or convex face) of the folded ring system.

■ This work was carried out at Hoechst in Frankfurt-am-Main by R. Henning and H. Urbach, *Tetrahedron Lett.*, 1983, **24**, 5339.

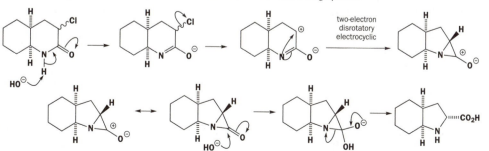

Suggested solutions for Chapter 38

38

Problem 1

Just to check your skill at finding fragmentations by numbers, draw a mechanism for each of these one-step fragmentations in basic solution (with an acidic work-up).

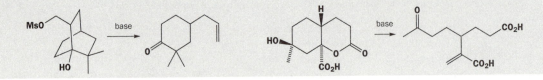

Purpose of the problem

As the problem says, to help you learn about simple fragmentations.

Suggested solution

We can identify the six-membered ring in both compounds – the sequence 1–6 is clearly the same in both with a side chain at C3. The fragmentation is easy enough too – the OH proton is removed

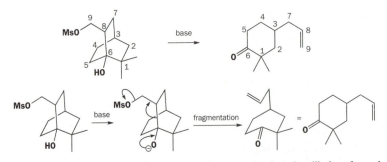

and the mesylate must be the leaving group so we have our 'push and pull' clear from the start.

The two CO_2H groups in the second product might cause a moment's concern but one is on a $-CH_2-CH_2-$ side chain and one is at a branchpoint and we can soon fill in the rest of the numbers. Clearly also, the OH proton is removed and one of the carboxyls is the leaving group. The stereochemistry disappears in the fragmentation but it is important as the conformational drawing shows. One lone pair on the O^-, the bond being fragmented, and the bond to the leaving group are all parallel (shown in thick lines).

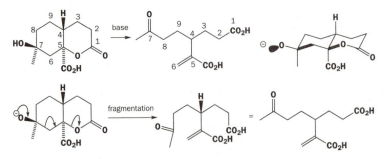

Problem 2

Treatment of this hydroxy-ketone with base followed by acid gives the enone shown. What is the structure of the intermediate A, how is it formed, and what is the mechanism of the formation of the final product?

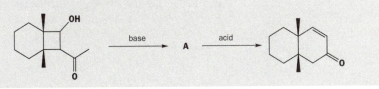

Purpose of the problem

Fragmentation may be followed by another reaction. Revision of Chapter 27.

Suggested solution

Removal of the hydroxyl proton by the base promotes a fragmentation that is really a reverse aldol reaction. It works because the bond being fragmented is in a four-membered ring. Then an acid-catalysed aldol reaction in the normal direction allows the formation of a much more stable six-membered ring.

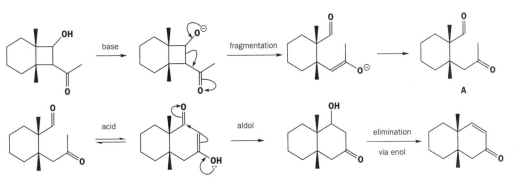

Problem 3

Suggest a mechanism for this reaction that involves a fragmentation as a key step.

Purpose of the problem

Revision of the mechanism of acetal hydrolysis plus an acid-catalysed fragmentation.

Suggested solution

Hydrolysis of the acetal gives a ketone whose enol can fragment. You would not expect such a fragmentation to occur unless, as here, the C–C bond being broken is in a strained ring. It is not even necessary to complete the hydrolysis of the acetal. Several intermediates in this reaction (p 344) have the required lone pair of electrons on oxygen and can initiate fragmentation.

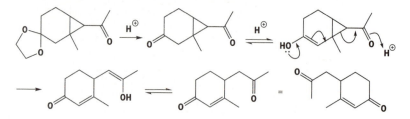

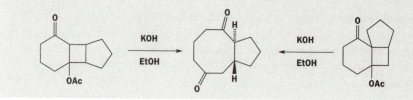

Problem 4

Explain why both these tricyclic ketones fragment to the same diastereoisomer of the same cyclo-octadione.

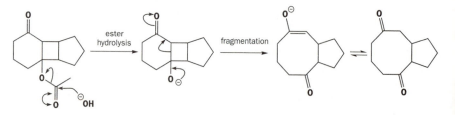

Purpose of the problem

Fragmentations linked to ester hydrolysis (contrast acetal hydrolysis in Problem 3) plus stereochemistry.

Suggested solution

It is obvious from the reactions that two features have disappeared from the starting materials: an ester group (OAc) and a four-membered ring. The ester can be hydrolysed with KOH and the four-membered ring disappears in the fragmentation. As usual, draw the mechanisms first and worry about the stereochemistry later. For the first compound, this sequence gives the enolate of a diketone and hence the diketone itself.

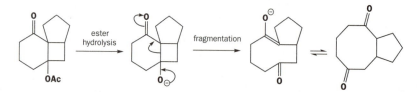

The second compound follows the same sequence and a different enolate emerges, but it is simply another enolate of the same diketone. Both compounds give the same basic structure.

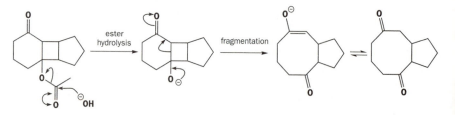

But what about stereochemistry? We are not told the stereochemistry of the starting materials, but we do know that the 5/4 fused rings must have *cis* ring junctions. This junction survives in the first compound so the stereochemistry looks wrong. The second compound gives us the clue. The fragmentation product is an enolate at the ring junction. When it tautomerizes to the ketone it will select the more stable *trans* 8/5 ring junction. In the same way, the enolate from the first sequence is in equilibrium under the reaction conditions with all the other enolates of the same diketone including those at the ring junction. This is a stereoselective reaction.

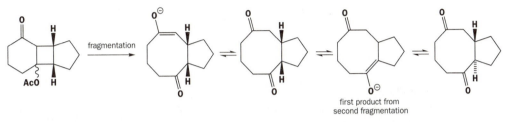

first product from
second fragmentation

Problem 5

Suggest a mechanism for this ring expansion in which fragmentation is one step.

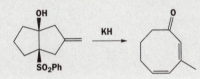

Purpose of the problem

A simple-looking mechanistic problem but it is not obvious what happens.

Suggested solution

Numbering is the best idea. There is no change necessary in the carbon skeleton other than the fragmentation of the C–C bond at the ring junction. The OH proton is obviously removed, the sulfonate group is lost, and the *exo*-methylene group becomes a methyl group. Evidently, other reactions than fragmentation must occur.

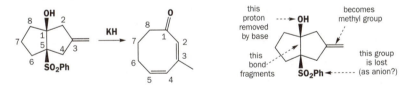

The C–C bond to be broken is not strained so a good reason for the fragmentation must be found. The base KH is easily strong enough to remove the OH proton entirely and sulfone is a good anion-stabilizing group.

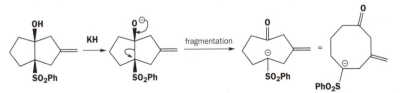

That gives us the carbon skeleton of the product. Now the sulfone-stabilized carbanion can remove a proton from either position next to the ketone to make an enolate. The Hs between the ketone and the double bond are the more acidic. Replacing this proton on the end of the double bond puts the methyl group in place. Finally, we repeat this process, removing a second hydrogen to make an extended enolate and eliminating the sulfone group as a sulfinate anion ($PhSO_2^-$) in an E1cB process. These proton transfers need not be intramolecular.

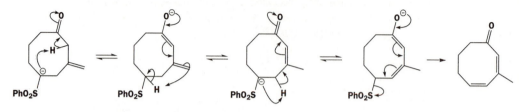

Problem 6

Suggest a mechanism for this fragmentation and explain the stereochemistry of the double bonds in the product. This is a tricky problem but find the mechanism and the stereochemistry will follow.

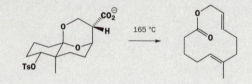

Purpose of the problem

Probably the most beautiful application of fragmentation yet by a true genius of chemistry, Albert Eschenmoser, *Angew. Chem., Int. Ed. Engl.*, 1979, **18**, 634, 636.

Suggested solution

The tosylate is obviously the leaving group, the two oxygen atoms in the rings must become the ester group, and the CO_2^- must leave as CO_2. All that remains is to trace an electron pathway from CO_2^- to OTs via one of the ring oxygens using parallel bonds. Though you can draw a mechanism for this double fragmentation in one step, it is not convincing. The only electrons antiparallel to the C–OTs bond are those in the ring junction bond and the equatorial lone pair on one of the ring oxygens. Marking these with heavy lines we carry out the first fragmentation.

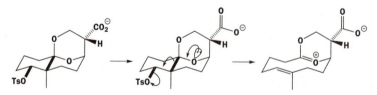

A conformational diagram shows that the geometry of this first alkene is predictable from the structure of the starting material. The second fragmentation is easier to see if we redraw the intermediate so that we can see which groups are antiparallel. A second conformational drawing shows that the second fragmentation also gives the correct alkene geometry.

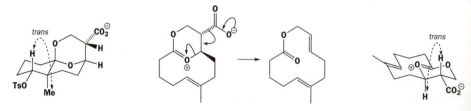

Problem 7

Suggest a mechanism for this reaction and explain why the molecule is prepared to abandon a stable six-membered ring for a larger ring.

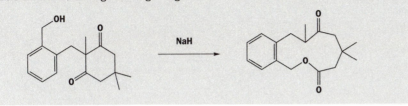

Purpose of the problem

A rather simpler exercise in fragmentation also used to form a medium-sized (11-membered) ring.

Suggested solution

The strong base removes the proton from the OH group and the oxyanion attacks one of the carbonyl groups (they are the same). This intermediate can decompose by the loss of the oxygen that has just come in but it can alternatively lose the enolate of the other ketone. The product is then an ester with its extra conjugation and protonation of the enolate completes the reaction. The eleven-membered ring is more stable than usual because of the benzene ring (compare the solution to Problem 3, Chapter 35) and because the ester oxygen does not suffer from interactions across the ring.

■ J. R. Mahajan and H. de Carvalho, *Synthesis*, 1979, 518.

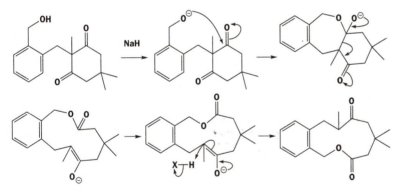

Problem 8

Give mechanisms for these reactions, commenting on the fragmentation.

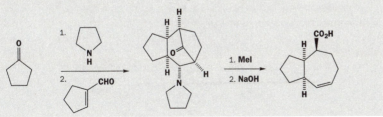

Purpose of the problem

Revision of conjugate addition of enols (Chapter 29), another ring expansion with an enolate as leaving group, and an interesting piece of stereochemistry.

Suggested solution

The first step is enamine formation and the second is conjugate addition. This produces an intermediate having the enamine of the ketone but the free aldehyde. The final product cannot be formed directly from this.

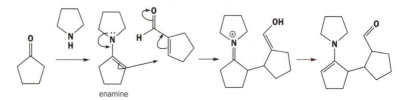

enamine

We need to use the enol of the ketone to attack the iminium derivative of the aldehyde. Some sort of enamine exchange must occur and there are various ways you could do this. We have simply exchanged the enamine of the ketone for that of the aldehyde.

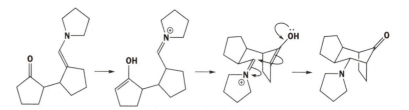

- As Hendrickson says; 'The reaction forms five asymmetric centres but its stereospecificity may be predicted from the presumption that the sequence of mechanistic steps includes no irreversible reactions. Hence the thermodynamically most stable product should result.'

- J. B. Hendrickson and R. K. Boeckman, *J. Am. Chem. Soc.*, 1971, **93**, 1307.

The last two diagrams show where the stereochemistry comes from. The final product has a chair six-membered ring. The 1,3-bridge on the bottom must be 1,3-diaxial so it is better to have the inevitable axial bond to the five-membered ring on the opposite, top, face. The pyrrolidine is equatorial. In other words, this is the most stable diastereoisomer of the product and that means reversible reactions and thermodynamic control. Finally, the fragmentation itself. Methylation of the amine provides the leaving group and addition of hydroxide to the ketone provides the electronic push. Notice that the C–N$^+$ bond, the C–C bond being fragmented, and a lone pair on the O$^-$ group are all parallel. Notice also that the hydrogen atoms on the *cis*-alkene in the product are already *cis* in the intermediate. The *cis* ring junction too is already there.

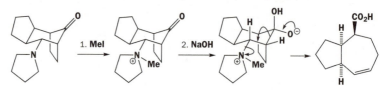

Problem 9

Propose mechanisms for the synthesis of the bicyclic intermediate and explain why only one diastereoisomer fragments (which one?).

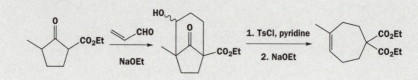

Purpose of the problem

More revision of conjugate addition of enols (Chapter 29), another ring expansion with an enolate as leaving group, and an interesting stereospecific reaction.

Suggested solution

The stable enolate of the keto-ester does conjugate addition even to an unsaturated aldehyde. The enolates of the products are in equilibrium and the ketone enolate can do a direct addition to the aldehyde to make a new six-membered ring without much selectivity at the OH group.

■ The ratio is in fact about 2:1 in favour of the useful equatorial alcohol (G. L. Buchanan and co-workers, *J. Chem. Soc., Perkin Trans. 1*, 1979, 1740).

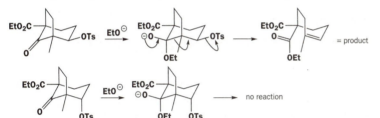

Tosylation provides a leaving group without changing the stereochemistry at the OH group and ethoxide adds to the ketone. One of these intermediates, the one with the equatorial tosylate, has the right stereochemistry for fragmentation with O⁻, C–C, and C–OTs all parallel.

Problem 10

Suggest mechanisms for these reactions, explaining the alkene geometry in the first case. Do you consider that they are fragmentations?

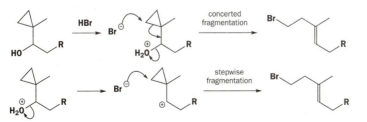

Purpose of the problem

Simple fragmentations involving the opening of three-membered rings.

Suggested solution

The first reaction is a fragmentation without any 'push' but that is all right because it is a bond in a three-membered ring that is being broken. You may have drawn a concerted mechanism or a stepwise one with a cation as an intermediate. Either may be correct. The stereochemistry of the alkene product is thermodynamically controlled.

■ W. S. Johnson and group, *J. Am. Chem. Soc.*, 1978, **100**, 4268.

■ K. Kondo *et al.*, *Tetrahedron Lett.*, 1978, 907.

The second reaction is base-catalysed and starts with the hydrolysis of the ester by NaOH. This fragmentation also needs 'push', though only a three-membered ring is being broken, because the leaving group is an enolate, nowhere near as electron-withdrawing as the water molecule or even the carbocation of the first example. Are they fragmentations? In both cases a C–C bond is broken but we would understand if you felt the first was not strictly a fragmentation, especially if it goes stepwise. Neither breaks the molecule into three pieces and it's all a matter of opinion.

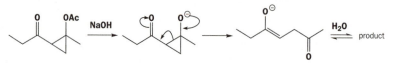

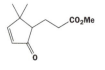

Problem 11

What steps would be necessary to carry out an Eschenmoser fragmentation on this ketone and what products would be formed?

Purpose of the problem

Revision of an important and complex reaction involving a fragmentation.

Suggested solution

The Eschenmoser fragmentation uses a tosyl hydrazone of an α,β-epoxy-ketone (p. 1008). The epoxide can be made with alkaline hydrogen peroxide and the tosylhydrazone needs just tosylhydrazine to form what is simply an imine. Then the fun can begin. The stereochemistry doesn't matter for once.

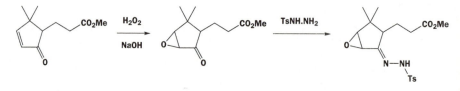

The fragmentation is initiated with base that removes a proton from the NHTs group. This anion fragments the molecule one way with the epoxide as leaving group and then the oxyanion fragments the molecule the other way with nitrogen gas and Ts⁻ as leaving groups. It comes as no surprise that the same Eschenmoser invented the other double fragmentation in Problem 6 of this chapter. The product is an acetylenic ketone or, as here, aldehyde.

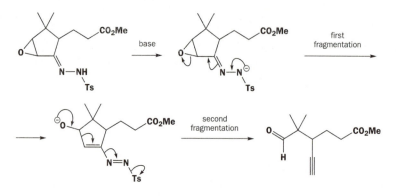

Problem 12

These related spirocyclic compounds give different naphthalenes when treated with sodium borohydride or with 5M HCl. Each reaction starts with a different fragmentation. Give mechanisms for the reactions and explain why the fragmentations are different. Treatment of the starting ketone with LiAlH₄ instead of NaBH₄ gives the alcohol below without fragmentation. Comment on the difference between the two reagents and the stereochemistry of the alcohol.

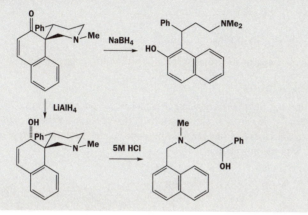

Purpose of the problem

A subtle effect of stereochemistry and reagent combined on two fragmentation pathways for the same compound.

Suggested solution

Sodium borohydride is a weak reducing agent and fragmentation (push by N and pull by C=O) is faster. The borohydride reduces the iminium salt to a methyl group and, at the other end of the molecule, the product is the 'keto' form of a phenol.

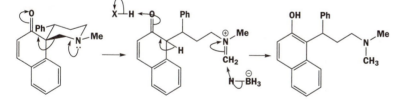

The second reaction starts with the rapid reduction of the ketone by the more powerful reducing agent LiAlH₄. Hydride is added from the less hindered face of the ketone opposite the phenyl group. Then, on treatment with acid, the alcohol does fragment but by a different pathway. The amine can no longer participate as it is protonated under the reaction conditions. It may look awkward to do the fragmentation with two separate cations as intermediates but both are stable and the stereochemistry is all wrong for anything concerted.

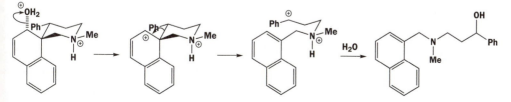

Problem 13

Revision content. Suggest mechanisms for these reactions explaining the stereochemistry.

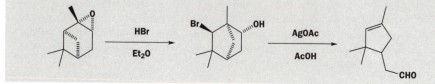

Purpose of the problem

Revision of rearrangements from Chapter 37 with a new fragmentation.

Suggested solution

The ring opening and the rearrangement cannot be all concerted because the group on the 'wrong' side of the molecule migrates. There must be a cationic intermediate. In contrast, attack of bromide occurs stereospecifically from the side opposite to the migrating group so this is presumably concerted with the migration.

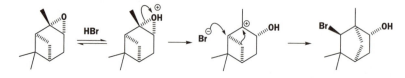

■ As it happens, the starting epoxide is that of natural α-pinene so it and the product are single enantiomers (P. H. Boyle *et al.*, *J. Chem. Soc., Chem. Commun.*, 1971, 395).

The second reaction is a fragmentation. Silver (I) is an excellent Lewis acid for halogens and probably produces a seco ndary carbocation intermediate. Push from the OH group completes the fragmentation.

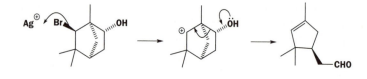

Suggested solutions for Chapter 39 **39**

Problem 1

In Chapter 33, Problem 13, we used a silylated ene-diol that was actually made in this way. Give a mechanism for the reaction and explain why the Me₃SiCl is necessary.

Purpose of the problem

Reminder of an important radical reaction and a modern development that makes it better.

Suggested solution

This is the acyloin condensation linking together radicals derived from esters by electron donation from a dissolving metal (here sodium). If the esters can form enolates, the addition of Me₃SiCl protects against that problem by removing the EtO⁻ by-product.

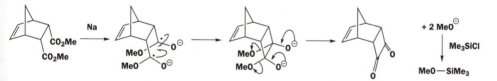

■ Details in B. M. Trost and group, *J. Org. Chem.*, 1978, **43**, 4559.

This first product is a very electrophilic 1,2-dione and it accepts electrons from sodium atoms even more readily that the original esters. The product is an ene diolate that is also silylated under the reaction conditions.

Problem 2

Heating the diazonium salt below in the presence of methyl acrylate gives a reasonable yield of a chloroacid. Why is this unlikely to be nucleophilic aromatic substitution by the S$_N$1 mechanism (Chapter 23)? Suggest an alternative mechanism that explains the regioselectivity.

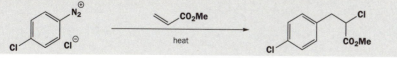

Purpose of the problem

Revision of nucleophilic aromatic substitution (p. 597) with diazonium salts and contrasting cations and radicals.

Suggested solution

The cation mechanism is perfectly reasonable as far as the diazonium salt is concerned but it will not do for the alkene. Conjugated esters are electrophilic and not nucleophilic alkenes. Even if it were to attack the aryl cation, we should find the reverse regioselectivity.

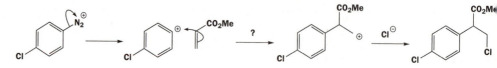

The only way to produce the observed product is to decompose the diazonium salt homolytically. To do this we must draw the 'salt' as a covalent compound or transfer one electron from the chloride ion to the diazonium ion. The other product would be a chlorine radical. Addition to the alkene gives the more stable radical, which abstracts chlorine from the diazocompound and keeps the chain going.

■ Notice that in the last step we have put in only half the mechanism – we shall generally do this from now on as it is clearer (p. 1022). There is nothing wrong with putting in another set of half-arrows going the other way if you want to.

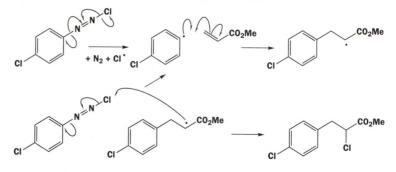

Problem 3

Suggest a mechanisms for this reaction and comment on the ring size formed. What is the minor product likely to be?

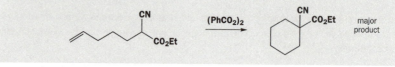

Purpose of the problem

A commonly used radical cyclization: to help you to realize that activated alkenes are unnecessary.

Suggested solution

The peroxide is a source of benzoyloxy radicals ($PhCO_2^•$) and these capture hydrogen atoms to give the most stable radical. The best radical is the tertiary one stabilized by both CN and CO_2Et. Cyclization on to the alkene gives a secondary radical on a six-membered ring and this abstracts a hydrogen atom from the starting material to complete the cycle.

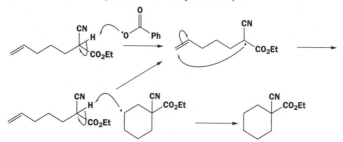

The alternative is to add to the more substituted end of the alkene. This gives a less stable primary radical but (p. 1050) this type of cyclization is often preferred because the orbital alignment is better. The minor product has a five-membered ring.

■ This topic is reviewed by Marc Julia in *Acc. Chem. Res.*, 1971, **4**, 386.

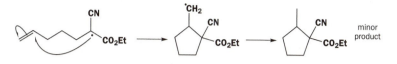

Problem 4

Treatment of this aromatic heterocycle with NBS (*N*-bromosuccinimide) and AIBN gives mainly one product but this is difficult to purify from minor impurities containing one or three bromine atoms. Further treatment with 10% aqueous NaOH gives one easily separable product in modest yield (50%). What are the mechanisms for the reactions? What might the minor products be?

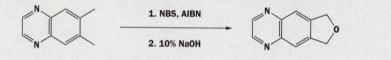

Purpose of the problem

An important radical reaction: bromination at benzylic and allylic positions by NBS plus an application.

Suggested solution

Two preliminary reactions need to take place: NBS is a source of a low concentration of bromine and AIBN initiates the radical chain by forming a nitrile-stabilized tertiary radical.

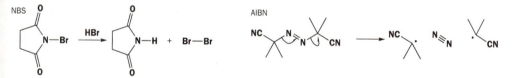

The new radical abstracts hydrogen atoms from the benzylic positions to make stable delocalized radicals. These react with bromine to give the benzylic bromide and release a bromine atom.

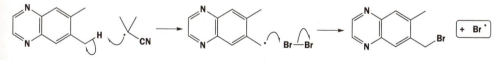

All subsequent hydrogen abstractions are carried out by the bromine atoms, either of the kind we have just seen or to remove a hydrogen atom from the other methyl group. This reaction provides the HBr that generates bromine from NBS.

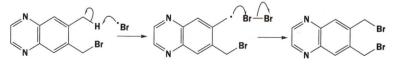

■ This product was used to make novel constrained amino acids by S. Kotha and co-workers, *Tetrahedron Lett.*, 1997, **38**, 9031.

Finally, the dibromide reacts with NaOH to give the heterocycle. Both S_N2 displacements are very easy at a benzylic centre, the second particularly so because it is intramolecular and forms a five-membered ring.

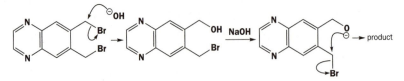

Problem 5

Propose a mechanism for this reaction accounting for the selectivity. Include a conformational drawing of the product.

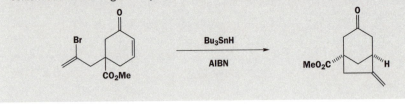

Purpose of the problem

Another important radical reaction: cyclization of alkyl bromides on to alkenes with Bu_3SnH as reagent.

Suggested solution

This time AIBN abstracts a hydrogen atom from the reagent and the $Bu_3Sn^\bullet$ radicals carry the chain along. First, they remove the bromine atom from the starting material to make a vinyl radical that cyclizes on to the unsaturated ester to give a radical stabilized by conjugation with the ketone group. The chain is completed by hydrogen atom abstraction from another molecule of Bu_3SnH.

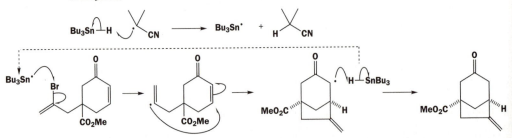

■ N. N. Marinovic and H. Ramanathan (*Tetrahedron Lett.*, 1983, **24**, 1871) developed this new chemistry.

The stereochemistry of the product comes from the requirement of a 1,3-bridge to be diaxial as this is the only way the bridge can reach across the ring. At the moment of cyclization, the vinyl radical side chain must also be in an axial position.

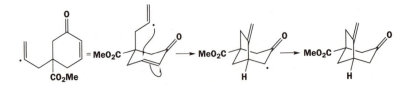

Problem 6

An ICI process for the manufacture of the diene used to make pyrethroid insecticides involves heating these compounds to 500°C in a flow system. Propose a radical chain mechanism for the reaction.

Purpose of the problem

Learning how to avoid a trap in writing radical mechanisms and to show you that radical reactions can be useful.

Suggested solution

The most likely initiation at 500°C is the cleavage of the C–Cl bond to release allyl and chloride radicals. The chloride radicals then attack the alkene and abstract a proton to give an allylic radical.

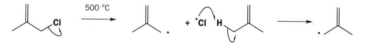

The trap is to form the product by dimerizing these allylic radicals. Dimerizing of radicals is known (in the acyloin reaction, for example, p. 1032), and these radicals will sometimes dimerize but it is a rare process.

Much more efficient is a chain reaction. If we add the allylic radical to the alkene of the allylic chloride, we can create a chain. The chloride radical released is used to abstract a hydrogen atom from the next molecule of alkene and the chain continues.

■ The original workers at ICI, D. Holland and D. J. Milner, suggested a different mechanism (*Chem. and Ind. (London)*, 1979, 707).

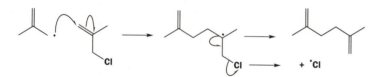

Problem 7

Heating this compound at 560°C gives two products with the spectroscopic data shown below. What are these products and how are they formed?

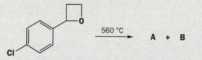

A has IR 1640 cm^{-1}; m/z 138 (100%), 140 (33%); δ_H (p.p.m.) 7.1 (4H, s), 6.5 (1H, dd, J 17, 11 Hz), 5.5 (1H, dd, J 17, 2 Hz), and 5.1 (1H, dd, J 11, 2 Hz).

B has IR 1700 cm^{-1}; m/z 111 (45%), 113 (15%), 139 (60%), 140 (100%), 141 (20%), and 142 (33%); δ_H (p.p.m.) 9.9 (1H, s), 7.75 (2H, d, J 9 Hz), and 7.43 (2H, d, J 9 Hz).

Purpose of the problem

Revision of structure determination and a radical reaction with a difference.

Suggested solution

Compound A contains chlorine (M 138/140 3:1) and that fits C_8H_7Cl, it still has the 1,4-disubstituted benzene ring (four aromatic Hs), and it is an alkene (IR 1640 cm^{-1}) with three hydrogen atoms on the double bond (δ_H 6.5, 5.5, and 5.1). We can write the structure immediately as there is no choice.

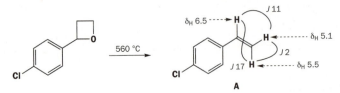

Compound B has M 140/142 (3:1) and a carbonyl group (IR 1700 cm^{-1}) and looks like C_7H_5ClO. It is an aldehyde (δ_H 9.9) and again still has the disubstituted benzene. The structure is even easier this time! We have included the fragmentation in the mass spectrum just to remind you that this starts with a radical reaction too. Occasionally in the mass spectrum (gas phase, high vacuum) you get the very rare loss of H$^\bullet$, a reaction that never occurs in solution. Here it does so because of the great stability of the acylium ion (139/140) (p. 554).

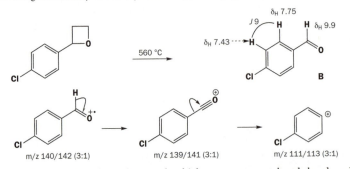

So how are these products formed? At such a high temperature, σ-bonds break and the weakest bonds in the molecule are the C–C and C–O bonds in the four-membered ring next to the benzene ring. Breaking these bonds releases strain and allows one of the radical products to be secondary and delocalized.

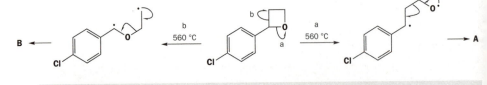

Problem 8

Treatment of methylcyclopropane with peroxides at very low temperature (–150°C) gives an unstable species whose ESR spectrum consists of a triplet with coupling 20.7 gauss and fine splitting showing dtt coupling of 2.0, 2.6, and 3.0 gauss. Warming to a mere –90°C gives a new species whose ESR spectrum consists of a triplet of triplets with coupling 22.2 and 28.5 gauss and fine splitting showing small ddd coupling of less than 1 gauss.

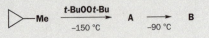

If methylcyclopropane is treated with *t*-BuOCl, various products are obtained, but the two major products are C and D. At lower temperatures more of C is formed and at higher temperatures more of D.

C 28% yield   D 20% yield

Treatment of the more highly substituted cyclopropane with PhSH and AIBN gives a single product in quantitative yield. Account for all of these reactions, identifying A and B and explaining the differences between the various experiments.

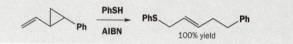

100% yield

Purpose of the problem

Working out the consequence of an important substituent effect on radical reactions: the cyclopropyl group.

Suggested solution

The peroxide is a source of *t*-BuO• radicals and these abstract a hydrogen atom from the methyl group of the hydrocarbon. The first spectrum is that of the cyclopropylmethyl radical. The odd electron is in a p orbital represented by the circle and the plane of the $CH_2^•$ group is orthogonal to the plane of the ring but the two H^as are the same because of rapid rotation. The odd electron has a large coupling to the two hydrogens (H^a) on the same carbon, a smaller doublet coupling to H^b, and small couplings to the two H^cs and two H^ds. The coupling to H^b is small because the p orbital containing the odd electron is orthogonal to the C–H^b bond.

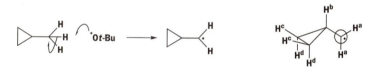

Warming to –90°C causes decomposition to an open-chain radical. The odd electron is coupled to the two hydrogens on its own carbon (H^a) and those on the next carbon (H^b), each giving a triplet (22.2 and 28.5). Coupling to the more remote hydrogen atoms is small.

■ These data come from J. K. Kochi's discovery of the details of this process (*J. Am. Chem. Soc.*, 1969, **91**, 1877).

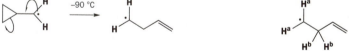

Decomposition of the same hydrocarbon with *t*-BuOCl produces the same sequence of radicals but now they can be intercepted by the chlorine atom of the reagent releasing more *t*-BuO• radicals and a radical chain started. At lower temperatures the rate of ring opening is less so more of the cyclopropane is captured.

■ C. S. Walling and P. S. Fredericks, *J. Am. Chem. Soc.*, 1962, **84**, 3326.

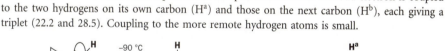

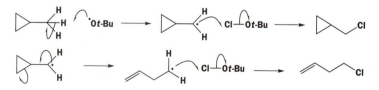

■ Ring opening of cyclopropanes is now a standard way of detecting radicals (J. W. Wilt in J. K. Kochi, *Free radicals*, Wiley, 1973, pp. 333–502).

The last example also produces a radical next to a cyclopropane ring but this time it can decompose very easily to give a stable secondary benzylic radical. This captures a hydrogen atom from PhSH releasing PhS• and maintaining an efficient radical chain.

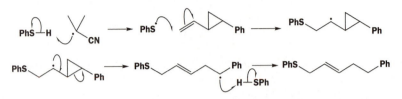

Problem 9

The last few stages of Corey's epibatidine synthesis are shown here. Give mechanisms for the first two reactions and suggest a reagent for the last step.

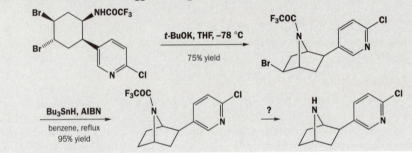

Purpose of the problem

Application of radical reactions in an important sequence plus revision of conformation and stereochemistry.

Suggested solution

The first step involves deprotonation of the rather acidic amide (the CF$_3$ group helps) and the displacement of the only possible bromide – the one on the opposite face of the six-membered ring as the S$_N$2 must take place with inversion.

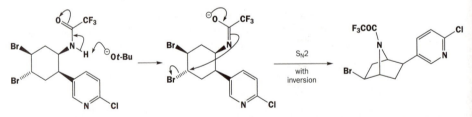

The second step is a standard dehalogenation by Bu$_3$SnH. AIBN generates Bu$_3$Sn• by hydrogen abstraction from the reagent and this removes the bromine. The only tricky thing is to make sure you complete the chain and don't use H• at any point.

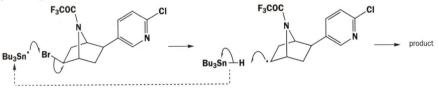

Finally, we need to hydrolyse the amide. This normally requires strong acid or alkali but the trifluoroacetyl group is significantly more electrophilic than most acyl groups and milder conditions can be used. In this final step of his epibatidine synthesis, E. J. Corey actually used sodium methoxide (NaOMe) in methanol at 13°C for two hours and the yield was 96%. Any reasonable conditions would be nearly as good.

■ E. J. Corey and group, *J. Org. Chem.*, 1993, **58**, 5600.

Problem 10

How would you make the starting material for this sequence of reactions? Give a mechanism for the first reaction that explains its regio- and stereoselectivity. Your answer should include a conformational drawing of the product. What is the mechanism of the last step? Attempts to carry out this last step by iodine-lithium exchange and reaction with allyl bromide fail. Why? Why is the reaction sequence here successful?

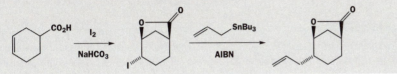

Purpose of the problem

Application of radical reactions when the alternative ionic reactions fail.

Suggested solution

The starting material is an obvious Diels–Alder product as it is a cyclohexene with a carbonyl group outside the ring on the opposite side. The first step is iodolactonization. Iodine attacks the alkene reversibly on both sides but ,when it attacks opposite the carboxylate anion, the lactone ring snaps shut.

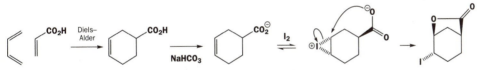

The question asks for a conformational drawing of the product and indeed that is necessary. The 1,3 lactone bridge must be diaxial as that is the only way to reach across and the carboxylate group must attack the iodonium ion from an axial position too. You can regard this as the formation of the *trans* diaxial product as in the opening of a cyclohexene oxide with a nucleophile (pp. 468–70).

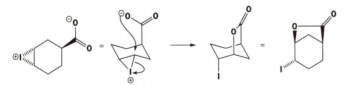

The last step is initiated by AIBN, which removes an iodine atom from the compound to make a secondary radical. This attacks the allyl stannane and the intermediate loses Bu₃Sn• and that takes over the job of removing iodine atoms to keep the chain going. The radical intermediate has no stereochemistry at the planar radical carbon. Attack occurs on the bottom face to avoid the 1,3-diaxial bridge.

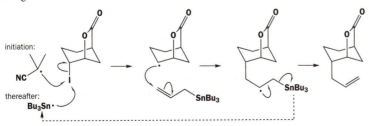

Anionic reactions cannot be used for this allylation. If the iodine were metallated, the organometallic compound would immediately expel the carboxylate anion as a good leaving group. The radical is stable because the C–O bond is strong and not easily cleaved in radical reactions.

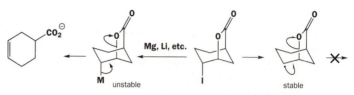

Problem 11

Suggest a mechanism for this reaction explaining why a mixture of diastereoisomers of the starting material gives a single diastereoisomer of the product. Is there any other form of selectivity?

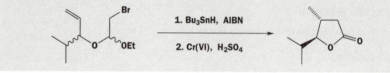

Purpose of the problem

A radical ring-closing reaction with a curious stereochemical outcome.

Suggested solution

The abstraction of bromine, at first by AIBN but thereafter by $Bu_3Sn^\bullet$, produces a radical that again does not eliminate but adds to an alkene. A five-membered ring is formed (this is usually the more favourable closure) by attack on the alkene on the opposite face to that occupied by the *i*-Pr group. The product is a mixture of diastereoisomers as no change occurs at the acetal centre.

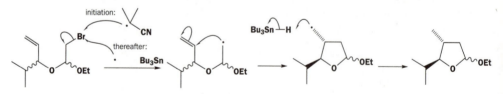

■ This reaction was a new method to make lactones discovered by Gilbert Stork and his group (*J. Am. Chem. Soc.*, 1983, **105**, 3741).

Acid-catalysed oxidation first hydrolyses the acetal and then oxidizes either the hemiacetal or the aldehyde to the lactone. Now the molecule is one compound (one diastereoisomer) as the ambiguous centre is planar. The other form of selectivity is the regioselectivity in the closure of the ring. That is referred to in Chapter 39 (p. 1049) but properly dealt with in Chapter 42 (pp. 1140–3).

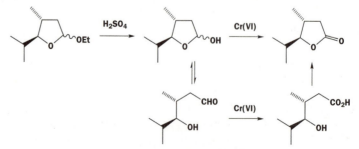

Problem 12

On the other hand, why does a single diastereoisomer of this organomercury compound give a mixture of diastereoisomers (68:32) on reduction with borohydride in the presence of acrylonitrile?

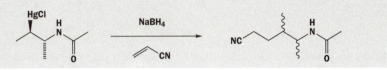

Purpose of the problem

Rehearsal of a useful method to make radicals and further exploration of stereochemistry.

Suggested solution

Borohydride reduces the mercury to unstable –HgH. This decomposes to provide an alkyl radical for conjugate addition to acrylonitrile. Though the starting material has stereochemistry, the alkyl radical is flat and, in an open-chain compound, has no reason to react one side rather than the other. Compare with the previous problem.

■ This reaction nevertheless provided a new approach to nitrogen heterocycles pioneered by A. P. Kozikowski and J. Scripko, *Tetrahedron Lett.*, 1983, **24**, 2051.

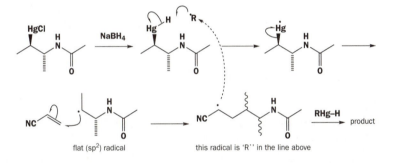

flat (sp²) radical   this radical is 'R'' in the line above

Problem 13

Reaction of this carboxylic acid ($C_5H_8O_2$) with bromine in the presence of dibenzoyl peroxide gives an unstable compound A ($C_5H_6Br_2O_2$) that gives a stable compound B ($C_5H_5BrO_2$) on treatment with base. The stable compound has IR 1735 and 1645 cm^{-1} and NMR δ_H (p.p.m.) 6.18 (1H, s), 5.00 (2H, s), and 4.18 (2H, s). What is the structure of the stable product? Deduce the structure of the unstable compound and mechanisms for the reactions.

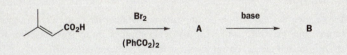

Purpose of the problem

Revision of structural analysis in combination with an important radical functionalization.

Suggested solution

The starting material is $C_5H_8O_2$ so the stable compound B has gained a bromine and lost three hydrogens. There must be an extra double bond equivalent (DBE) in B. The IR spectrum shows that the OH has gone and suggests a carbonyl compound, possibly an ester because of the high frequency, and an alkene. The NMR shows that both methyl groups have gone and been replaced by

CH$_2$ groups. The bromine must be on one of them and the ester oxygen on the other. The extra DBE is a ring.

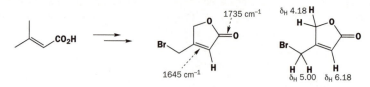

Since both methyl groups are functionalized, the unstable A must have one Br on each methyl group. The peroxide produces benzoate radicals that abstract a proton from allylic positions to give stabilized radicals and these attack bromine molecules providing bromine atoms to continue the chain reaction. In base, the carboxylate ion cyclizes on to the *cis* CH$_2$Br group.

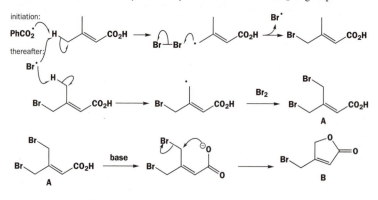

Problem 14

The product formed in Problem 9 of Chapter 20 was actually used to make this cyclic ether. What is the mechanism?

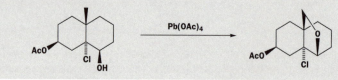

Purpose of the problem

A slightly unusual method to make radicals and a remote functionalization reaction.

Suggested solution

Lead tetraacetate provides radicals on heating (maybe AcO$^{\bullet}$ or lead radicals) and these abstract a hydrogen atom from the OH group. The oxygen radical is in an ideal position to abstract a hydrogen atom from the methyl group (conformational diagram essential).

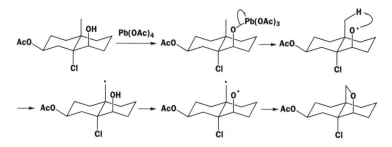

Problem 1

Suggest mechanisms for these reactions.

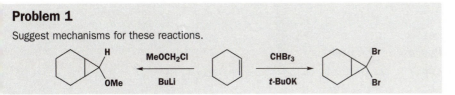

Purpose of the problem

Two simple carbene reactions initiated by base.

Suggested solution

Going to the right we must remove the rather acidic proton from $CHBr_3$ to give the carbanion. This loses bromide ion to give dibromocarbene and insertion into cyclohexene gives the product.

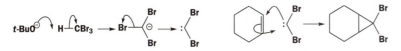

The second reaction is very similar. α-Elimination of HCl gives a carbene that inserts into the double bond. These are the simplest types of carbene reactions and are very common.

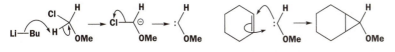

Problem 2

Suggest a mechanism and explain the stereochemistry of this reaction.

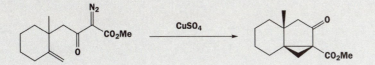

Purpose of the problem

Another important carbene method used in the synthesis of a natural antibiotic.

Suggested solution

The diazo-compound decomposes to gaseous nitrogen and a carbene under catalysis by Cu(II). Insertion into the exposed alkene gives the three-membered ring. The stereochemistry comes partly from the 'tether' – the linkage between the carbene and the rest of the molecule that delivers the carbene to the bottom face of the alkene. The rest comes from the inevitable *cis* fusion between the six- and three-membered rings.

■ This reaction established the skeleton of cycloeudesmol (margin) and was carried out by E. Y. Chen, *Tetrahedron Lett.*, 1982, **23**, 4769, at Sandoz.

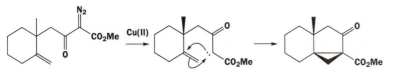

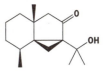

cycloeudesmol

Problem 3

Comment on the selectivity shown in these two reactions.

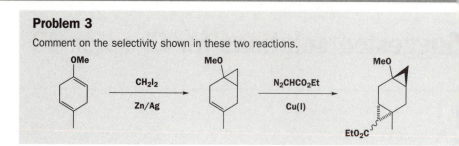

Purpose of the problem

A study in chemoselectivity during carbene insertion into alkenes.

Suggested solution

The first reagent is a variation on the Simmons-Smith cyclopropanation. Though strictly a carbenoid rather than a carbene, it delivers a CH_2 group from an organozinc compound bound to an oxygen atom, in this case the OMe group. Only that alkene reacts. We follow the description in the chapter (pp. 1067–8).

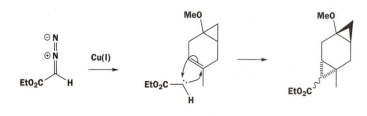

■ There is little selectivity for the stereochemistry of the CO_2Et group but this fortunately did not matter in the synthesis of a natural defence substance from a sponge by G. A. Schiehser and J. D. White, *J. Org. Chem.*, 1980, **45**, 1864.

The second cyclopropanation occurs at the only remaining alkene with a carbene generated from a diazoester. The stereoselectivity comes from attack on the opposite side of the ring to the three-membered ring already present.

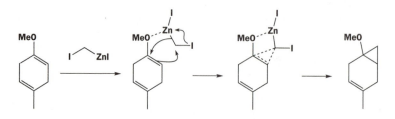

Problem 4

Suggest a mechanism for this ring contraction.

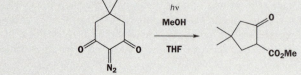

Purpose of the problem

Drawing mechanisms for a rearrangement involving a carbene formed photochemically.

Suggested solution

The carbene formed by loss of nitrogen from the diazoketone rearranges with the migration of either CC bond to give a ketene picked up by the methanol.

■ This product was the starting material in Problem 33.9 (J. Froborg and G. Magnusson, *J. Am. Chem. Soc.*, 1978, **100**, 6728).

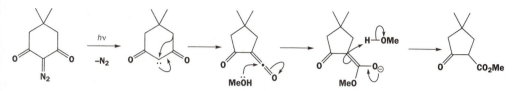

Problem 5

Suggest a mechanism for the formation of this cyclopropane.

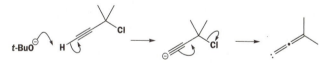

Purpose of the problem

An unusual type of carbene but it behaves normally.

Suggested solution

There is no doubt that *t*-BuO⁻ is a base, but which proton does it remove? The OH proton, perhaps, but that doesn't lead to a carbene. The proton from the alkyne? That molecule has a leaving group (chloride) but is it too far away?

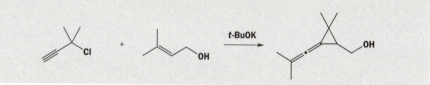

Not if you push the electrons though the molecule in a γ-elimination. The carbene is an allenyl carbene with no substituent at the carbene centre. It inserts into the alkene in the other molecule. This sequence is the start of a synthesis of chrysanthemic acid.

■ Normal elimination to give an alkene is β-elimination. Both α- and γ-elimination can produce carbenes.

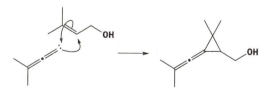

Problem 6

Problem 4 in Chapter 31 (p. 821) asked: 'Decomposition of this diazo compound in methanol gives an alkene A ($C_8H_{14}O$) whose NMR spectrum contains two signals in the alkene region: δ_H (p.p.m.) 3.50 (3H, s), 5.50 (1H, dd, J 17.9, 7.9 Hz), 5.80 (1H, ddd, J 17.9, 9.2, 4.3 Hz), 4.20 (1H, m), and 1.3–2.7 (8H, m). What is its structure and geometry?'

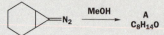

In order to work out the mechanism of the reaction you might like to take these additional facts into account. Compound A is unstable and even at 20°C isomerizes to B. If the diazo compound is decomposed in methanol containing a diene, compound A is trapped as an adduct. Account for all of these reactions.

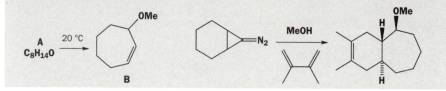

Purpose of the problem

Revision of structural analysis, double bond geometry, and cycloaddition reactions with carbenes as a mechanistic link.

Suggested solution

The structure of A, as deduced in Problem 31.4, was a *trans*-alkene in a seven-membered ring. It is obviously formed from the cyclopropyl carbene released when nitrogen is lost from the diazo compound.

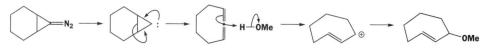

■ H. Jendralla, *Angew. Chem., Int. Ed. Engl.*, 1980, **19**, 1032.

The twisted alkene of a *trans*-cycloheptene is unstable and rotates to the much more stable *cis*-alkene even at 20°. It can rotate because the overlap between the p orbitals is very weak as they are not parallel. Trapping in a Diels–Alder reaction preserves the *trans* stereochemistry.

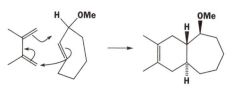

Problem 7

Give a mechanism for the formation of the three-membered ring in the first of these reactions and suggest how the ester might be converted into the amine with retention of configuration.

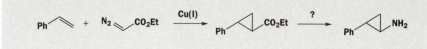

Purpose of the problem

Reminder of nitrenes as analogues of carbenes after a routine carbene insertion.

Suggested solution

The diazoester gives the carbene under Cu(I) catalysis, and insertion into the alkene follows its usual course. The only extra is the stereoselectivity: the insertion happens more easily if the two large groups (Ph and CO_2Et) stay as far apart as possible.

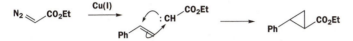

Conversion of acid derivatives into amines with the loss of the carbonyl group can be done in various ways. In Chapter 40 we recommended either the Curtius rearrangement or the Hofmann degradation (p. 1073). The Hofmann degradation is easier starting with an ester. We convert into the amide with ammonia and treat with bromine in basic solution. The N-bromo derivative forms a nitrene by α-elimination that rearranges to an isocyanate.

■ The product is an antidepressant discovered by A. Burger and W. L. Yost, *J. Am. Chem. Soc.*, 1948, **70**, 2198.

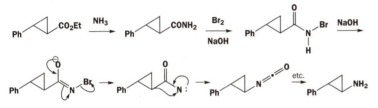

Problem 8

Explain how this highly strained ketone is produced, albeit in very low yield, by these reactions. How would you attempt to make the starting material?

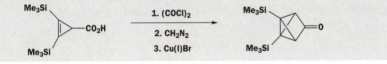

Purpose of the problem

To show that intramolecular carbene insertion is a powerful way to prepare cage compounds.

Suggested solution

Oxalyl chloride makes the acid chloride (details on p. 296) and diazomethane converts this into the diazoketone (details on p. 1057).

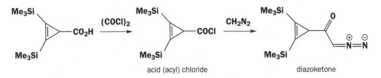

acid (acyl) chloride   diazoketone

Now the carbene chemistry. Treatment with Cu(I) removes gaseous nitrogen and forms the carbene. Incredibly, this is able to reach across the molecule and insert into the alkene forming one three- and two new four-membered rings in one step. You will not be surprised that the yield in this reaction is rather low (about 10%).

■ This very strained ketone was used in a vain attempt to make tetrahedrane by G. Maier and group, *Angew. Chem., Int. Ed. Engl.*, 1983, **22**, 990.

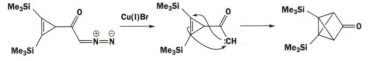

■ G. Maier and group, *Angew. Chem., Int. Ed. Engl.*, 1982, **21**, 437.

How would you attempt to make the starting material? The original workers used another carbene reaction – the Cu(I)-catalysed insertion of a diazoester into *bis* trimethylsilyl acetylene.

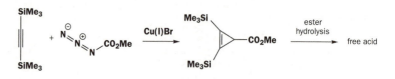

Problem 9

Attempts to prepare compound A by a phase-transfer-catalysed cyclization required a solvent immiscible with water. When chloroform (CHCl₃) was used, compound B was formed instead and it was necessary to use the more toxic CCl₄ for success. What went wrong?

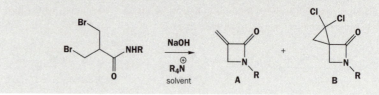

Purpose of the problem

Carbene chemistry is not always what is wanted! How do you avoid it?

Suggested solution

■ This chemistry was used to make new β-lactams at the then ICI by S. R. Fletcher and I. T. Kay, *J. Chem. Soc., Chem. Commun.*, 1978, 903.

Product B is clearly an adduct of A and dichlorocarbene, which must be formed by attack of NaOH on chloroform. The good news is that A was evidently formed in the basic reaction mixture so, if we simply avoid the solvent that makes the carbene, all is well.

■ Did you notice that the starting material is a derivative of a compound made in Problem 39.13?

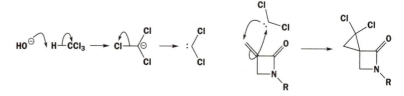

Problem 10

Revision content. How would you carry out the first step in this sequence? Propose mechanisms for the remaining steps, explaining any selectivity.

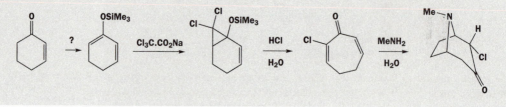

Purpose of the problem

Revision of specific enol formation (Chapter 21, pp. 538–9), rearrangement reactions (Chapter 38), or even electrocyclic reactions (Chapter 36), and conjugate addition (Chapter 10) plus the usual carbene chemistry.

Suggested solution

The first step requires a specific enol from an enone. Treatment with LDA achieves kinetic enolate formation by removing one of the more acidic hydrogens immediately next to the carbonyl group. The other enolate would be the thermodynamic product. The lithium enolate is trapped with Me₃SiCl to give the silyl enol ether.

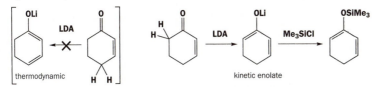

The next step is dichlorocarbene insertion into the more nucleophilic of the two alkenes. Dichlorocarbene is an electrophilic carbene so the main interaction is between the HOMO (π) of the alkene and the empty p orbital of the carbene. The carbene is formed by decarboxylation, a process that needs no strong base.

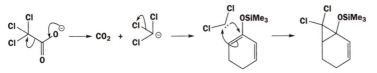

You can draw the ring expansion in a number of ways. All start with the removal of the Me₃Si group with water. You might then simply produce the product in one step by mechanism (a) but an electrocyclic process via the cyclopropyl cation (b) might be better. This is allowed since the inevitably *cis* ring junction requires H and OH to rotate outwards.

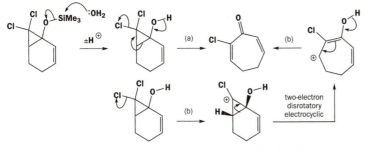

Finally, a double conjugate addition of MeNH₂ to the dienone forms the bicyclic amine. Conjugate addition probably occurs first on the more electrophilic chloroenone, though it doesn't much matter. There is some stereoselectivity in that the remaining Cl prefers the equatorial position on the new six-membered ring but this is thermodynamic control as that position is easily enolized.

■ The product has the skeleton of the tropane alkaloids (pp. 1416–18) and this chemistry allowed T. L. Macdonald and R. Dolan (*J. Org. Chem.*, 1979, **44**, 4973) to make a number of these natural products.

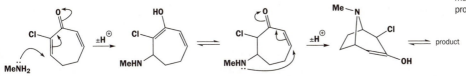

Problem 11

How would you attempt to make these alkenes by metathesis?

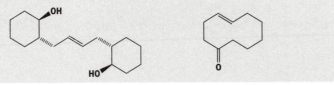

powerful method.

Suggested solution

Metathesis is usually *E*-selective and these are both *E*-alkenes so prospects are good. We must split each alkene down the middle of the double bond and add something to the end of each, probably just CH_2 as the other product will then be volatile ethylene.

Each starting material must now be made. The stereochemistry of the first tells us that we should add an allyl metal compound to an epoxide. The metathesis catalyst can be any of those mentioned

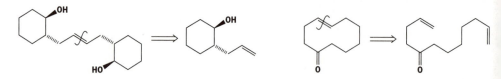

in the chapter (pp. 1074–5).

The second molecule is not symmetrical but this is all right as it is an intramolecular (ring-closing) metathesis so we can expect few cross-products. There are many ways to make the required

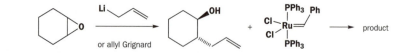

starting material: alkylation of a simple ketone is probably the simplest though conjugate addition would have its advantages. If you are not sure about your method, check with someone who knows. The same catalyst can be used and only about 2% would be needed.

Purpose of the problem

Problem 12

Heating this acyl azide in dry toluene under reflux for 3 hours gives a 90% yield of a heterocyclic product. Suggest a mechanism, emphasizing the involvement of any reactive intermediates.

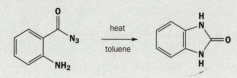

Demonstrating the practical nature of nitrene chemistry in the context of heterocyclic synthesis.

Suggested solution

Heating an azide liberates nitrogen gas and forms a nitrene. In this case rearrangement to an isocyanate is followed by intramolecular nucleophilic attack by the *ortho* amino group.

■ A. O. Fitton and R. K. Smalley, *Practical heterocyclic chemistry*, Academic Press, London, 1968, p. 43.

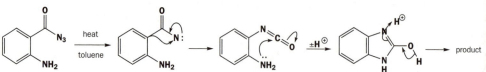

Problem 13

Give mechanisms for the steps in this conversion of a five- to a six-membered aromatic heterocycle.

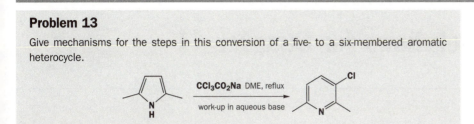

Purpose of the problem

It is the turn of carbene chemistry to show its usefulness in that most practical of all subjects – heterocyclic synthesis.

Suggested solution

Decomposition of trichloroacetate ion releases the Cl_3C^- carbanion. Loss of chloride gives dichlorocarbene and addition to one of the double bonds in the pyrrole gives a bicyclic intermediate.

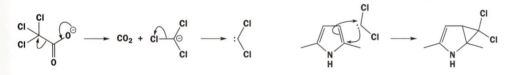

Ring expansion can be drawn in various ways. There is a direct route from the neutral amine, or its anion, that doesn't look very convincing, or you can ionize one of the chlorides first and open the cyclopropyl cation in an electrocyclic process. This is a good method to make 3-substituted pyridines.

■ A. O. Fitton and R. K. Smalley, *Practical heterocyclic chemistry*, Academic Press, London, 1968, p. 67.

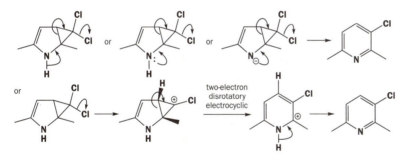

Suggested solutions for Chapter 41

41

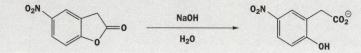

Purpose of the problem

Investigating a reaction where there are serious doubts about the mechanism.

Suggested solution

The reaction is an ester hydrolysis so the obvious mechanism is to attack the carbonyl group with hydroxide (p. 291). Notice that we draw out each stage and do not use any summary or shorthand mechanisms.

mechanism 1: normal ester hydrolysis

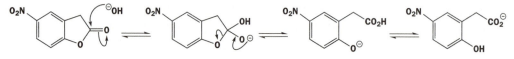

But the ester group is attached to an aromatic ring with a *para* nitro group. Nucleophilic substitution on the aromatic ring (p. 591) would give the same product.

mechanism 2: nucleophilic aromatic substitution

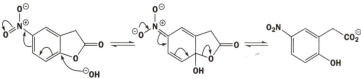

Finally, the CH_2 group between the carbonyl group and the benzene ring is acidic and can react with hydroxide as base to form an enolate. Elimination gives a ketene that can be attacked by hydroxide acting as a nucleophile to give the product.

mechanism 3: enolate elimination to give a ketene

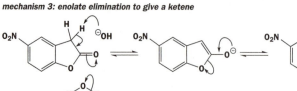

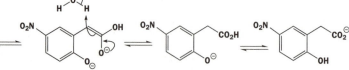

Mechanism 3 requires the exchange of at least one hydrogen atom with the solvent so, if D_2O were used as the solvent or, better, deuterated starting material were used, the exchange of one whole deuterium atom would indicate mechanism 3 while no exchange or minor amounts from the inevitable enolization would show mechanisms 1 or 2. In mechanisms 1 and 3, the added OH ends up in the acid group but in mechanism 2 it ends up as the phenol. Using labelled $H_2^{18}O$ or, better, labelling the ester oxygen as ^{18}O would separate mechanisms 1 and 3 from 2.

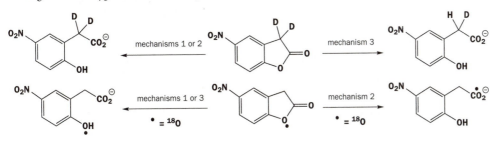

Other experiments we might do could include trying to trap the ketene in a [2 + 2] cycloaddition, perhaps with dihydropyran as in Problem 35.1, studying the reaction by UV, hoping to see the release of *p*-nitrophenolate in the slow step, changing the structure of the starting material so that one or other of the mechanisms would become more difficult, even measuring the kinetics and studying the effects on the rate of aryl substituents (pp. 1090–100), or looking for a deuterium isotope effect in the labelled lactone.

Problem 2

Explain the stereochemistry and labelling pattern in this reaction.

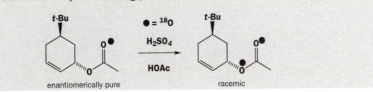

Purpose of the problem

A combination of labelling and stereochemistry reveals the details of an interesting rearrangement.

Suggested solution

The randomization of the label and the racemization suggest that the carboxylate falls off the allyl cation and comes back again at either end. While they are detached, the distinction between the two ends of both cation and anion disappears as they are delocalized.

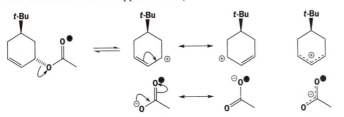

The product is racemic because the two intermediates have planes of symmetry and are achiral. The retention of *relative* stereochemistry (formation of *trans* product from *trans* starting material) could result from stereoselective recombination (the two faces of the allylic cation are not the same) or from the two ions sticking together as an ion pair so that the acetate anion slides across the cation

■ This question is based on more complex chemistry described by H. L. Goering *et al.*, *J. Am. Chem. Soc.*, 1964, **86**, 1951.

on the same face. It cannot be explained by a [3,3]-sigmatropic rearrangement (Chapter 36) as this would not randomize the label in the same way.

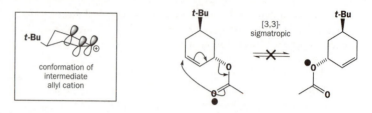

Problem 3

The Hammett ρ value for migrating aryl groups in the acid-catalysed Beckmann rearrangement is –2.0. What does this tell us about the rate-determining step?

Purpose of the problem

The Hammett relationship gives an intimate picture of the Beckmann rearrangement.

Suggested solution

The normal mechanism for the Beckmann rearrangement involves protonation at OH and migration of the group *anti* to the N–O bond (pp. 997–9), in this case the substituted benzene ring.

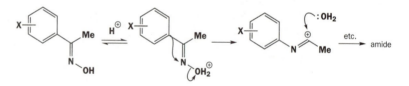

If this mechanism is correct, we should expect the migration itself to be the slow step. The first step is just a proton transfer to oxygen and must be fast. The steps after the migration involve attack of water on a carbocation and proton transfers between O and N. These must all be fast. The migration forms and breaks a C–C bond and creates an unstable cation. But does this agree with the evidence? Starting material and product in the migration step are cations so the transition state must be a cation too. Any contribution to cation stability made by the migrating group will be welcome so we should expect electron-donating groups to migrate faster. This is what we see: a ρ value of –2.0 shows a modest acceleration by electron-donating groups (pp. 1093–8).

In the Beckmann rearrangement, the *anti* group migrates, but in other rearrangements the migrating group is chosen for a very different reason: it is normally the group best able to stabilize a positive charge and benzene rings can do this by π participation (pp. 973–5). This would be the participation mechanism.

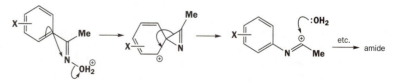

The Hammett ρ value of –2.0 gives very definite evidence that π participa tion does not occur. If it did, the closure of the unstable three-membered ring would undoubtedly be the slow step and a positive charge would be formed inside the benzene ring itself. This would give a much larger ρ value of something like –5.0 or more. One reason participation does not occur is that the starting material is planar and the p orbitals in the benzene ring cannot point in the right direction to interact with the σ* orbital of the N–O bond. They are orthogonal to it.

■ This is early work of D. E. Pearson and group, *J. Org. Chem.*, 1952, **17**, 1511; 1954, **19**, 957.

Problem 4

Between pH 2 and 7, the rate of hydrolysis of this thiol ester is independent of pH. At pH 5 the rate is proportional to the concentration of acetate ion [AcO⁻] in the solution and the reaction goes twice as fast in H_2O as in D_2O. Suggest a mechanism for the pH-independent hydrolysis. Above pH 7, the rate increases with pH. What kind of change is this?

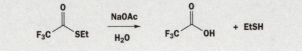

Purpose of the problem

Time for you to try your skill at interpreting pH-rate profiles.

■ You might have started your answer by sketching the pH-rate profile (p. 1104).

Suggested solution

The second part of the question is easily dealt with. In alkaline solution the rate of hydrolysis of this ester simply increases with pH and we have the normal specific base-catalysed reaction in which hydroxide ion attacks the carbonyl group.

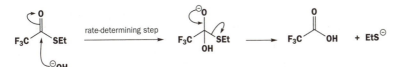

But this is no ordinary ester. The leaving group is a thiol (pK_a about 8) not the usual alcohol (pK_a about 16) and so the thiolate anion is a much better leaving group than EtO⁻. Also. the CF₃ group is very electron-withdrawing so nucleophilic attack on the carbonyl group will be unusually fast. This is why there is a region of pH-independent hydrolysis not found with EtOAc (p. 1104). You could have suggested that acetate was a nucleophilic or a general base catalyst but the solvent deuterium isotope effect suggests that it is a general base.

■ Fast in comparison with attack on CH_3CO_2Et but still the slow step.

■ This is a much simplified description of a series of experiments described by L. R. Fedor and T. C. Bruice, *J. Am. Chem. Soc.*, 1965, **87**, 4138.

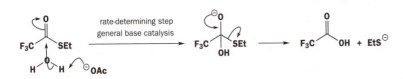

Problem 5

In acid solution, the hydrolysis of this carbodiimide has a Hammett ρ value of –0.8. What mechanism might account for this?

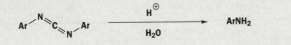

Purpose of the problem

Interpretation of a small Hammett ρ value.

Suggested solution

The most obvious explanation for a small Hammett ρ value, that the aromatic ring is too far away from the reaction site, will not wash here as the aromatic rings are joined directly to the reacting nitrogen atoms of the carbodiimide. The reaction must surely start with the protonation of one of these nitrogen atoms. This cannot be the slow step and it would in any case have a large negative ρ value. The small observed ρ value suggests that the rate-determining step must have a positive ρ value that nearly cancels out the large negative ρ value for the first step. Attack by water on the protonated carbodiimide looks right.

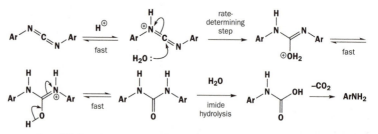

■ The hydrolysis of carbodiimides in acid and base solution was studied by S. Hünig *et al.*, *Liebig's Annalen*, 1953, **579**, 87.

The expected equilibrium Hammett ρ value for the protonation would be about -2.5 to -3 so the kinetic Hammett ρ value for the attack of water would have to be about $+2$ to give a net Hammett ρ value of -0.8. This looks fine. The rest of the mechanism involves proton transfers, hydrolysis of an imide, and decarboxylation.

Problem 6

Explain the difference between these Hammett ρ values by mechanisms for the two reactions. In both cases the ring marked with the substituent 'X' is varied. When R = H, $\rho = -0.3$ but, when R = Ph, $\rho = -5.1$.

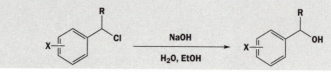

Purpose of the problem

Interpretation of a variation in Hammett ρ value with another structural variation.

Suggested solution

The reaction is obviously a nucleophilic substitution at the benzylic centre so we are immediately expecting either the S_N1 or the S_N2 mechanism. When R = H, the reaction occurs at a primary alkyl group and S_N2 is expected. When R = Ph, the reaction occurs at a secondary benzylic centre and S_N1 is expected.

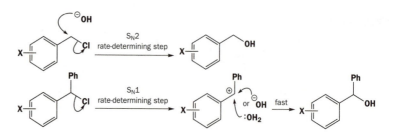

Since S_N1 produces a carbocation delocalized around the benzene ring in the slow step, a large negative Hammett ρ value is reasonable. It is not obvious what sign the Hammett ρ value would have in the S_N2 mechanism as there is no build-up of charge near the benzene ring in the transition state so a small value is reasonable. The actual value -0.3 is very small indeed but, if we can read anything into it at all, it does suggest a slightly loose transition state with a very small (δ^+?) charge on the carbon atom.

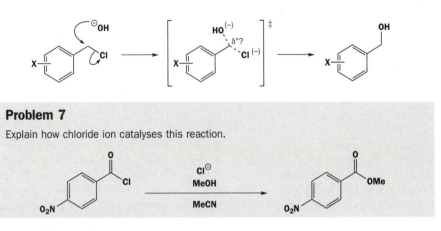

Problem 7

Explain how chloride ion catalyses this reaction.

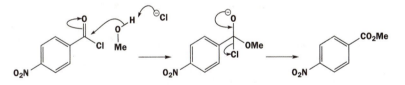

Purpose of the problem

An extreme example of surprising catalysis.

Suggested solution

At first you might ask how chloride can catalyse anything at all. It is a weak base and not a very good nucleophile for the carbonyl group. However, in polar aprotic solvents like acetonitrile (MeCN), chloride is not solvated and is both more basic and more nucleophilic. In this reaction it cannot be a nucleophilic catalyst as attack on the carbonyl group simply regenerates the starting material. It cannot be a specific base as it is too weak, even in MeCN, to deprotonate methanol. But it can act as a general base. Notice how we have arrived at general species catalysis as a last resort after eliminating the other possibilities.

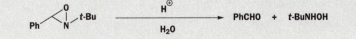

Problem 8

The hydrolysis of this oxaziridine in 0.1M sulfuric acid has $k(H_2O)/k(D_2O) = 0.7$ and an entropy of activation of $\Delta S^{\ddagger} = -76$ J mol^{-1}K^{-1}. Suggest a mechanism for the reaction.

$$\text{Ph} \diagup \overset{\displaystyle O}{\underset{N}{\triangle}} \text{-} t\text{-Bu} \quad \xrightarrow[\text{H}_2\text{O}]{\text{H}^{\oplus}} \quad \text{PhCHO} \quad + \quad t\text{-BuNHOH}$$

Purpose of the problem

Deducing a mechanism from isotope effects and entropy of activation.

■ You might have considered that the ring-opening step is electrocyclic.

Suggested solution

The inverse solvent deuterium isotope effect indicates specific acid catalysis (pp. 1102–4) and the modest negative entropy of activation suggests some bimolecular involvement. There are various mechanisms you might propose but the right one must surely involve cleavage of the three-membered ring in the protonated amine. The second or possibly the third step could be rate-determining.

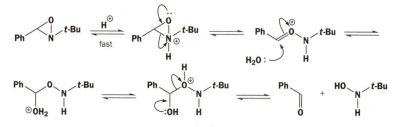

Once the three-membered ring is opened, the rest of the reaction amounts to acid-catalysed hemiacetal hydrolysis. The alcohol is a hydroxylamine and the carbonyl compound is benzaldehyde. An alternative mechanism, favoured by the original authors (J. H. Fendler and group, *J. Chem. Soc., Perkin Trans. 2*, 1973, 1744), starts with protonation of the oxygen atom and ends up with the hydrolysis of an imine. Again, the second or the third step could be rate-determining.

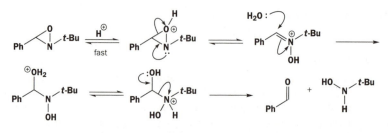

Problem 9

Explain how both methyl groups in the product of this reaction come to be labelled. If the starting material is re-isolated at 50% reaction, its methyl group is also labelled.

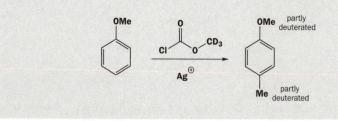

Purpose of the problem

Exploring a mechanism with labelling.

Suggested solution

The role of silver ion (Ag^+) is the removal of halide to give an acylium ion that reacts, not at the carbonyl group, but at the methyl group to give CO_2 and a methylated benzene ring. The simple Friedel–Crafts route cannot be the whole story: it explains how the added methyl group is labelled but not why it is only partly labelled nor how label gets into the MeO group on the benzene ring.

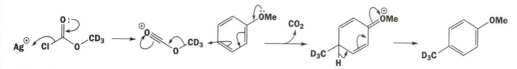

The only way in which we can explain those extra features is to suggest that the methylation initially occurs on the oxygen atom and that methyl groups are transferred from there to the benzene ring. We should never have detected this detail without labelling studies. The alkylation on oxygen provides an alkylating agent that can transfer CH_3 or CD_3 and also forms the necessary trideuterio-toluene.

■ We hope you did *not* suggest a methyl cation as an intermediate!

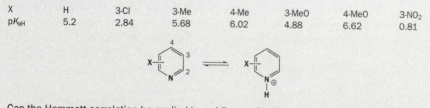

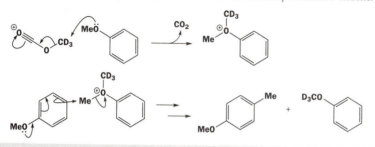

Problem 10

The pK_{aH} values of some substituted pyridines are as follows.

X	H	3-Cl	3-Me	4-Me	3-MeO	4-MeO	3-NO₂
pK_{aH}	5.2	2.84	5.68	6.02	4.88	6.62	0.81

Can the Hammett correlation be applied to pyridines using the σ values for benzenes? What equilibrium ρ value does it give and how do you interpret it? Why are no 2-substituted pyridines included in the list?

Purpose of the problem

Making sure you understand the ideas behind the Hammett relationship.

Suggested solution

The obvious thing to do is to plot the pK_{aH} values of the substituted pyridines against the σ values for the substituents using *meta* values for the 3-substituted and *para* values for the 4-substituted compounds (table on p. 1092). This gives quite a good straight line and we get a slope (Hammett ρ value) of +5.9. The sign is, of course, positive as the same electronic effects that make benzoic acids more acidic will also make pyridinium ions more acidic. The large ρ value may have surprised you, but reflect. Ionization of benzoic acids occurs outside the ring and the change isn't delocalized round the ring. Deprotonation of pyridinium ions occurs on the ring and the charge (positive this time) is delocalized round the ring.

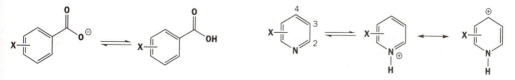

■ This work was done to apply the Hammett relationship to the reaction of pyridines with acid chlorides (R. B. Moody and group, *J. Chem. Soc., Perkin Trans 2*, 1976, 68).

There are no 2-substituted pyridines in the list since they are like *ortho*-substituted benzenes and cannot be expected to give a good correlation because of steric effects.

Problem 11

These two reactions of diazo compounds with carboxylic acids give gaseous nitrogen and esters as products. In both cases the rate of the reaction is proportional to [diazo compound] · [RCO₂H]. Use the data for each reaction to suggest mechanisms and comment on the difference between them.

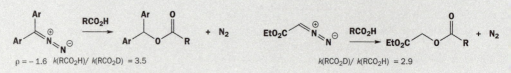

$\rho = -1.6$ $k(RCO_2H)/\ k(RCO_2D) = 3.5$ $k(RCO_2D)/\ k(RCO_2H) = 2.9$

Purpose of the problem

Application of contrasted isotope effects to detailed mechanistic analysis.

Suggested solution

The first reaction has a normal kinetic isotope effect (RCO_2H reacts *faster* than RCO_2D) while the second has an inverse deuterium isotope effect (RCO_2H reacts *slower* than RCO_2D). This suggests that there is rate-determining proton transfer in the first reaction but specific acid catalysis (i.e. fast equilibrium proton transfer followed by slow reaction of a protonated species) in the second. Protonation must occur at carbon to give the products and this can be slow.

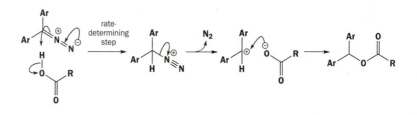

■ This is just a small sample of work on the acid-catalysed decomposition of diazo compounds reviewed by R. A. More O'Ferrall, *Adv. Phys. Org. Chem.*, 1967, **5**, 331.

Loss of nitrogen is fast because it gives a stable secondary benzylic cation. You might correctly have called the slow protonation at carbon general acid catalysis. This mechanism fits a negative Hammett ρ value as electrons are being donated out of the site of attachment of the aromatic rings and a small value because the electron donation from the diazo group is more important. If the loss of nitrogen had been the rate-determining step, the Hammett ρ value would have been large and negative.

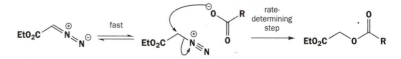

The second reaction follows much the same pathway except that loss of nitrogen is now difficult because the cation would be very unstable (primary and next to a carbonyl group) and the second step is S_N2 rather than S_N1 and is rate-determining.

Problem 12

Suggest mechanisms for these reactions and comment on their relevance to the Favorskii family of mechanisms.

(a)   (b)

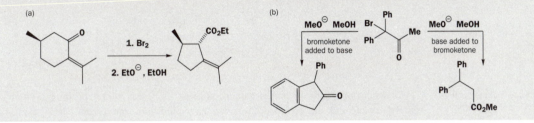

Purpose of the problem

Extension of a section of the chapter (p. 1111) into new reactions with internal trapping of intermediates.

Suggested solution

In reaction (a) the bromination must occur on the alkene to give a dibromide. We cannot suggest stereochemistry at this stage and it is better to continue with the standard Favorskii mechanism and see what happens. Everything is the same until the very last step when the opening of the cyclopropane provide electrons in just the right place to displace the second bromide and put the double bond back where it was. This alternative behaviour of a proposed intermediate gives us confidence that the intermediate really is involved.

■ The description of the ring closure as an electrocylic process is explained on p. 1112.

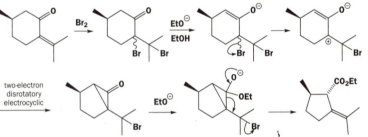

The stereochemistry of the initial bromination turns out to be irrelevant as it disappears when the oxyallyl cation is formed. We know the stereochemistry of the final product so we know the stereochemistry of the cyclopropanone: it must be on the opposite face of the five-membered ring to the methyl group. The disrotatory closure of the oxyallyl cation goes preferentially one way with the H and CMe₂Br substituents going upwards and the carbonyl group going down.

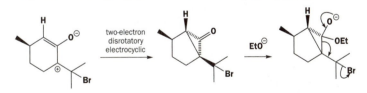

Reaction (b) is a normal Favorskii reaction going to the right in the diagram. The only point of interest is the way in which the three-membered ring breaks up. The more stable carbanion is the doubly benzylic one so the right product is formed.

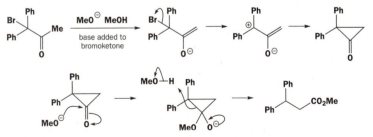

■ This is rather like the Nazarov reaction described in Chapter 36 (p. 962).

The reaction with excess bromoketone starts in the same way but the oxyallyl cation is captured by one of the benzene rings in a four-electron (and therefore conrotatory) electrocyclic closure to give the new five-membered ring.

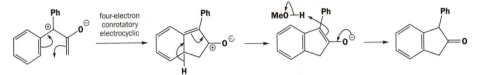

■ This work was part of a thorough investigation into the mechanism of the Favorskii reaction by F. G. Bordwell and group, *J. Am. Chem. Soc.*, 1970, **92**, 2172.

You may wonder how the presence of excess MeO⁻ stops this from happening. It doesn't. The oxyallyl cation and the cyclopropanone are in equilibrium and excess MeO⁻ captures the cyclopropanone and drives the normal Favorskii reaction onwards. If there is no excess MeO⁻, the oxyallyl cation lasts long enough for the five-membered ring to be the main product.

Problem 13

Propose mechanisms for the two reactions at the start of the chapter. The other product in the first reaction is the imine PhCH=NSO₂Ph.

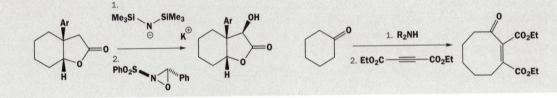

Purpose of the problem

Following up two interesting and unusual reactions from p. 1080.

Suggested solution

We put these reactions into the chapter just to remind you that there are lots of reactions you haven't met yet. They were chosen deliberately to have less than obvious mechanisms so, if you got reasonable answers, you should be pleased. The first is an example of the oxidation of an enolate by an 'oxaziridine' – one of a family introduced by F. A. Davis and B. C. Chen, *Chem. Rev.*, 1992, **92**, 919. The weak N–O bond breaks under attack by the enolate and the sulfonyl group makes the nitrogen atom a good leaving group. The electrophilic oxygen atom adds to the top face as that is the outside (*exo* or convex face) of a folded bicyclic system (pp. 865–70).

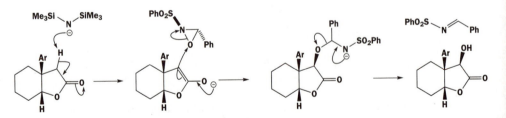

The second reaction is a kind of cycloaddition followed by a ring opening with the sequence of intermediates shown here. It leads to a ring expansion by two carbon atoms and allows us to make eight-membered rings easily.

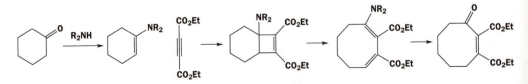

The problem is that the thermal [2 + 2] cycloaddition is not allowed and the electrocyclic ring opening would have to be conrotatory and that would put a *trans* double bond in the eight-membered ring. You may not have seen these difficulties or you may have escaped from them by using ionic rather than pericyclic mechanisms.

■ This chemistry is to be found in K. C. Brannock *et al., J. Org. Chem.,* 1963, **28**, 1464.

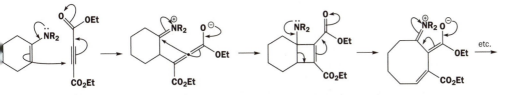

Problem 14

A typical Darzens reaction involves the base-catalysed formation of an epoxide from an α-haloketone and an aldehyde. Suggest a mechanism for the Darzens reaction consistent with the results shown below.

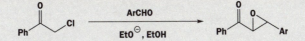

(a) The rate expression is: rate = k_3[PhCOCH$_2$Cl][ArCHO][EtO$^-$]
(b) When Ar is varied, the Hammett ρ value is +2.5.
(c) The following attempted Darzens reactions produced unexpected products.

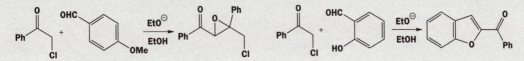

Purpose of the problem

Trying for a complete picture of a reaction using physical data and structural variation.

Suggested solution

The ethoxide is not incorporated into the product but appears in the rate expression. Its role must be as base and there is only one set of enolizable protons. We start by making the enolate of the chloroketone. Then we can attack the aldehyde with the enolate and finally close the epoxide ring with a nucleophilic displacement of chloride ion.

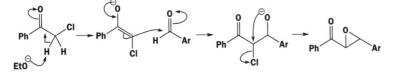

If this mechanism is right, the kinetic data show that the second step is rate-determining so the first becomes a pre-equilibrium. We can write

$$\text{rate} = k_2[\text{enolate}][\text{ArCHO}]$$

But we know from the pre-equilibrium that

$$K_1 = \frac{[\text{enolate}]}{[\text{PhCOCH}_2\text{Cl}][\text{EtO}^-]}$$

And so the rate expression becomes (substituting for unknown [enolate])

$$\text{rate} = k_2 K_1[\text{PhCOCH}_2\text{Cl}][\text{EtO}^-][\text{ArCHO}]$$

and this matches the given rate expression though the apparently third-order rate constant is revealed as an equilibrium constant (K_1) multiplied by a second-order rate constant (k_2).

The Hammett ρ value shows a modest gain of electrons in the transition state. We must not take the pre-equilibrium into account in this as ArCHO is not involved in that step. In fact a Hammett ρ value of +2.5 is typical of nucleophilic attack on a carbonyl group conjugated with the aromatic ring (see p. 1095 for an example).

The unexpected products come from variations in this mechanism. *para*-Methoxy-benzaldehyde is conjugated and unreactive so that the chloroketone ignores it and does a Darzens reaction on itself, rather like the self-condensations we met in Chapter 21.

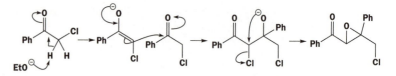

With salicylaldehyde, the second example, the OH group will exist as an oxyanion under the reaction conditions. Alkylation with the chloroketone allows enolate formation from the product leading to an intramolecular aldol reaction.

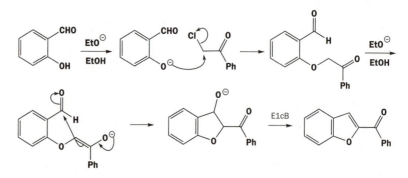

Problem 15

If you believed that this reaction went by elimination followed by conjugate addition, what experiment would you carry out to try and prove that the enone is an intermediate?

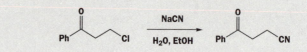

Purpose of the problem

Turning the usual question backwards: what evidence do you want, rather than how to interpret what you are given.

Suggested solution

The suggested mechanism of elimination followed by conjugate addition might be contrasted with the direct S$_N$2 displacement to see what evidence is needed.

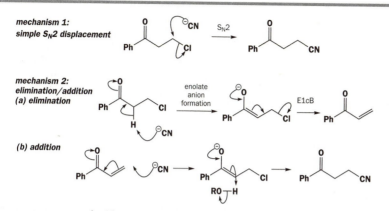

mechanism 1:
simple S$_N$2 displacement

mechanism 2:
elimination/addition
(a) elimination

enolate
anion
formation

(b) addition

There are many types of evidence you could have considered including those below. There are many others.

- Exchange of protons in D$_2$O/EtOD is necessary for the elimination/addition mechanism.
- Kinetic evidence (tricky as you cannot be sure what is the rate-determining step).
- A Hammett plot with substituted benzene rings. Again the result depends on which step is slow in mechanism 2, but the simple S$_N$2 reaction would give a very small ρ value as the benzene ring is a long way from the scene of action.
- Base catalysis. Mechanism 2 is base-catalysed; mechanism 1 is not.
- Kinetic isotope effects might be found in mechanism 2.
- Stereochemistry. If a substituent were added to make the terminal carbon chiral, inversion would be expected for mechanism 1 but racemization for mechanism 2 as the enone is achiral.

S$_N$2 inversion   elimination/addition   achiral   racemic

Problem 16

This question is about three related acid-catalysed reactions: (a) the isomerization of Z-cinnamic acids to E-cinnamic acids; (b) the dehydration of the related hydroxy-acids; (c) the racemization of the same hydroxy-acids. You should be able to use the information provided to build up a complete picture of the interaction of the various compounds and the intermediates in the reactions.

(a) Data determined for the acid-catalysed isomerization of Z-cinnamic acids in water include the following.

   (i)   The rate is faster in H$_2$O than in D$_2$O: $k(H_2O)/k(D_2O) = 2.5$.

  (ii)   The product contains about 80% D at C2.

 (iii)   The Hammett ρ value is −5.

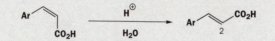

Suggest a mechanism for the reaction that explains the data.

(b) The dehydration of the related hydroxy-acids also gives E-cinnamic acids at a greater rate under the same conditions but the data for the reaction are rather different.

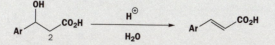

   (i)   Hydroxy-acid deuterated at C2 shows a kinetic isotope effect: $k_H/k_D = 2.5$.

(c) If the dehydration reaction is stopped after about 10% conversion to products, the remaining starting material is completely racemized. Data for the *racemization* reaction include:

(i)   The rate i s slower in H_2O than in D_2O.
(ii) Hydroxy-acid deuterated at C2 shows practically no kinetic isotope effect.
(iii) The Hammett ρ value is –4.5.

What conclusions can you draw about the dehydration?

Recalling that the dehydration goes faster than the isomerization, what would be present in the reaction mixture if the isomerization were stopped at 50% completion?

Purpose of the problem

Interpretation of data for three closely related reactions where results from each support the others

Suggested solution

Our first impressions of the *E/Z* isomerization are that it is a simple reaction. Protonation obviousl occurs at C2 in the rate-determining step and a cation delocalized round the benzene ring must b formed in the step too. The slow step is proton transfer to carbon. There is more than 50% deuteration at C2 because, although only one deuterium atom is added, the last step will lose H rather than D because of the kinetic isotope effect. A ratio of 4:1 is reasonable.

the E/Z isomerization mechanism

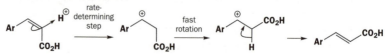

The dehydration reaction shows a normal kinetic deuterium isotope effect indicating loss of the C2 proton as the rate-determining step. This fits with fast protonation at oxygen followed by fast formation of the same cation with the rest of the mechanism the same. The dehydration is faster than the isomerization so loss of water is faster than protonation of the alkene. We have rate determining loss of proton from carbon.

the dehydration mechanism

The racemization doesn't include that last step, rate-determining in the dehydration. Now we see an inverse solvent deuterium isotope effect indicating specific acid catalysis. The loss of water to form the same cation is now the slow step. The large Hammett ρ value also suggests the formation of a cation delocalized around the ring in the rate-determining step.

the racemization mechanism

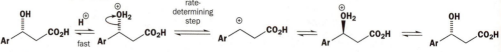

■ This is a difficult problem! Don't be too disappointed if you failed to see the whole picture.

Note that the racemization tells us that water is adding to the cation and falling off again faster than any other process. This means that during the *E/Z* isomerization, any amount, perhaps a majority, of the material in the reaction vessel may be hydroxy-acid. This is the true picture.

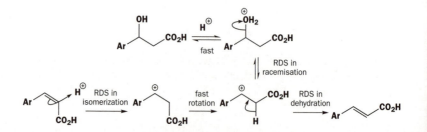

Suggested solutions for Chapter 42

42

Problem 1

Predict the most favourable conformations of these insect pheromones.

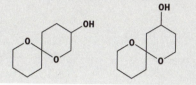

Purpose of the problem

Just to check that you can draw the conformations of spirocyclic acetals.

Suggested solution

There are many good ways to draw these conformations and yet more not quite so good. Make sure you have each acetal oxygen atom axial on the other ring to enjoy the full anomeric effect (pp. 1128–31). We show three ways to draw each compound. The natural products are, in fact, the compounds in the frames according to R. Baker *et al., J. Chem. Soc., Chem. Commun.*, 1982, 601.

■ You get extra credit if you noticed that these compounds can each exist as two diastereoisomers and each diastereoisomer as two enantiomers.

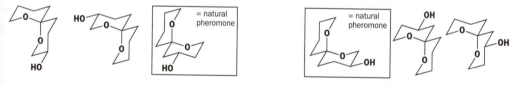

Problem 2

Refluxing cyclohexanone with ethanolamine in toluene with a Dean Stark separator to remove the water gives an excellent yield of this spirocycle. What is the mechanism, and why is acid catalysis (or any other kind) unnecessary?

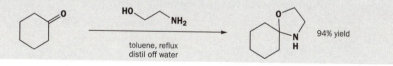

Purpose of the problem

Revision of the mechanism of a reaction related to acetal and imine formation (Chapter 14).

Suggested solution

The nitrogen atom is more nucleophilic than the oxygen atom so we should start with the amine attacking the carbonyl group. The first product will be an imine and addition of the alcohol to that gives the product. Imine formation is acid-catalysed but occurs quite rapidly at neutral pHs and the intramolecular second step is fast. The whole system is under equilibrium but the other product, water, is driven off as an azeotrope with benzene so the equilibrium is driven over to give the

■ Although the closure of the ring is formally a 5-*endo-trig* cyclization, it is very like the familiar and allowed acetal formation (frame on p. 1142).

heterocyclic product in 77–94% yield according to A. C. Cope and E. M. Hancock, *J. Am. Chem. Soc.,* 1942, **64**, 1503.

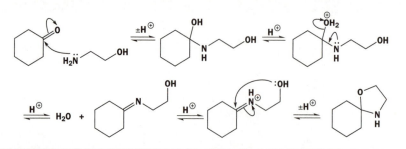

Problem 3

What is A in the following reaction scheme and how does it react to give the final product?

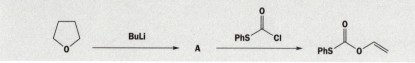

Purpose of the problem

Revision of reverse cycloadditions (Chapter 35) and an exercise in lithiation of heterocycles.

Suggested solution

The only sensible place to lithiate the starting material (which is, of course, the popular solvent THF) is next to the oxygen atom. A reverse cycloaddition then eliminates ethylene (ethene) and gives the lithium enolate of acetaldehyde. It is easier to see the reverse cycloaddition if we write the lithium derivative as an anion.

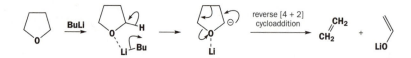

■ This chemistry was used to make a new protecting group (A. J. Duggan and F. E. Roberts, *Tetrahedron Lett.,* 1979, 595).

The intermediate A is the lithium enolate of acetaldehyde formed in the absence of any acetaldehyde (see p. 707 for the importance of this). It acylates on oxygen with the acid chloride in what is evidently a charge-controlled interaction between a hard nucleophile (the oxygen atom of the enolate) and a hard electrophile (the acid chloride).

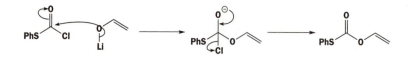

Problem 4

Give mechanisms for the formation of this *spiro* heterocycle. Why is the product not formed simply on reacting the starting materials in acid solution without Me₃Al?

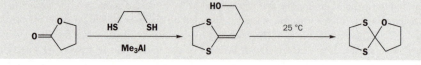

Purpose of the problem

Analysis of the advantages of heterocycle formation.

Suggested solution

Organoaluminium compounds, like those of magnesium or lithium, react rapidly with acidic protons like those in SH groups so the first thing that happens is evolution of two molecules of methane as S–Al bonds are formed.

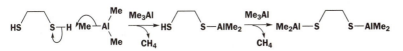

The Al–S bond is rather like, say, the O–Li bond and provides a nucleophilic sulfur atom. Probably the lactone first adds to the aluminium to give a tetrahedral aluminium anion but, if you drew a direct attack releasing Me_2Al^+, that is good enough.

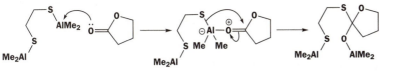

This first product is a kind of tetrahedral intermediate and eliminates the alkoxide leaving group. You may object that S^- is more stable than O^- but the Al–O bond is stronger than the Al–S bond and that decides which group leaves.

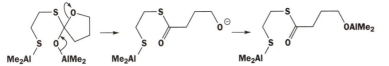

Now the second sulfur atom repeats the process and the loss of the second oxygen atom requires an elimination into the carbon skeleton as this time a carbonyl group cannot be formed.

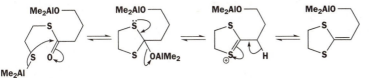

Finally, on work-up, the Al–O bond is hydrolysed and cyclization to the final product occurs, probably by protonation of the very nucleophilic alkene.

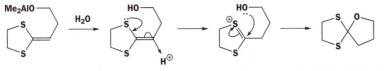

■ This orthoester-like functional group was introduced by E. J. Corey and D. J. Beames, *J. Am. Chem. Soc.*, 1975, **95**, 5829, as a new protecting group for esters and lactones.

Problem 5

The *Lolium* alkaloids have a striking skeleton of saturated heterocycles. One way to make this skeleton is shown below. Explain both the mechanism and the stereochemistry.

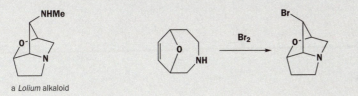

a *Lolium* alkaloid

Purpose of the problem

Analysis of a reaction to make a double heterocycle stereospecifically.

Suggested solution

■ This particular reaction was used by S. R. Wilson *et al.*, *J. Org. Chem.*, 1981, **46**, 3887, to help establish the correct structure of the *Lolium* alkaloids.

Bromine, of course, attacks the alkene to form the bromonium ion. If it has the right stereochemistry, it cyclizes but, if it doesn't, it reverts to starting material. This reaction might remind you of halolactonization (p. 872).

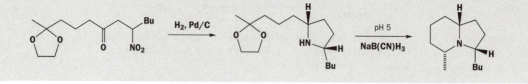

Problem 6

Explain the stereochemical control in this synthesis of a fused bicyclic saturated heterocycle – the trail pheromone of an ant.

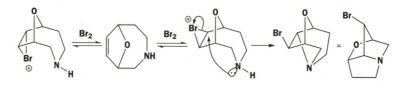

Purpose of the problem

Analysis of a reaction to make a double heterocycle stereoselectively.

Suggested solution

The first stage involves reduction of the nitro group to an amine, condensation with the ketone to give an imine, and then the interesting step, reduction of the imine to the secondary amine. The imine binds to the catalyst on the side away from the butyl group – the only group not in the plane. The product is shown as one enantiomer but it is, of course, racemic if the starting material is racemic.

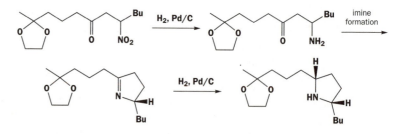

■ This monomorine synthesis is by R. V. Stevens and A. W. M. Lee, *J. Chem. Soc., Chem. Commun.*, 1982, 102.

The acetal is hydrolysed at pH 5 to give a second ketone: condensation with the secondary amine gives an iminium salt that is reduced by the cyanoborohydride. The reagent approaches from the less hindered side, that is, opposite the two groups already there. The result is that all three hydrogen atoms end up on the same side of the ring.

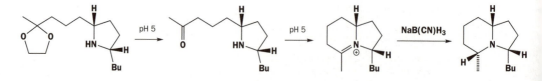

Problem 7

In Chapter 31, one of the questions asked you to comment on the difference between these two reactions. Now would you like to comment again and add comments on the way we drew the starting materials.

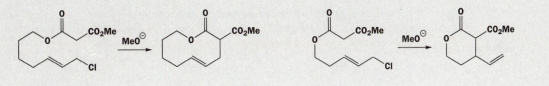

Purpose of the problem

Stereoelectronics comes to the fore in the synthesis of medium-sized heterocyclic rings.

Suggested solution

The mechanisms are easy enough to draw – the only question is, why are the reactions different?

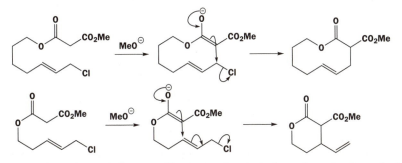

As you now know (p. 1134) all esters and lactones prefer to be in the conformation shown for the first compound so that they can enjoy stabilization by the anomeric effect. The second ester would also prefer to be like this too (first diagram below) but in this conformation it cannot cyclize at all. Even in the less favourable conformation it uses for reaction, it prefers conjugate to direct substitution as the latter would give a strained *trans*-cycloheptene. The first compound can enjoy both the anomeric effect and conjugate addition as a *trans*-alkene in a ten-membered ring is fine. It is better than fine in the heterocyclic product as the lactone also enjoys the anomeric effect.

■ This is an adapted version of a macrolide synthesis reported by J. Tsuji and group, *J. Am. Chem. Soc.*, 1978, **100**, 7424.

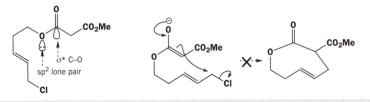

Problem 8

In Chapter 32, Problem 3, we asked you to work out the stereochemistry of a sugar. One of the sugar components in the antibiotic kijanimycin has the gross structure (shown in the margin) and NMR spectrum (shown below). What is its stereochemistry?

δ_H (p.p.m.) 1.33 (3H, d, J 6 Hz), 1.61* (1H, broad s), 1.87 (1H, ddd, J 14, 3, 3.5 Hz), 2.21 (1H, ddd, J 14, 3, 1.5 Hz), 2.87 (1H, dd, J 10, 3 Hz), 3.40 (3H, s), 3.47 (3H, s), 3.99 (1H, dq, J 10, 6 Hz), 1.33 (3H, d, J 6 Hz), 4.24 (1H, ddd, J 3, 3, 3.5 Hz), and 4.79 (1H, dd, J 3.5, 1.5 Hz). The signal marked * exchanges with D_2O.

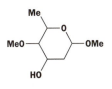

When you did this problem, you probably thought about the conformation but now draw it and say why you think the molecule prefers that conformation.

Purpose of the problem

Stereoelectronics decides the choice of conformation in a cyclic acetal.

Suggested solution

The deduction of the stereochemistry (and the conformation) is given in full in the solution to Problem 32.3. The molecule has the configuration shown here and there are two chair conformations, A or B, it could adopt.

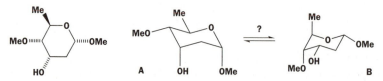

■ A. K. Mallams *et al.*, *J. Am. Chem. Soc.*, 1981, **103**, 3938.

There seems at first sight to be little to choose between A and B: each has two axial and two equatorial substituents. In fact, the NMR shows clearly that the molecule exists as A. The acetal prefers the OMe group axial because of the anomeric effect so that in A three groups have their preferred conformation.

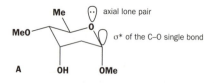

Problem 9

Revision of Chapters 35 and 37. Give mechanisms for these reactions, commenting on the formation of that particular saturated heterocycle in the first reaction. What is the alternative product from the migration and why is it not formed?

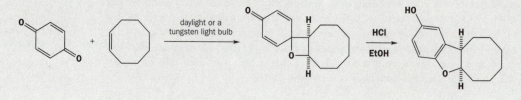

Purpose of the problem

An example of the formation and rearrangement of a four-membered heterocycle.

Suggested solution

The first reaction is a photochemically allowed [2 + 2] cycloaddition (pp. 927–9). The alternative product would come from the alkene in the quinone being used instead of the carbonyl group. This regiochemistry is not obvious but the light is absorbed by the quinone and the coefficient in the SOMO must be larger in the carbonyl groups. The stereochemistry simply results from the concerted reaction: those hydrogens were *cis* in cyclooctene and remain *cis* in the product.

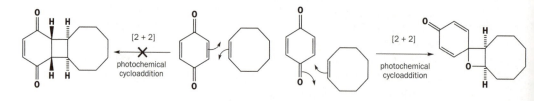

The second step is the dienone-phenol rearrangement (pp. 988–9). The oxygen atoms in the product still have a *para* relationship so it must be the carbon atom in the ring that migrates. Notice that the carbon atom migrates with retention of configuration, as expected from other migrations, so the *cis* ring junction is preserved in the product.

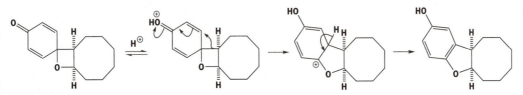

But why does carbon migrate rather than oxygen? Oxygen is good at migrating because it can participate (pp. 969–73) and a very reasonable product could be formed.

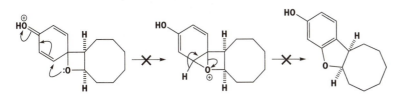

It is true that the stereoelectronics for both the first and second step are not good but the main reasons are probably that the intermediate is very strained with fused three- and four-membered rings and that oxygen is also very good at staying behind and encouraging other groups to migrate. We should redraw our first mechanism slightly so that the oxygen atom can still use a lone pair.

■ These reactions formed part of an extensive investigation at Reading University into photochemistry in solution by D. Bryce-Smith *et al.*, *J. Chem. Soc. (C)*, 1967, 383.

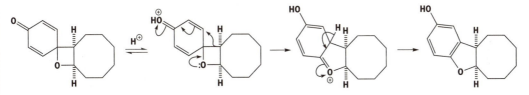

Problem 10

Though the anion of dithiolane decomposes as described in the chapter (p. 1127) and cannot be used as a d^1 reagent, the example shown here works well without any decomposition. Explain and comment on the regioselectivity of the reaction. Anions of dithianes are notorious for preferring direct to conjugate addition.

Purpose of the problem

Encouraging you to think rather than just apply rules.

Suggested solution

We should first draw a mechanism for what does happen and then consider the reasons. However, as we draw the mechanism we see the reason: the anion formed from this dithioacetal is an enolate.

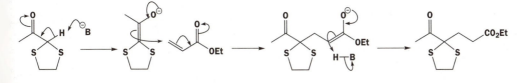

■ This reaction was used in a synthesis of the natural product isocomene by W. G. Dauben and D. M. Walker, *J. Org. Chem.*, 1981, **46, 1103.**

The decomposition of the dithiolane anion is a reverse cycloaddition and occurs because the stabilization given to the anion by two sulfur atoms is not very great. This anion is an enolate as well so it is much more stable. Conjugate addition is preferred (p. 751) with more stable enolates and this anion is an enolate additionally stabilized by two sulfur atoms. The three groups (ester and two S atoms) combine to make a stable anion good at conjugate additions.

Problem 11

Propose a mechanism for this reaction. It does not occur in the absence of an *ortho-* or *para-* OH group.

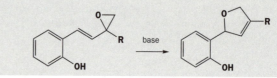

Purpose of the problem

A mechanistic exercise in the transformation of a three-membered heterocycle into a five-membered one.

Suggested solution

■ This reaction was discovered by P. Bravo and C. Ticozzi at Milan, *J. Chem. Soc., Chem. Commun.*, 1979, 438.

The most acidic proton is that of the phenol so we should remove that first. Now we can push electrons through the molecule to break open the reactive epoxide ring. The released oxyanion returns in a conjugate addition to give the more stable five-membered ring.

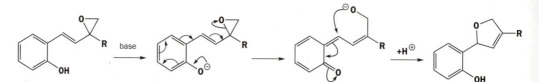

Problem 12

Explain why this cyclization gives a preponderance (3:1) of the oxetane though the tetrahydrofuran is much more stable.

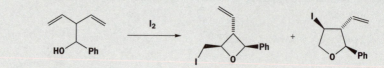

Purpose of the problem

A reminder that Baldwin's rules may apply in any cyclization.

Suggested solution

Clearly, iodine attacks an alkene and the OH group adds to the intermediate iodonium ion. Let us draw this first without any stereochemistry and see what happens.

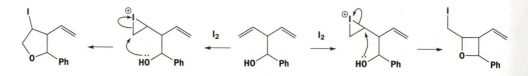

Whether the oxetane or the THF is formed depends on which end of the iodonium ion is attacked by the OH group. In terms of Baldwin's rules (pp. 1140–4), oxetane formation is a simple 4-*exo-tet* reaction and is favoured. THF formation is slightly more complicated. It is 5-*exo-tet* as far as the new ring is concerned but it is 6-*endo-tet* as far as the S$_N$2 reaction is concerned. In the transition state the nucleophile, the carbon atom under attack, and the leaving group are part of the same six-membered ring. It is very difficult to get the two dotted lines in the transition state diagram at the required 180° to each other.

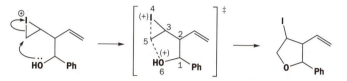

Now, what about the stereochemistry. Did you notice that each product has all-*trans* stereochemistry? And what about that second alkene? The two alkenes are, in fact, diastereotopic and which one is attacked by iodine as well as on which face determines the stereochemistry of the product. This is rather like iodolactonization. Iodine adds randomly and reversibly to both faces of both alkenes. Only when cyclization gives the all-*trans* product does it continue.

■ This important reaction was used to make antiviral compounds by M. E. Jung and C. J. Nichols, *Tetrahedron Lett.*, 1996, **37**, 7667.

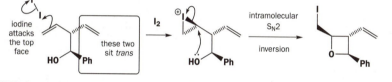

Problem 13

Reduction of this keto-ester with LiAlH$_4$ gives a mixture of diastereoisomers of the diol. Treatment with TsCl and pyridine at –25°C gives a monotosylate from each. Treatment of these with base leads to the two very different products shown. Explain.

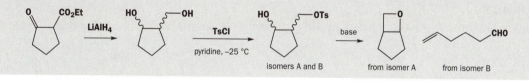

Purpose of the problem

Is it chemo-, regio-, or stereoselectivity at work here?

Suggested solution

Reduction of both ester and ketone is straightforward and a low-temperature tosylation favours the less hindered primary alcohol. We shall start with the two tosylates. The oxetane had to have a *cis* ring junction so isomer A must be the *cis* isomer. That leaves isomer B as *trans* and its fate is fragmentation (Chapter 38). Notice in the fragmentation that one lone pair on the oxyanion, the C–C bond being fragmented, and the C–O bond to the leaving group are all *anti*-planar.

■ This remarkable difference between two diastereoisomers was discovered by G. Kinast and L. Tietze, *Chem. Ber.*, 1976, **109**, 3626.

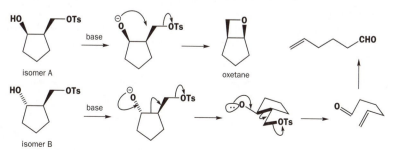

Problem 14

Draw a mechanism for the following multistep reaction. Do the cyclization steps follow Baldwin's rules? What other stereoelectronic effects are involved?

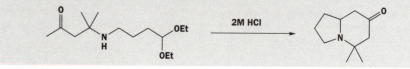

Purpose of the problem

Baldwin's rules at work in the synthesis of a bicyclic heterocycle with one nitrogen atom in both rings

Suggested solution

■ This new synthesis of a bicyclic amine was reported by F. D. King, *Tetrahedron Lett.*, 1983, **24**, 3281.

Hydrolysis of the acetal releases an aldehyde and Mannich-style condensation leads to the product. The cyclization step in which the enol attacks the iminium ion is *endo* in both components (6-*endo-trig* for both the electrophile and the nucleophile) and thus difficult to do. However, by folding the molecule in a chair (frame in margin) a reasonable overlap between the required p orbitals is possible.

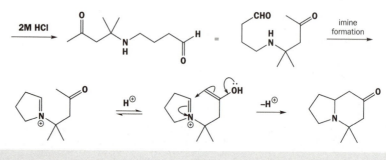

Problem 15

Consider the question of Baldwin's rules for each of these reactions. Why do you think they are successful?

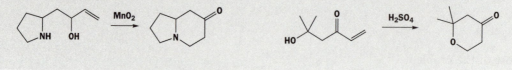

Purpose of the problem

Developing judgement in using Baldwin's rules in the synthesis of heterocycle compounds.

Suggested solution

■ J. J. Tufariello and R. C. Gatrone, *Tetrahedron Lett.*, 1978, 2753.

The first ring system is the same as the one we have just been considering but the route to it is different and is more demanding in terms of Baldwin's rules, though we should still describe it as 6-*endo-trig*. Manganese dioxide is a specific oxidant for allylic alcohols and conjugate addition of the amine to the enone gives the bicyclic amine. This works because *endo* reactions are just about all right in a six-membered ring and because conjugate addition is under thermodynamic control: as long as *some* of the reaction occurs, the product is the most stable compound in the mixture.

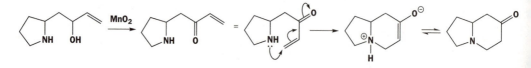

The second example is again *6-endo-trig* but it is acid-catalysed and this makes a big difference as the protonation increases the reactivity of the enone system and reduces the rigidity of the enone system. Both these *6-endo-trig* reactions go through chair-like transition states rather like the example in Problem 14 but these are not *endo* as far as the nucleophile is concerned, only for the electrophile.

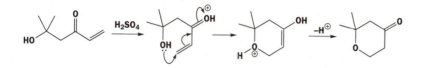

Suggested solutions for Chapter 43   **43**

Problem 1

For each of the following reactions: (a) State what kind of substitution is suggested; (b) suggest what product might be formed if monosubstitution occurs.

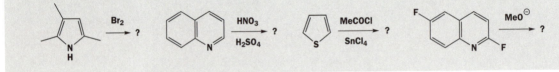

Purpose of the problem

A simple exercise in aromatic substitution with heterocycles.

Suggested solution

The first three reactions are all electrophilic substitution: a bromination of a pyrrole, the nitration of quinoline, and a Friedel–Crafts acylation of thiophene. Bromination occurs on the pyrrole at the only remaining site. Nitration of quinoline occurs on the benzene rather than the quinoline ring (actually giving a mixture of 5- and 8-nitroquinolines, but don't worry if you didn't see that, p. 1174), and the acylation would occur next to sulfur.

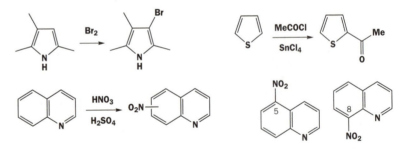

The last reaction is an aromatic nucleophilic substitution: here the pyridine is more reactive than the benzene as the negative charge in the intermediate can be delocalized on to the pyridine nitrogen atom.

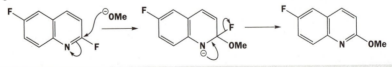

Problem 2

Give a mechanism for this side-chain extension of a pyridine.

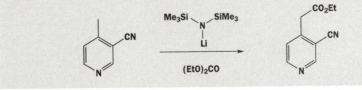

Purpose of the problem

An exercise on side-chain extension with aromatic heterocycles.

Suggested solution

The strong base (LHMDS or Lithium HexaMethyl DiSilazide) removes a proton from the methyl group so that the charge is delocalized on both nitrogen atoms and acylation follows. If you drew the lithium atom covalently bound to one of the nitrogen atoms, your answer is better than ours.

■ This sort of chemistry was introduced by D. J. Sheffield and K. R. H. Wooldridge, *J. Chem. Soc., Perkin 1*, 1972, 2506 and used by A. S. Kende and T. P. Demuth, *Tetrahedron Lett.*, 1980, **21**, 715 in a synthesis of the antileukaemic sesbanine featured in Problem 6, Chapter 33.

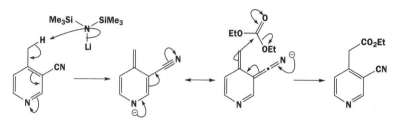

Problem 3

Give a mechanism for this reaction, commenting on the position on the furan ring that reacts.

Purpose of the problem

An unusual electrophilic substitution of a furan with interesting regioselectivity.

Suggested solution

The allylic alcohol evidently gets protonated and forms an allylic cation. We can work out the mechanism just from the structure of the product.

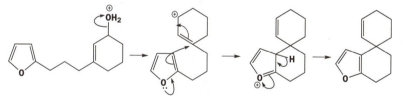

Furans normally prefer substitution at the α positions (2 and 5) but one α position is already blocked and the other is too far away to reach the allylic cation. Attack at the other end of the allylic cation would give an eight-membered ring with a *trans* alkene in it. That would theoretically be possible but closure of six-membered rings is much faster. We could summarize both those reasons by saying that electrophile and nucleophile are *tethered*.

■ This new way to make spirocyclic compounds was invented by S. P. Tanis *et al., Tetrahedron Lett.*, 1985, **26**, 5347.

Problem 4

Suggest which product might be formed in each of these reactions and justify your choices.

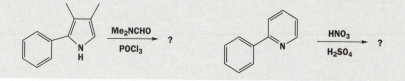

Purpose of the problem

Regioselectivity test with contrasted electrophilic aromatic substitution.

Suggested solution

In each case we have a choice between an aromatic heterocycle and a benzene ring. The pyrrole is more reactive than benzene and the pyridine less. The pyrrole does a Vilsmeier reaction (p. 1158) at the remaining free position while the pyridine acts as a deactivating and *meta*-directing substituent on the benzene ring (Chapter 22).

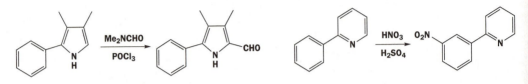

Problem 5

Comment on the mechanism and selectivity of this reaction of a pyrrole.

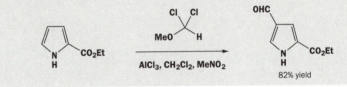

Purpose of the problem

Can you explain this unusual regioselectivity in electrophilic substitution on a pyrrole?

Suggested solution

This is clearly some sort of Friedel–Crafts reaction with an alkyl halide and AlCl₃ as the reagents. The alkyl halide forms a cation very easily because of the MeO group. Now we can draw the mechanism and we won't worry about the regioselectivity until we have.

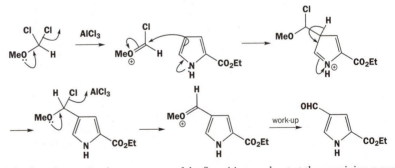

■ This chemistry was discovered by P. Fournari, M. Farnier, and C. Fournier at Dijon, *Bull. Soc. Chim. France*, 1972, 283 and used in a synthesis of the coenzyme methoxatin (p. 48) by J. B. Hendrickson and J. G. deVries, *J. Org. Chem.*, 1982, **47**, 1150.

But why does the reaction happen at one of the β positions and not at the remaining α position? The ester group deactivates the remaining α position but that is not enough. It must be the additional effect of a complex formed with the ester group and the Lewis acid that deactivates the remaining positions. We show this by the delocalization of charge.

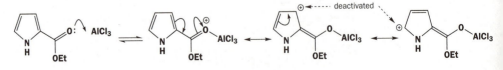

Problem 6

Explain the formation of the product in this Friedel-Crafts alkylation of an indole.

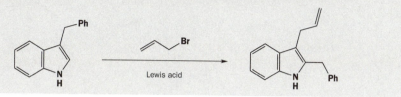

Purpose of the problem

Checking up on your understanding of indole chemistry.

Suggested solution

The Lewis acid combines with allyl bromide to give either the allyl cation or, maybe, the complex we show here. In either case, electrophilic attack occurs on the 3-position, as is almost always the case with indole (pp. 1170–1), and the benzyl group migrates to the 2-position where there is a hydrogen atom that can be lost to restore the aromaticity.

■ This reaction was discovered at Parma by G. Casnati *et al.*, *Tetrahedron Lett.*, 1972, 5277.

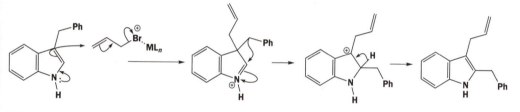

Problem 7

Explain the order of events and choice of bases in this sequence.

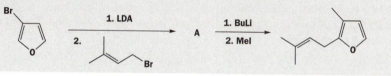

Purpose of the problem

A clever use of selective lithiation in furan chemistry.

Suggested solution

The allyl group evidently goes into the 2-position so lithiation must occur there first of all, helped by both the Br atom and the oxygen in the ring (pp. 1161–2). The second electrophile (MeI) goes in place of the bromine so the second lithiation occurs there. It is not surprising that LDA acts as a base to remove a proton (*ortho*-lithiation) whereas BuLi exchanges with a bromine atom.

■ The product is related to a constituent of the perfume of roses and was made by N. D. Ly and M. Schlosser, *Helv. Chim. Acta,* 1977, **60**, 2085.

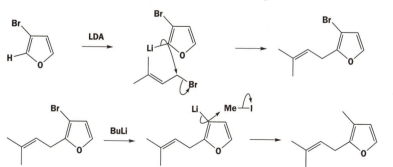

Problem 8

Explain the difference between these two pyridine reductions.

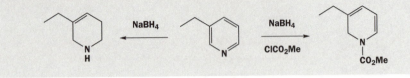

Purpose of the problem

Can you see why substitution at the nitrogen atom of pyridine affects the extent of reduction?

Suggested solution

The reducing agent is the same in both reactions ($NaBH_4$): the only difference is the acylation at the pyridine nitrogen atom. The reaction to the right looks straightforward so we'll start with that. Acylation at nitrogen gives an electrophilic pyridinium ion that accepts hydride from BH_4^- to give the product. Steric hindrance is not an issue with this small reagent and the ethyl group both deactivates the other α position and stabilizes the diene product.

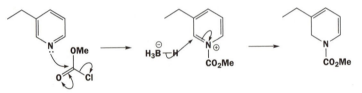

■ This chemistry was part of a synthesis of the alkaloid catharanthine by S. Raucher and R. F. Lawrence, *Tetrahedron Lett.*, 1983, **24**, 2927.

The second reduction starts in the same way, though without the preliminary acylation, but the product is a simple enamine and lone pair electrons on nitrogen are now available for protonation in enamine style. A second reduction can then occur. Notice that the ethyl group keeps the alkene to itself: the weak σ conjugation makes a trisubstituted alkene more stable than the disubstituted alternatives. The mysterious but necessary weak acid 'HX' could be solvent (EtOH) or protons from the boron or from the NH.

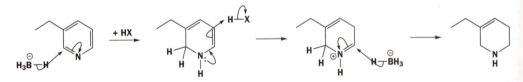

Problem 9

Why can this furan not be made by the direct route from available 2-benzylfuran?

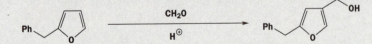

The same furan can be made by the route described below. Suggest mechanisms for the first and the last step. What is the other product of the last step?

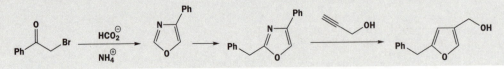

Purpose of the problem

Revision of cycloadditions from Chapter 35 and further exploration of the chemistry of furan.

Suggested solution

Even if the proposed electrophilic substitution by that notoriously erratic reagent formaldehyde (methanal, CH_2O) gave a clean product, it would not be the one required as furan reacts more readily in the α positions so we should get 2-benzyl-4-furanol instead. The route that works starts with the synthesis of a heterocycle known as an oxazole in a strange reaction called the 'Bredereck reaction'. It probably goes like the scheme below, but the order of events is uncertain. If you reacted the ammonia with the ketone first, or even with the formate to give formamide ($HCONH_2$), you might well be right. The product is aromatic (four electrons from the double bonds and two from the oxygen).

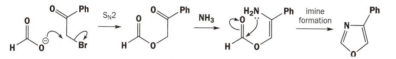

The last step involves a Diels-Alder reaction between the oxazole as diene and the alkyne as dienophile followed by a reverse Diels-Alder reaction to give the required furan. The other product turns out to be benzonitrile PhCN!

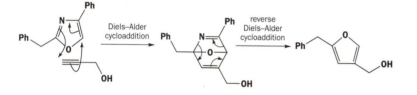

Problem 10

What aromatic system might be based on the skeleton given in the margin? What sort of reactivity might it display?

Purpose of the problem

A chance for you to be creative about aromatic compounds.

Suggested solution

The aromatic system has the poetic name 'pyrrocoline' and you will have found it by trial and error. One ring looks like a pyridine and one like a pyrrole but counting the electrons should have made you realize that we need the lone pair of the nitrogen atom to get a 10-electron aromatic system (eight from the four double bonds and two from the nitrogen) so it is a pyrrole-like nitrogen. We can therefore expect it to be good at electrophilic substitution on the five-membered ring and that is exactly what it does. The easiest pyrrocolines to make happen to have alkyl groups at position 3 and these compounds are nitrated to give mainly the 4-nitro compound. Acylation under Friedel–Crafts conditions gives the same isomer.

■ There is an old review from Glaxo at Greenford by E. T. Borrows and D. O. Holland in *Chem. Rev.*, 1948, **42**, 611 showing that this system has been known for a long time.

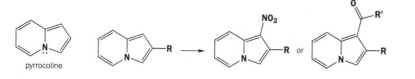

pyrrocoline

Problem 11

The reactions outlined in the chart below describe the early steps in a synthesis of an antiviral drug by the Parke-Davis company.

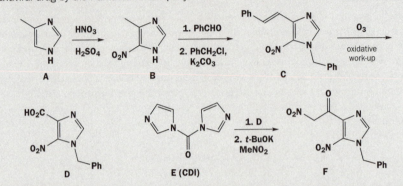

Consider how the reactivity of imidazoles is illustrated in these reactions, which involve not only the skeleton of the molecule but also the reagent E. You will need to draw mechanisms for the reactions and explain how they are influenced by the heterocycles.

Purpose of the problem

An exploration of the chemistry of imidazole beyond that considered in the chapter (pp. 1165–7).

Suggested solution

The first reaction is the nitration of imidazole in one of only two free positions. The position next to only one nitrogen is more reactive than the one between the two nitrogens. This shows that imidazole, with one pyrrole-like and one pyridine-like nitrogen, is more reactive than pyridine and more controlled than pyrrole.

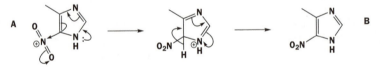

■ *ortho*-Nitro toluene would not react like this.

The second reaction is like an aldol condensation between the methyl group on the imidazole proving the 'enol' and benzaldehyde as the electrophile. The nitro group provides the main stabilization for the 'enol' but that would not be enough without the imidazole. The elimination also is E1cB-like, going through a similar 'enol' intermediate.

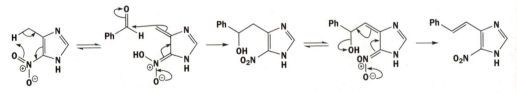

Next, alkylation on one of the nitrogen atoms of the imidazole. To get reaction on either nitrogen, we need the anion of the heterocycle and alkylation on the lower nitrogen preserves the longest conjugated chain in the product (from the nitrogen lone pair to the nitro group).

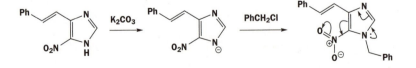

Ozonolysis of the alkene frees the carboxylic acid, which reacts with carbonyl diimidazole (CDI) in a nucleophilic substitution at the carbonyl group. The relatively stable imidazole anion is the leaving group. The anion of nitromethane displaces the second imidazole anion and completes the sequence.

■ This compound was used in the synthesis of pentostatin, an antiviral and antitumour drug, by D. C. Baker and S. R. Putt (*J. Am. Chem. Soc.*, 1979, **101**, 6127) at the Warner–Lambert/Parke-Davis research laboratories at Ann Arbor, Michigan.

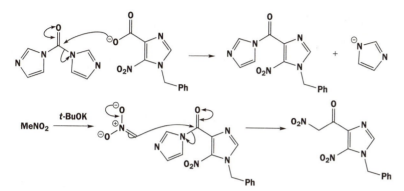

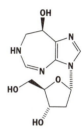

pentostatin

Problem 12

The synthesis of DMAP, the useful acylation catalyst mentioned in Chapters 8 and 12, is carried out by initial attack of thionyl chloride (SOCl₂) on pyridine. Suggest how the reactions might proceed.

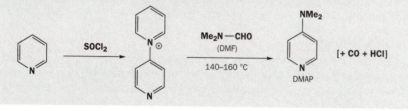

Purpose of the problem

Finding a slightly unusual mechanism for a reaction on a pyridine ring.

Suggested solution

Thionyl chloride (SOCl₂) is an electrophile. It cannot attack the pyridine ring but it could attack the pyridine nitrogen atom. Let's see where that leads us. It at least allows us to add the second pyridine in the right position and then maybe restore the aromaticity of the first pyridine by the loss of SO and HCl.

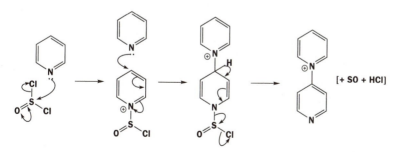

Now we can do a relatively normal nucleophilic aromatic substitution on one pyridine with the other acting as leaving group. The exact order of the later steps is open to conjecture but the

■ This reaction was described in the original paper describing DMAP (G. Höfle and W. Steglich, *Angew. Chem., Int. Ed. Engl.*, 1978, **17**, 569).

decomposition of the DMF adduct to carbon monoxide could take place as we have shown or by attack of chloride to give 'HCOCl', which is known to decompose to CO and HCl. The chlorine comes from the counterion of the intermediate.

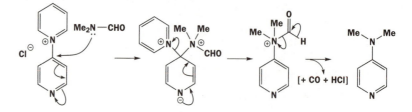

Problem 13

Suggest what the products of these nucleophilic substitutions might be.

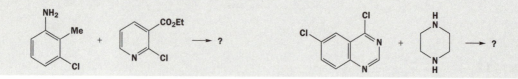

Purpose of the problem

Checking your understanding of nucleophilic aromatic substitution with chemo- and regioselectivity decisions.

Suggested solution

Each compound has potential nucleophilic and electrophilic sites. In the first case the benzene ring is not activated towards nucleophilic substitution but the pyridine is, both by the pyridine nitrogen atom and by the ester group. The NH_2 group on the benzene ring is much more nucleophilic than the pyridine nitrogen atom.

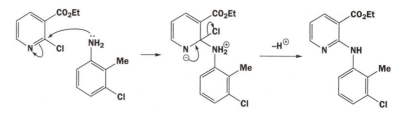

In the second case, the chlorine on the heterocyclic ring is much more reactive towards nucleophilic substitution as the intermediate is stabilized by both nitrogen atoms and the benzene ring is not disturbed. The saturated heterocycle (a piperazine) can be made to react once only as the product under the reaction conditions is strictly the hydrochloride of the unreacted amino group. This is much more basic than the one that has reacted as its lone pair is conjugated with the heterocyclic ring.

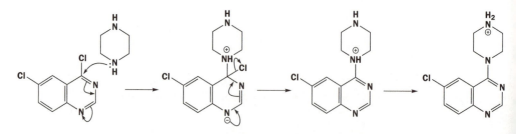

Problem 14

Suggest how 2-pyridone might be converted into the amine shown. This amine undergoes mononitration to give compound A with the NMR spectrum given. What is the structure of A? Why is this isomer formed?

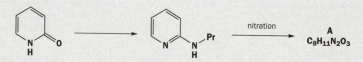

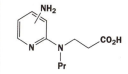

δ_H (p.p.m.) 1.0 (3H, t, J 7 Hz), 1.7 (2H, sextet, J 7 Hz), 3.3 (2H, t, J 7 Hz), 5.9 (1H, broad s), 6.4 (1H, d, J 8 Hz), 8.1 (1H, dd, J 8 and 2 Hz), and 8.9 (1H, d, J 2 Hz).

Compound A was needed for conversion into the potential enzyme inhibitor shown in the margin. How might this be achieved?

Purpose of the problem

Revision of proof of structure together with electrophilic and nucleophilic substitution on pyridines.

Suggested solution

The first step requires nucleophilic substitution so we could convert the pyridone into a chloropyridine and displace the chloride with the amine.

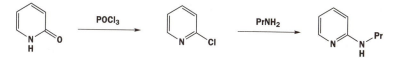

The nitration occurs only because this pyridine is activated by the extra amino group so you could have started by predicting which compound might be made. Alternatively, you could work out the structure from the spectrum. The key points are that A has only three aromatic protons so nitration has occurred on the ring, that there is only one coupling constant large enough to be *ortho* (8 Hz), and there is a proton that has only *meta* coupling (2 Hz) at large shift. The pyridine nitrogen causes large shifts at positions 2, 4, and 6, the nitro group causes large *ortho* shifts, and the amino group causes upfield *ortho* shifts to smaller δ. All this fits the structure and mechanism shown. This isomer is formed because the amino group directs *ortho, para* and the *para* position is preferred sterically.

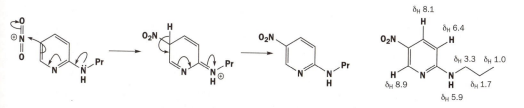

To get the potential enzyme inhibitor, we need to reduce the nitro group to an amine and add the new chain to the other amine. This conjugate addition must be done first while there is only one nucleophilic amine. The ester is probably the best acid derivative to use but you may have chosen another.

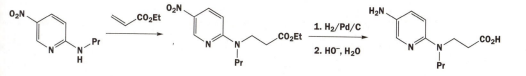

Suggested solutions for Chapter 44

44

Problem 1

In this pyridine synthesis, give a structure for A and mechanisms for the reactions. Why is hydroxylamine used instead of ammonia in the last step?

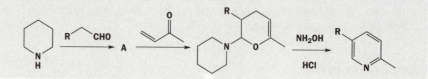

Purpose of the problem

A simple exercise on the synthesis of a pyridine from a masked dicarbonyl compound.

Suggested solution

■ The cyclization of the enamine onto the enone could be drawn as a Diels–Alder reaction with two heteroatoms (N and O) instead of the ionic mechanism we have drawn.

Compound A is clearly the enamine from the aldehyde and piperidine and its reaction with an unsaturated ketone would be expected to lead to conjugate addition (Chapter 29). Evidently, in this

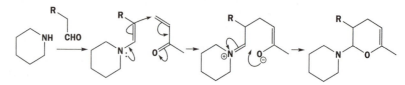

■ This chemistry was developed by G. Botteghi *et al.*, *Synth. Commun.*, 1979, **9**, 69.

case the intermediate is trapped.

The acid conditions for the next reaction release the piperidine and addition of hydroxylamine eventually produced an *N*-hydroxy-dihydropiperidine rather like the pyridine synthesis described in the chapter (pp. 1193–4). Loss of water gives the aromatic pyridine. The exact order of events shown below may well not be correct – there are other ways to get to the product – but the last intermediate is definitely formed. Hydroxylamine is used instead of ammonia so that the last step is

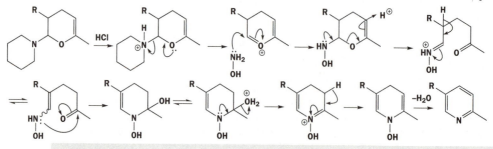

Problem 2

Suggest a mechanism for this synthesis of a tricyclic aromatic heterocycle.

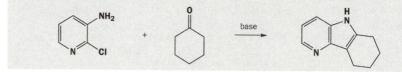

a simple elimination of water rather than the oxidation required with ammonia.

Purpose of the problem

A simple exercise on the synthesis of a pyridine fused to a pyrrole (or an indole with an extra nitrogen atom).

■ This compound was used in a study of radical reactions by R. Beugelmans *et al.*, *Tetrahedron Lett.*, 1980, **21**, 1943.

Suggested solution

The first step must be the formation of an enamine between the primary amine and the ketone. Now, because we have a pyridine and not a benzene ring, nucleophilic aromatic substitution can occur. These 'aza-indoles' are more easily formed than indoles.

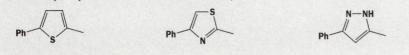

Problem 3

How would you synthesize these aromatic heterocycles?

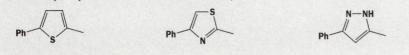

Purpose of the problem

Your first chance to devise syntheses for five-membered aromatic heterocycles with one or two heteroatoms.

Suggested solution

These compounds all look much the same, but the strategies needed for each are rather different. Removing the heteroatom from the thiophene reveals a 1,4-diketone to be made by one of the methods in Chapter 30 (pp. 800–1). We have chosen to use an enamine and an α-bromoketone though there are many other good choices.

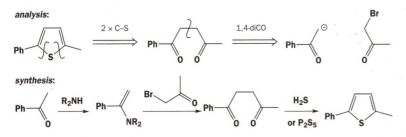

The second compound is a thiazole and we want to use a thioamide to make it (pp. 1199–200). We should disconnect C–N and C–S bonds to give the thioamide and another α-bromoketone, remembering to let the nucleophiles exercise their natural preferences: sulfur displaces the halide and nitrogen attacks the carbonyl group.

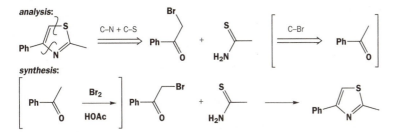

The third has the two heteroatoms joined together and we do best to keep them that way so we disconnect the C–N bonds outside the hidden molecule of hydrazine (NH_2NH_2). We need a 1,3-diketone so we are into Claisen ester chemistry (pp. 791–4 and 1188).

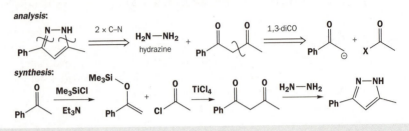

Problem 4

Is the heterocyclic ring created in this reaction aromatic? How does the reaction proceed? Comment on the selectivity of the cyclization.

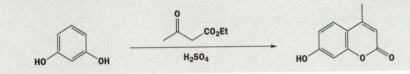

Purpose of the problem

An oxygen heterocycle and the question of selectivity in its synthesis.

Suggested solution

The left-hand ring is obviously aromatic as it is a benzene ring. The right-hand ring has four electrons from the double bonds and can have a lone pair from the oxygen in the ring, making six in all. Alternatively, the whole system can be considered as a $4n + 2$ system with 10 electrons. Either way, it is aromatic. This is more obvious in a delocalized form.

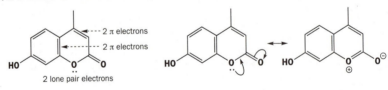

■ This is a very old reaction discovered by H. von Pechmann and C. Duisberg, *Ber.*, 1883, **16**, 2119.

The first step in the reaction is transesterification and the cyclization then occurs in an *ortho* position. The selectivity is regioselectivity for one *ortho* position and it looks as though steric hindrance is important. The other OH group is also *ortho*, *para*-directing and the position chosen is *para*.

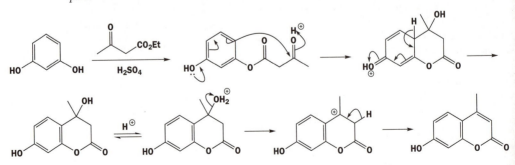

Problem 5

Suggest mechanisms for this unusual indole synthesis. How does the second reaction relate to electrophilic substitution at indoles as discussed in Chapter 33?

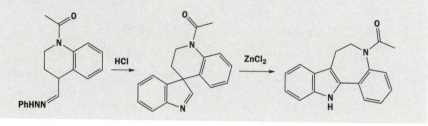

Purpose of the problem

A combination of a Fischer indole synthesis with revision of a bit of indole chemistry from the last chapter.

Suggested solution

The first step starts off as a normal Fischer indole; you just have to draw the molecules carefully to show the *spiro* ring system, and you have to stop before an indole is formed as the quaternary centre prevents aromatization.

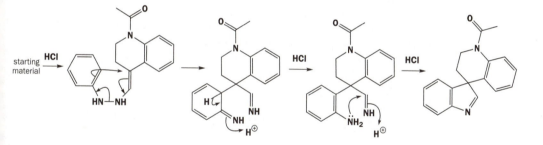

Treatment with a Lewis acid initiates a rearrangement very like those occurring when 3-substituted indoles are attacked by electrophiles (pp. 1170–1). The aromatic ring is a better migrating group than the primary alkyl alternative and an indole can finally be formed.

■ The new seven-membered heterocycle (an azepine) is found in some tranquillizers (T. S. T. Wang, *Tetrahedron Lett.*, 1975, 1637).

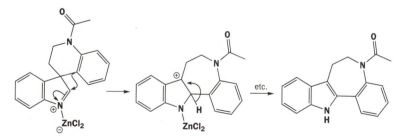

Problem 6

Explain the reactions in this partial synthesis of methoxatin, the coenzyme of bacteria living on methanol.

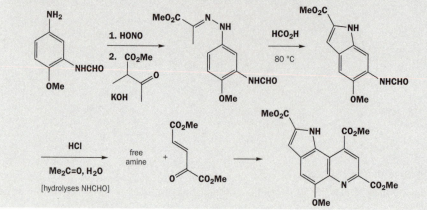

Purpose of the problem

A combination of a Fischer indole synthesis with revision of a bit of indole chemistry from the last chapter.

Suggested solution

■ This methoxatin synthesis is described by E. J. Corey and A. Tramontano, *J. Am. Chem. Soc.,* 1981, **103**, 5599.

There is clearly a Fischer indole synthesis in the second step but the first step makes the usual hydrazone in a most unusual way. The first reaction is diazotization so we have to combine the diazonium salt with the enolate of the keto-ester. That creates a quaternary centre and the KOH deacylates it to give the aryl hydrazone needed for the Fischer reaction.

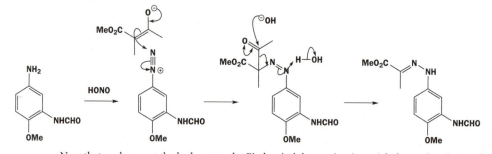

Now that we have got the hydrazone, the Fischer indole reaction is straightforward and gives the indole-2-carboxylic acid derivative. There is only one site for the enamine, and the indole is formed away from the substituents already on the ring.

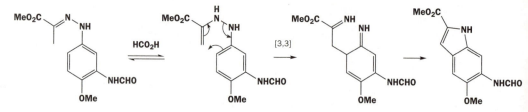

The next stage obviously involves the primary amine as nucleophile and the conjugated keto-diester as electrophile. You may well have done direct addition to the ketone as that gives the product by a very reasonable mechanism. In fact, it is known that conjugate addition occurs as the tertiary alcohol A can be isolated. The dehydration of this alcohol is obviously acid-catalysed and the oxidation by air to the aromatic heterocycle is also acid-catalysed.

■ A can be oxidized to the product with Ce(IV) if required.

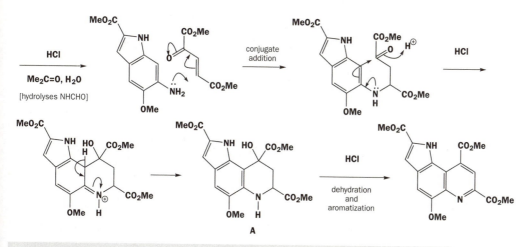

Problem 7

Suggest a synthesis of fentiazac, a nonsteroidal anti-inflammatory drug. The analysis is in the chapter but you need to explain why you need these particular starting materials as well as how you would make them.

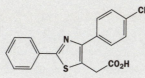

Purpose of the problem

Further exploration of concepts and chemistry from the chapter.

Suggested solution

The analysis in the chapter (p. 1200 reproduced below) shows that we need the thioamide of benzoic acid (thiobenzamide available from Aldrich) and an α-bromoketone. We need these particular starting materials because the soft sulfur atom will displace the bromide in a frontier-orbital-controlled S_N2 reaction whilst the hard amino group will attack the ketone in a charge-controlled nucleophilic attack on a carbonyl group.

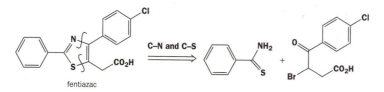

We still need to make the bromoketone and that will come from the bromination of the parent keto-acid. This in turn, being a 1,4-dicarbonyl compound, comes from chemistry discussed in Chapter 30 (pp. 800–1) and in Problem 3 above. A selection of possible disconnections is given here. There is no problem about the bromination as the ketone is more

easily enolized than the acid and it can be brominated only on one side. The aromatic ring i
deactivated by the carbonyl group.

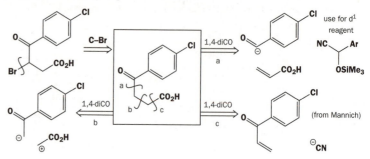

The route reported by J. Cavalla and group, *J. Med Chem.*, 1974, **17**, 1177, is given below.

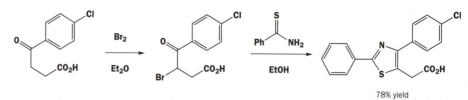

78% yield

Problem 8

Explain why these two quinoline syntheses from the same starting materials give (mainly)
different products.

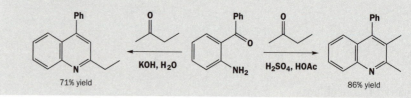

71% yield   86% yield

Purpose of the problem

An exercise in regioselectivity of heterocyclic synthesis controlled by pH.

Suggested solution

■ This selective route to quinolines
by the Friedländer synthesis was
discovered by E. A. Fehnel, *J. Org.
Chem.*, 1966, **31**, 2899.

You have a choice here: either you first form an enol(ate) from butanone and do an aldol reaction
with the aromatic ketone or you first make an imine and then form enamines from that. In either
case, you would expect enol or enamine formation on the more substituted side in acid but on the
less substituted side in base.

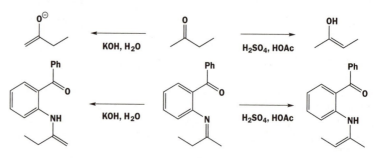

Problem 9

Give mechanisms for these reactions used to prepare a fused pyridine. Why is it necessary to use a protecting group?

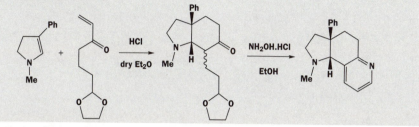

Purpose of the problem

Saturated and aromatic heterocycles combined with stereochemistry make a neat synthesis.

Suggested solution

The first starting material is a stable cyclic enamine and conjugate addition is what we should expect with an enone. Of course, if the aldehyde were unprotected, direct addition might occur there as well as carbonyl condensations. The product is in equilibrium with both its enols, one of which can cyclize to form the new six-membered ring. The enol must attack the five-membered ring in a *cis* fashion as the tether is too short to reach the other side. There is no control over one stereogenic centre but that is unimportant as it is soon to disappear.

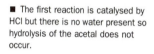

■ The first reaction is catalysed by HCl but there is no water present so hydrolysis of the acetal does not occur.

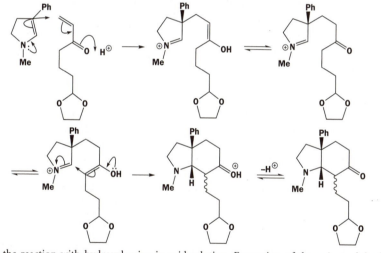

Now the reaction with hydroxylamine in acid solution. Formation of the oxime of the ketone (p. 348) produces one molecule of water – just enough to hydrolyse the acetal – and the pyridine synthesis can be completed by cyclization and a double dehydration (p. 1194).

■ This chemistry was used to make the alkaloids from *Sceletium* species (C. P. Forbes *et al.*, *Tetrahedron*, 1978, **34**, 487).

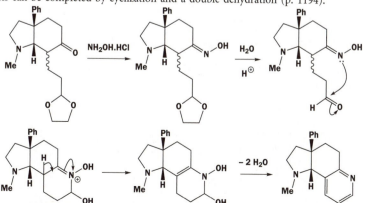

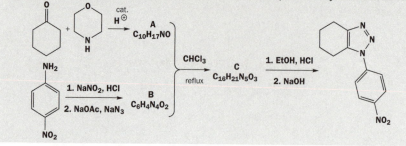

Problem 10

Identify the intermediates and give mechanisms for the steps in this synthesis of a triazole.

Purpose of the problem

Revision of aromatic nucleophilic substitution and a chance to unravel an interesting mechanism

Suggested solution

The first reaction forms A, just the enamine from the ketone and the secondary amine (morpholine). Below we have diazotization of an aromatic amine and replacement by azide. This nucleophilic aromatic substitution (Chapter 23) could occur by the addition-elimination mechanism, activated by the nitro group, or by the S_N1 mechanism.

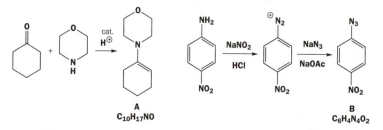

■ An alternative mechanism is a 1,3-dipolar cycloaddition of the azide on to the enamine.

Now comes the interesting bit. The two reagents A and B combine without losing anything – it is evident that the enamine must be the nucleophile and so the azide must be the electrophile. We can see from the final product that the enamine attacks one end or the other of the azide. Trial and error takes over! Here is one possible solution with the side chains of the intermediate abbreviated for clarity. This product C can be isolated but its stereochemistry is not known.

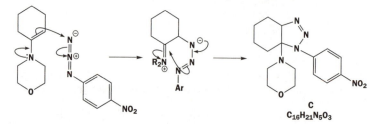

■ R. Fusco et al., Gazz. Chim. Ital., 1961, **91**, 849.

Finally, the new aromatic system (a triazole) can be formed by elimination of the aminal. Protonation of the most basic nitrogen is followed by expulsion of morpholine and aromatization by deprotonation. This synthesis of 1,2,3-triazoles was discovered in Milan during a mechanistic study of the reactions of azides and enamines.

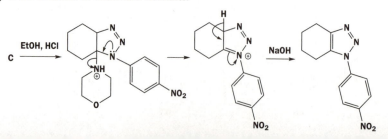

Problem 11

Give detailed mechanisms for this pyridine synthesis. The first part revises Chapters 27 and 29.

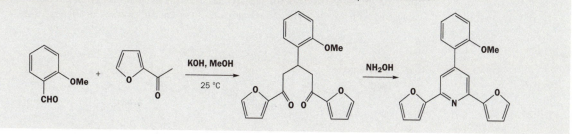

Purpose of the problem

Revision of aldol and conjugate addition reactions of enol(ate)s and a synthesis involving one pyridine and two furans.

Suggested solution

The first reactions are aldol condensation and conjugate addition, summarized below. Make sure you can draw the full mechanism for both. The last step is a standard pyridine synthesis already used in Problems 1 and 9.

■ This chemistry was actually used to make some *spiro* compounds related to pyridines (D. D. Weller and G. R. Luellen, *Tetrahedron Lett.*, 1981, **22**, 4381).

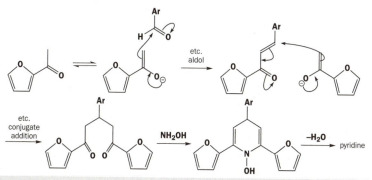

Problem 12

This question revises a number of previous chapters, especially Chapters 24–26 and 39. Give mechanisms for the reactions in this synthesis of a furan and comment on the choice of reagents for the various steps.

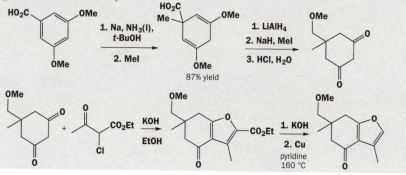

Purpose of the problem

Revision, drawing mechanisms, discussing selectivity, and furan synthesis.

Suggested solution

This question is long so a quick answer is best. Addition of two electrons and one proton in a Birch reduction (pp. 628–9) gives a unique structure with the dianion on oxygen and the two alkenes conjugated with the OMe groups. Methylation of this enolate gives the first intermediate.

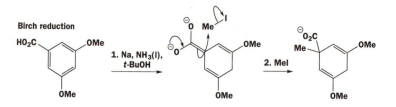

The next few steps need no comment but the double hydrolysis of the enol ethers may. This was considered in Chapter 21 (pp. 540–1) and is summarized here.

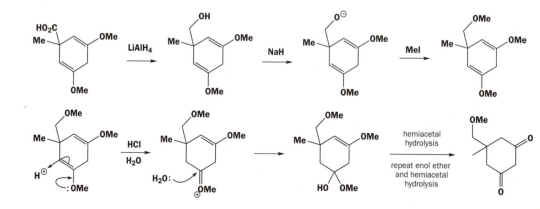

■ This sequence was used to make linderalactone (A. Gopalan and P. Magnus, *J. Am. Chem. Soc.*, 1980, **102**, 1756).

Finally, the furan synthesis, which takes place in the same reaction as the alkylation of the enolate. The only surprise is that the aldol reaction occurs first and O-alkylation afterwards. Maybe the chlorine is very crowded.

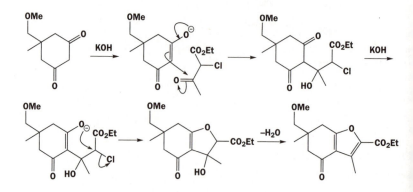

Problem 13

Suggest syntheses for this compound, explaining why you choose this particular approach.

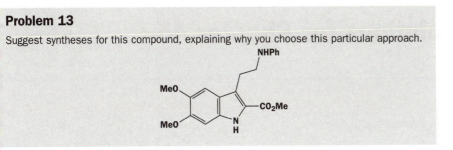

Purpose of the problem

The synthesis of an indole with a slight twist.

Suggested solution

This looks very much like a perfect subject for the Fischer indole synthesis. Let us see.

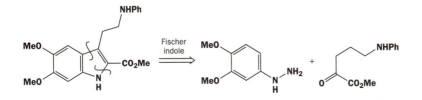

This looks fine, though we may wonder how we are going to have an amino group in that position on the keto-ester. Surely it will cyclize on to the ester to form a lactam? One solution would be to protect it with something like a Boc group but the solution found by the Sterling Drug company was partly motivated by a desire to make a variety of compounds with different amines on the side chain. They chose an OH group as a generally replaceable group and accepted that the starting material would exist as a lactone. They made it like this.

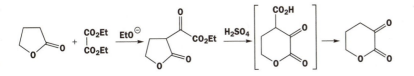

The first step is a typical Claisen ester condensation (pp. 728–31) and the second is a thermodynamically controlled reorganization to give the more stable six-membered lactone with decarboxylation of the intermediate in brackets. Now the Fischer indole works fine and work-up with dry HCl in methanol gives the required alkyl chloride. Displacement with aniline, or any other amine if needed, gives a series of antidepressants.

■ This chemistry is in the patent literature alone, but see S. Archer, *Chem. Abstr.*, 1971, **75**, 29442.

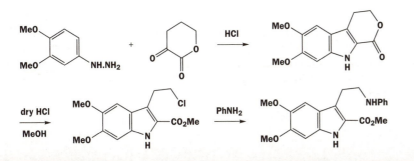

Suggested solutions for Chapter 45

45

■ This method was invented by
B. W. Bycroft and G. R. Lee, *J. Chem.
Soc., Chem. Commun.*, 1975, 988.

Problem 1

Explain how this asymmetric synthesis of amino acids, starting with natural proline, works. Explain the stereoselectivity of each reaction.

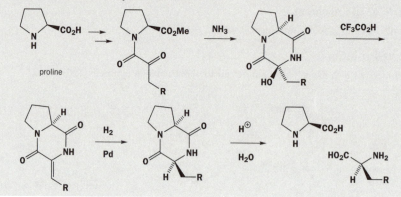

Purpose of the problem

A simple exercise in the creation of a new stereogenic centre via a cyclic intermediate.

Suggested solution

Nothing exciting happens until the hydrogenation step. The stereoselectivity in the reaction with ammonia is interesting but not of any consequence as that stereochemistry disappears in the elimination. This gives the *E*-enone as expected since the alkene and the carbonyl are in the same plane.

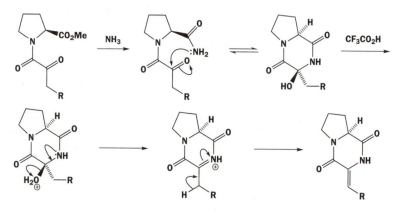

The new centre is formed in the hydrogenation step. The molecule is slightly folded and the catalyst interacts best with the outside (*exo*, convex) face so that it adds hydrogen from the same face as the ring junction hydrogen atom. All that remains is to hydrolyse the compound without racemization. Did you notice that the configuration of the new amino acid is (*S*), the same as that of the natural amino acids?

Problem 2

This is a synthesis of the racemic drug tazadolene. If the enantiomers of the drug are to be evaluated for biological activity, they must be separated. At which stage would you advocate separating the enantiomers, and how would you do it?

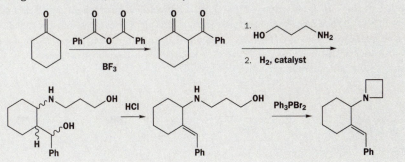

Purpose of the problem

First steps in planning an asymmetric synthesis by resolution.

Suggested solution

You need to ask: which is the first chiral intermediate? Can it be conveniently resolved? Will the chirality survive subsequent steps? The first intermediate is chiral but it enolizes very readily and the enol is achiral so that's no good. The second compound is definitely chiral but it has three chiral centres and these are evidently not controlled. We should have to separate diastereoisomers before resolution and that would be a waste of time since all the diastereoisomers give the next intermediate.

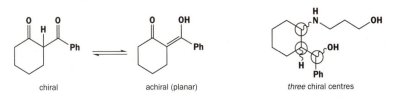

chiral	achiral (planar)	three chiral centres

The next intermediate, the amino alcohol, is ideal: it has only one chiral centre and that is not affected by the last reaction. It has two 'handles' for resolution – the amine and the alcohol. We might make a salt of the amine with tartaric acid or an ester of the alcohol with some chiral acid. Alternatively, we could resolve tazadolene itself: it still has an amino group so we could again form a salt with a suitable acid.

■ This synthesis is from Upjohn Company and is in only the patent literature (*Chem. Abstr.*, 1984, **100**, 6311).

Problem 3

How would you make enantiomerically enriched samples of these compounds (either enantiomer)?

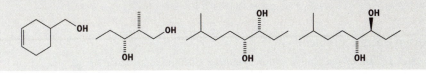

Purpose of the problem

First steps in planning an asymmetric synthesis.

Suggested solution

There are many possible answers here. What we had in mind was some sort of asymmetric Diel
Alder reaction for the first (pp. 1228–31), an asymmetric aldol reaction for the second or e
opening an epoxide made by the Sharpless epoxidation (pp. 1239–41), asymmetric dihydroxylatic
on an *E*-alkene for the third (pp. 1241–3), and perhaps asymmetric dihydroxylation on a *Z*-alke
for the fourth. Of course, you might have used resolution or catalytic hydrogenation, or the chi
pool, or any other strategy you could devise.

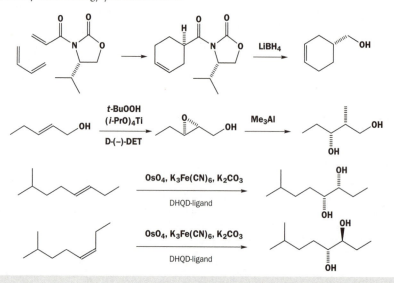

Problem 4

What is happening in stereochemical terms in this sequence of reactions? What is the other product from the crystallisation
from hexane? The product is one enantiomer of a phosphine oxide. If you wanted the other enantiomer, what would you do?
Revision. This phosphine oxide is used in the synthesis of DIPAMP, the chiral ligand for asymmetric catalytic hydrogenation

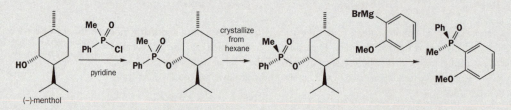

mentioned in the chapter. What are the various reagents doing in the conversion into DIPAMP?

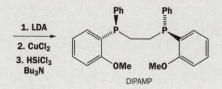

Purpose of the problem

Detailed analysis of a resolution and the synthesis of an important chiral ligand.

Suggested solution

The whole process is a resolution with the difference that the resolving agent (menthol) becomes the leaving group in a constructive nucleophilic displacement that builds an essential C–P bond. Racemic phosphinoyl chloride combines with a single enantiomer of menthol to give a separable mixture of diastereoisomers.

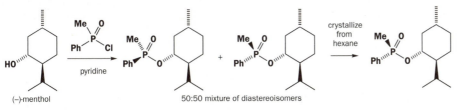

(–)-menthol   50:50 mixture of diastereoisomers

Reaction of the phosphinate ester with a Grignard reagent displaces the alkoxide leaving group, regenerating menthol, the resolving agent, and forms a phosphine oxide. This reaction is stereospecific with inversion. If you wanted the other enantiomer, you would have to separate the other diastereoisomer of the intermediate from the solution used in the crystallization, purify it, and react it with the same Grignard reagent.

■ This is the route to single enantiomers of phosphorus compounds devised by K. Mislow and group, *J. Am. Chem. Soc.*, 1968, **90**, 4842.

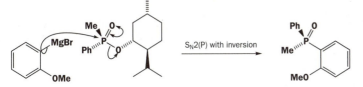

The dimerization to give DIPAMP is a radical reaction. LDA forms the lithium derivative and Cu (II) removes one electron from it to give a radical that dimerizes. You might have drawn this as the homolytic cleavage of a C–Cu bond. Finally, trichlorosilane reduces the phosphine oxide to the phosphine with inversion.

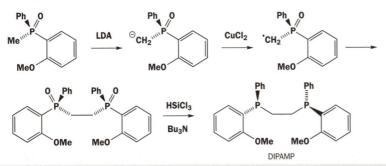

DIPAMP

Problem 5

An alternative to the Evans chiral auxiliary described in the chapter is this oxazolidinone, made from natural (S)-(–)-phenylalanine. What strategy is used for this synthesis and why are the conditions and mechanism of the reactions important?

synthesis of Evans's chiral auxiliary from (S)-phenylalanine

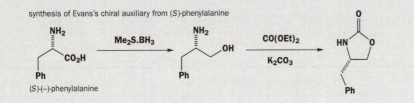

(S)-(–)-phenylalanine

Purpose of the problem

Detailed analysis of the synthesis of an important chiral auxiliary by the chiral pool strategy.

Suggested solution

This is the chiral pool strategy (pp. 1222–5): a cheap, readily available natural product (an amino acid from protein) is built into the product and provides its chirality. It is essential that no racemization occurs while the carbonyl group is still there so a reducing agent effective without strong acid or base is ideal. The cyclization with diethyl carbonate is not so critical as the chiral centre is now more secure but the conditions are again mild.

Problem 6

In the following reaction sequence, the chirality of mandelic acid is transmitted to a new hydroxy-acid by a sequence of stereochemically controlled reactions. Give mechanisms for the reactions and state whether each is stereospecific or stereoselective. Offer some rationalization for the creation of new stereogenic centres in the first and second reactions.

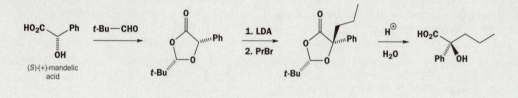

Purpose of the problem

Your chance to examine an ingenious method of asymmetric induction.

Suggested solution

The first reaction amounts to cyclic acetal formation except that one of the 'alcohols' is a carboxylic acid. The reaction is stereospecific (retention – no change) at the original chiral centre and stereoselective at the new one.

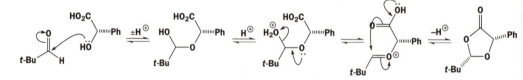

The second reaction creates a lithium enolate and alkylates it. It is again stereospecific at the unchanged chiral centre but stereoselective at the newly created quaternary centre. Finally, acetal hydrolysis preserves the new quaternary centre unchanged (stereospecific) by a mechanism that is the reverse of the first step

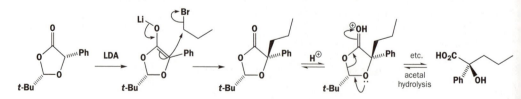

Now, as far as the rationalization is concerned, the first sequence is all reversible and therefore under thermodynamic control so the most stable product will be formed. It seems surprising that

this should be the *cis* compound, but the conformation of this chair-like five-membered ring prefers to have the two substituents di-*pseudo*-equatorial.

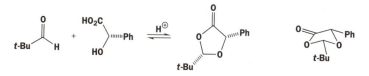

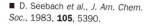

The alkylation is under kinetic control and, as lithium enolate has more or less a flat ring, the alkyl halide comes in on the opposite face to the *t*-butyl group. It has to approach roughly orthogonal to the ring (margin) as it must overlap with the p orbital of the enolate. This is Seebach's clever method of preserving the knowledge of a chiral centre while it is destroyed in a reaction. The original centre is used to create a temporary centre (at the *t*-butyl group); the original centre is then destroyed and the temporary centre used to re-create it.

■ D. Seebach *et al.*, *J. Am. Chem. Soc.*, 1983, **105**, 5390.

Problem 7

This reaction sequence can be used to make enantiomerically enriched amino acids. Which compound is the origin of the chirality and how is it made? Suggest why this particular enantiomer of the amino acid might be made. Suggest reagents for the last stages of the process. Would the enantiomerically enriched starting material be recovered?

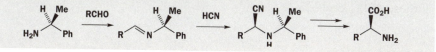

Purpose of the problem

Step-by-step discussion of a simple but useful sequence.

Suggested solution

The amine, phenylethylamine, is the origin of the chirality. It is easily made by resolution (Chapter 16), for example, by crystallizing the salt of the racemic amine with tartaric acid. This means that both enantiomers are equally available.

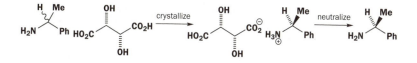

This particular enantiomer of the amino acid product is the natural (*S*)-enantiomer. It is, of course, true that the unnatural (*R*)-enantiomer would be valuable for different reasons. The last stages of the process require cleavage of one C–N bond and hydrolysis of the nitrile. It will be important to do this without racemizing the newly created centre.

The C–N bond can be reductively cleaved by hydrogenation as it is an *N*-benzyl bond. This would also hydrogenate the nitrile so that must be hydrolysed first using acid or base, as weak as possible.

■ There is a good example of the application of this method to a new amino acid by K. Q. Do *et al.*, *Helv. Chim. Acta*, 1979, **62**, 956.

The resulting carboxylic acid would not be reduced as carbonyl groups are very stable to hydrogenation. The starting material is not recovered and the chirality is lost as the by-product is just ethyl benzene, the nitrogen atom being incorporated in the main product.

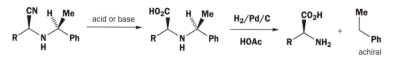

Problem 8

Submitting this racemic ester to hydrolysis by an enzyme found in pig pancreas leaves enantiomerically enriched ester with the absolute stereochemistry shown. What are the advantages and disadvantages of this method? Why is the ee not 100%?

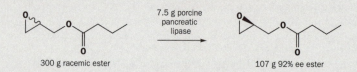

300 g racemic ester   107 g 92% ee ester

How could the same enantiomerically enriched compound be formed by chemical means? What are the advantages and disadvantages of this method?

Purpose of the problem

Comparison of two good methods.

Suggested solution

■ The other enantiomer of the alcohol is, of course, the product of the enzymatic reaction and is left behind in solution.

The advantages are that the enzymatic resolution is quick and efficient, that the catalyst is needed in only small amounts, and that the reagent is water. The disadvantages are that the yield can never be more than 50%, that the other enantiomer cannot be made, and that the enzyme is likely to be expensive. The ee is not 100% because this is not the natural substrate for the enzyme. It is remarkable that the enzyme does as well as this.

■ These methods are compared by W. E. Ladner and G. M. Whitesides, *J. Am. Chem. Soc.*, 1984, **106**, 7250.

Chemically, the obvious way is to use Sharpless epoxidation followed by esterification. The advantages of this method are that it can be used to make either enantiomer, that it is cheap and catalytic, and that an unskilled worker can do it. Only *t*-BuOOH is consumed.

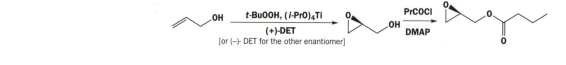

Problem 9

The BINAP-catalysed hydrogenations described in the chapter can also be applied to the reduction of ketones – the same ketones indeed as can be reduced by baker's yeast. Compare these results and comment on the differences between them.

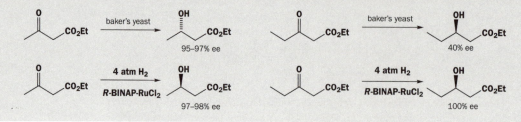

Purpose of the problem

Comparison of two more good methods.

Suggested solution

Baker's yeast is successful only if the compound is a good substrate for the reducing enzymes in the yeast. Clearly, the compound on the left is, and that on the right isn't. The BINAP-catalysed hydrogenation doesn't have such specific requirements and works well for both compounds. If it works, the yeast is a good and cheap if rather messy method while the Ru catalyst is expensive but more reliable.

■ The baker's yeast method is reviewed by C. J. Sih and C. Chen, *Angew. Chem., Int. Ed. Engl.*, 1984, **23**, 570, and the practicalities of the BINAP reduction are discussed by R. Noyori and group, *Org. Synth.*, 1993, **71**, 1.

Problem 10

Both of these bicyclic compounds readily undergo hydrogenation of the alkene to give the *syn* product. Explain why asymmetric hydrogenation of only one of the compounds would be of much value in synthesis.

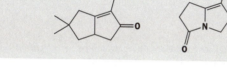

Purpose of the problem

Teaching you to avoid a trap in asymmetric synthesis.

Suggested solution

The two compounds look much the same in stereochemical terms and these are the very similar results of a simple hydrogenation.

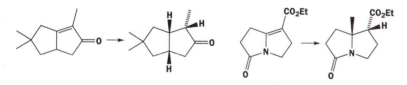

The difference is that there was already a chiral centre in the enone, but there wasn't in the amide because the nitrogen atom is flat. All the stereochemistry in the amide product is induced in the hydrogenation. For the enone, if the starting material was racemic, the product will be racemic too. If, on the other hand, the enone starting material was a single enantiomer, then we don't need asymmetric reduction. In either case, asymmetric reduction is too late.

■ The enone was hydrogenated by J. Froborg and G. Magnusson, *J. Am. Chem. Soc.*, 1978, **100**, 6728, and the amide by W. Flitsch and P. Wernsmann, *Tetrahedron Lett.*, 1981, **22**, 719. No asymmetric reduction was attempted.

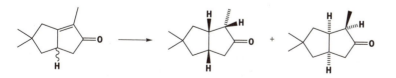

Problem 11

Explain the stereochemistry and mechanism in the synthesis of the chiral auxiliary 8-phenylmenthol from (+)-pulegone. After the reduction with Na in *i*-PrOH, what is the minor (13%) component of the mixture?

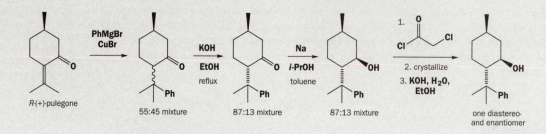

Purpose of the problem

A combination of conformational analysis, stereoselective reactions, and resolution to get a single enantiomer.

Suggested solution

The first reaction is a conjugate addition that evidently goes without any worthwhile stereoselectivity. The stereochemistry is not fixed in the addition but in the protonation of the enolate in the work-up. Equilibration of the mixture by reversible enolate formation gives mostly the all-equatorial compound.

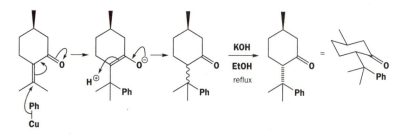

Reduction by that smallest reagent, electrons, gives the all-equatorial product. Since the ratio of products is the same as the ratio of starting materials (87:13), the reduction is totally stereoselective. The all-equatorial ketone gives 100% all-equatorial alcohol and the other isomer must give one other diastereoisomer (we cannot say which).

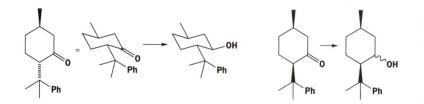

The mixture still has to be separated and, as it is a mixture of diastereoisomers, it can be separated by physical means. The chloroacetate is just a convenient crystalline compound. Two fractional crystallizations actually give about 50% of the required compound according to the details given by O. Ort, *Org. Synth. Coll.*, 1993, **VIII**, 552.

Problem 12

Describe the stereochemical happenings in these processes. You should use terms like diastereoselective and diastereotopic where needed. If you wanted to make single enantiomers of the products by these routes, at what stage would you introduce the asymmetry? (You are *not* expected to say *how* you would induce asymmetry!)

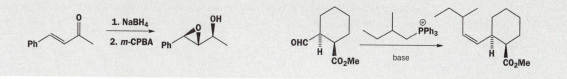

Purpose of the problem

Making sure you understand some of these stereochemical terms and concepts.

Suggested solution

First we should see what is happening at each stage of the reaction.

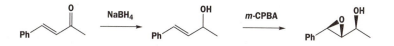

The faces of the ketone are enantiotopic (the carbonyl group lies in a plane of symmetry) so the reduction must give a 50:50 mixture of enantiomers, that is, the alcohol is racemic. Now the faces of the alkene are diastereotopic (there is no longer a plane of symmetry) and the epoxidation takes place stereoselectively on one face because the OH group hydrogen bonds to the reagent (p. 897). Asymmetry must be introduced at the first stage to create a single enantiomer of the alcohol. Then the epoxidation will automatically create a single enantiomer of the epoxide.

The second reaction is a simple Wittig process. It stereoselectively gives the *cis* (*Z*-) alkene as it is a non-stabilized ylid (pp. 814–18). The three-dimensional stereochemistry is entirely controlled by the starting material. If we want a single enantiomer of the product, we must start with a single enantiomer of the aldehyde-ester. Asymmetry must be introduced before this reaction sequence.

Problem 13

The unsaturated amine A, a useful intermediate in the synthesis of the *amaryllidaceae* (daffodil) alkaloids, can be made from the three starting materials shown below. What kind of chemistry is required in each case? Which is best adapted for asymmetric synthesis? Outline your chosen synthesis.

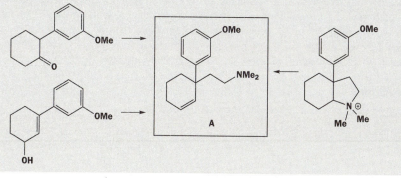

■ We give the published solutions but there are many others – yours, if different, may be as good.

Purpose of the problem

Devising syntheses when asymmetry is an issue.

Suggested solution

The more stable enolate, conjugated with the aromatic ring, can be made by a variety of methods and alkylated by various electrophiles with groups suitable for transformation into the required side chain. It is easiest to go straight there. Reduction of the ketone gives an alcohol that can eliminate only in the direction we want.

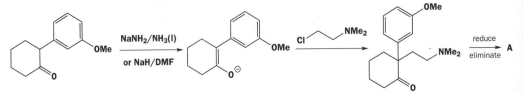

The allylic alcohol is designed for use in a [3,3]-sigmatropic rearrangement (pp. 945–6). Again, a variety of acid derivatives could be used but it is easiest to use the one we want. This route is shorter and more direct.

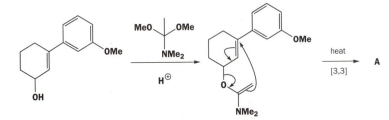

The last route depends on the stereochemistry of the starting material. If there is a *cis* ring junction, the conformation can be just right for an elimination (pp. 484–5) in base to give A directly. If it has a *trans* ring junction, this conformation is much less favourable.

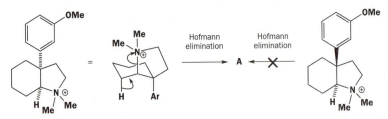

■ Compound A is an analgesic made by the methods outlined here (though in racemic form) by H. Bruderer and K. Bernauer, *Helv. Chim. Acta*, 1983, **66**, 570.

In all three cases the starting materials are already chiral. If the allylic alcohol or the cyclic ammonium salt were prepared, by resolution or by asymmetric synthesis, as a single enantiomer, then A too would be a single enantiomer. However, the ketone used in the first route, though also chiral, has an achiral enolate so asymmetry could be introduced at that point. It isn't possible to use the chiral enolates described in the chapter (p. 1230). Probably the easiest approach is to make the allylic alcohol by asymmetric reduction, say by the CBS reducing agent (p. 1234).

Problem 14

Suggest syntheses for single enantiomers of these compounds.

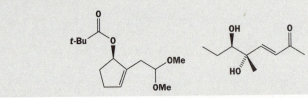

Purpose of the problem

Devising your own asymmetric syntheses.

Suggested solution

The first compound is an ester derived from a cyclic secondary alcohol that should be made from the corresponding enone by asymmetric reduction.

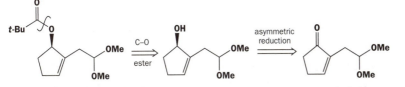

Reduction with Corey's CBS reducing agent (margin and p. 1234) gave the alcohol in 93% ee and this compound was used to make a compound from the gingko tree by E. J. Corey and A. V. Gaval, *Tetrahedron Lett.*, 1988, **29**, 3201.

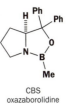

CBS
oxazaborolidine

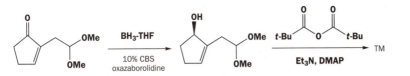

The second compound could be made by a Wittig reaction with a stabilized ylid and the required diol aldehyde derived from an epoxy-alcohol and hence from an allylic alcohol by Sharpless epoxidation.

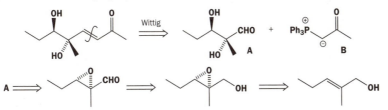

The first part of the synthesis was a demonstration that the Sharpless epoxidation was useful (K. B. Sharpless and group, *J. Am. Chem. Soc.*, 1981, **103**, 464) as it gave an intermediate that had already been used (S. Masamune *et al.*, *J. Am. Chem. Soc.*, 1975, **97**, 3512) in the synthesis of the antibiotic methymycin.

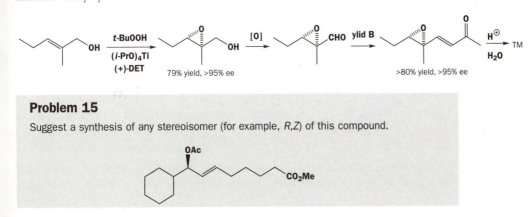

Problem 15

Suggest a synthesis of any stereoisomer (for example, *R,Z*) of this compound.

Purpose of the problem

Devising your own asymmetric synthesis with no strings attached.

■ FGI stands for functional group interconversion.

Suggested solution

This time there are so many possibilities that we don't want to push any one solution. We had in mind the asymmetric reduction of an acetylenic ketone by CBS or some other asymmetric reducing agent and then using the alkyne to make the *trans* alkene. You also have to consider at what stage to introduce the ester group. This open-ended problem cannot have a more defined solution.

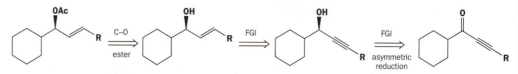

Problem 16

Revision. Give mechanisms for the steps in the synthesis of tazadolene in Problem 2.

Purpose of the problem

Revision of Chapters 28, 14, 24, 19, and 17.

Suggested solution

The first step is a Lewis acid-catalysed acylation of an enol (Chapter 28).

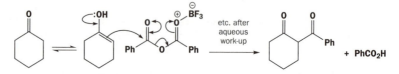

Next, we have a reductive amination: the imine is formed (Chapter 14) from the more reactive aliphatic ketone rather than from the conjugated aryl ketone and the hydrogenation reduces both the imine and the remaining ketone. This is unusual (Chapter 24) but there are catalysts that will do the job.

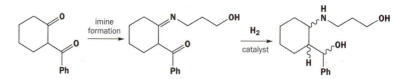

Now comes an acid-catalysed elimination (Chapter 19), probably by the E1 mechanism as the alcohol is secondary and benzylic. Each step is reversible so thermodynamic control ensures that the *E*-enone is formed. The *Z*-enone would have a steric clash between the carbonyl oxygen atom and the phenyl group.

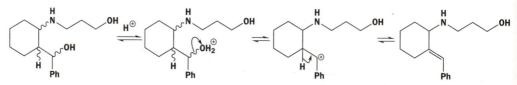

■ This synthesis is from Upjohn Company and is in only the patent literature (*Chem. Abstr., 1984,* **100**, 6311).

Now the most interesting step: a Mitsunobu-like reaction (Chapter 17) turning the OH group into a leaving group so that an intramolecular S_N2 reaction by the amine makes the four-membered ring.

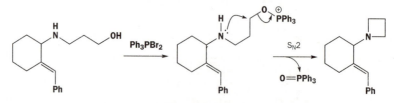

Suggested solutions for Chapter 46

46

Problem 1

Suggest structures for intermediates A and B and mechanisms for the reactions.

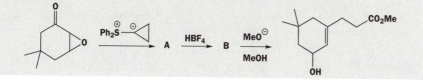

Purpose of the problem

A simple exercise in sulfur chemistry with some remarkable results.

Suggested solution

Incredible as it may seem, this ylid with an extra three-membered ring, still does the epoxide formation mentioned in the chapter (pp. 1258–60). The very strained intermediate rearranges in acid and finally fragments in a way quite like the Eschenmoser fragmentation (p. 1008).

■ This method was called 'secoalkylation' by its inventors B. M. Trost and M. J. Bogdanowicz, *J. Am. Chem. Soc.*, 1972, **94**, 4777.

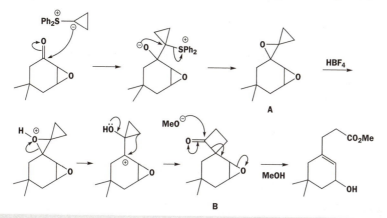

Problem 2

Suggest a mechanism for this reaction, commenting on the selectivity and the stereochemistry.

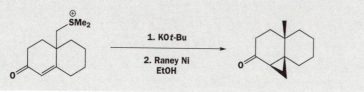

Purpose of the problem

An intramolecular version of an important reaction – what does it mean?

Suggested solution

■ R. S. Matthews and T. E. Meteyer, *J. Chem. Soc., Chem. Commun.,* 1971, 1536.

The ylid forms in the usual way but can't reach across the ring to attack the carbonyl group directly so it has to do conjugate addition instead. It also has to attack from the top face as it is tethered there. Completion of the cyclopropane-forming reaction leaves the sulfur still attached to the angular methyl group, from which it is freed by Raney nickel. This reaction shows that simple sulfonium ylids can do conjugate addition – they just prefer not to.

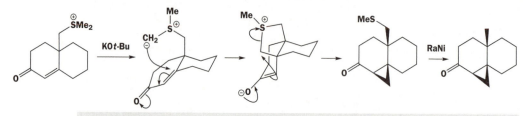

Problem 3

The product X of the following reaction has δ_H (p.p.m.) 1.28 (6H, s), 1.63 (3H, d, *J* 4.5 Hz), 2.45 (6H, s), 4.22 (1H, s), 5.41 (1H, d, *J* 15 Hz), and 5.63 (1H, dq, *J* 15, 4.5 Hz). Suggest a structure for X and a mechanism for its formation.

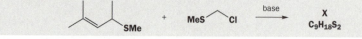

Purpose of the problem

Revision of NMR, structural identification, and sigmatropic rearrangements.

Suggested solution

The product still has the CMe_2 group, though no longer on a double bond; it has a *trans* alkene with a methyl group at one end, two SMe groups (6H singlet at 2.45), and hydrogen at 4.22 that might be next to two sulfur atoms. All that adds up to the structure shown and a [2,3]-sigmatropic shift of a sulfonium ylid.

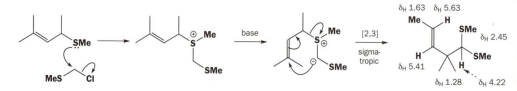

Problem 4

The thermal elimination of sulfoxides (example below) is a first-order reaction with almost no rate dependence on substituent at sulfur (Ar) and a modest negative entropy of activation. It is accelerated if R is a carbonyl group (that is, R = COR'). The reaction is (slightly) faster in less polar solvents. Explain.

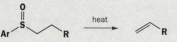

Explain the stereochemistry of the first reaction in the following scheme and the position of the double bond in the final product.

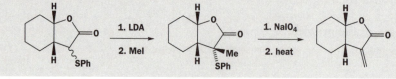

Purpose of the problem

Checking that you understand the mechanism and stereochemistry of sulfoxide eliminations and revision of Chapter 41.

Suggested solution

The mechanism given in the chapter (pp. 1269–70) is a reverse cycloaddition of a sulfenic acid and an alkene. This can be drawn in two ways (they are really the same) depending on whether you draw the sulfoxide with S=O or as an ylid. This fits the evidence quite well. The substituent on sulfur is not involved but that on carbon (R) will accelerate the reaction if it makes the proton that is removed more acidic. We might expect a positive entropy of activation as two molecules are being formed from one, but pericyclic reactions are notorious for large negative entropies of activation because of the high order of the transition state. A modest negative value seems fair. The sulfoxide is polar but the products are not so the transition state is presumably slightly less polar than the starting materials; hence the solvent effect.

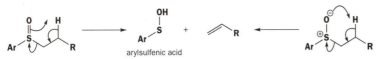

arylsulfenic acid

The first reaction is the alkylation of an enolate. The molecule is folded so the planar enolate will add the electrophile from the outside (*exo-* or convex) face.

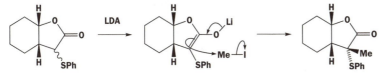

Sulfoxide elimination is a *cis* reaction so the oxygen atom cannot reach the hydrogen at the ring junction and has to eliminate into the side chain methyl to give the unstable *exo*-methylene compound. We know that the elimination into the ring is preferred because the other diastereoisomer, which has a free choice, prefers to do just that.

■ P. A. Grieco and J. J. Reap, *Tetrahedron Lett.*, 1974, 1097.

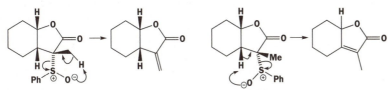

Problem 5

Revision content. Explain the reactions and the stereochemistry in these first steps in a synthesis of the B vitamin biotin.

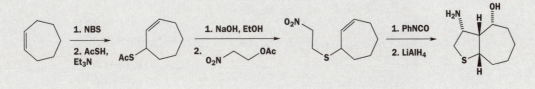

Purpose of the problem

Reminding you of radical bromination (p. 1039), how to make a thiol, that thiols are good at conjugate addition, and revision of 1,3-dipolar cycloadditions and their stereochemistry (pp. 932–6) when sulfur acts as a tether. This chemistry is described in outline on those pages.

Suggested solution

The first step is a typical allylic bromination with NBS, fully described on p. 1039. The second step is an excellent S_N2 reaction with thiolacetate as a nucleophile displacing bromide from an allylic position.

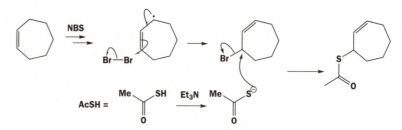

The reagents for the next step are both disguised and are both unmasked by NaOH. The thiolester is hydrolysed to the free thiolate and the nitrocompound eliminates acetate in an E1cB reaction. An excellent conjugate addition follows: both nucleophile (thiolate) and Michael acceptor (nitroalkene) are ideal.

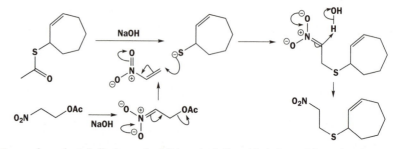

■ This synthesis of biotin is described by P. N. Confalone *et al.*, *J. Am. Chem. Soc.*, 1980, **102**, 1954.

Now we have the 1,3-dipolar cycloaddition. A nitrile oxide is formed from the nitro compound by dehydration with PhNCO (the full mechanism is on p. 934) and cycloaddition can occur only with the regio- and stereoselectivity shown because of the short tether. We have shown the nitrile oxide approaching from the bottom face of the alkene so that all three hydrogens are up. Reduction of the C=N bond then occurs from the top (outer, *exo*-, convex face) of the bowl-shaped molecule.

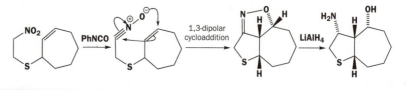

Problem 6

Explain the regio- and stereoselectivity of this reaction.

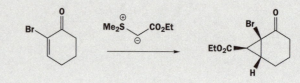

Purpose of the problem

Exploration of sulfur ylid chemistry.

Suggested solution

The ylid is stabilized by conjugation of the anion with the ester group – it is also an enolate. We can expect reversible addition to the carbonyl group and hence conjugate addition by thermodynamic control. The stereochemistry of the ring junction is inevitable: only a *cis* ring junction can be made. The only interesting centre is the ester on the three-membered ring. It too is in the more stable arrangement: on the outside of the folded molecule. The intermediate is probably a mixture of diastereoisomers so either the conjugate addition is reversible and only that diastereoisomer cyclizes that can give the more stable product or, less likely, the product epimerizes by enolization.

■ This is the first step in a synthesis of a series of natural products by P. A. Wender *et al.*, *Tetrahedron Lett.*, 1980, **21**, 2205.

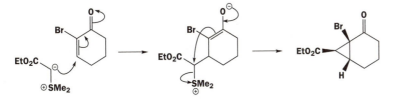

Problem 7

Draw mechanisms for these reactions of a sulfonium ylid and the rearrangement of the first product. Why is BF_4^- chosen as the counterion?

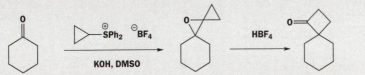

The intermediate may alternatively be reacted with a selenium compound in this sequence of reactions. Explain what is happening commenting on the regioselectivity. Why is the intermediate in square brackets not usually isolated?

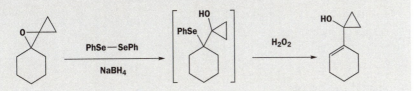

Purpose of the problem

Some sulfonium ylid chemistry, revision on rearrangements (Chapter 37), and some simple selenium reactions.

Suggested solution

The first step has already been discussed in Problem 1 so we just give the mechanism, registering again our surprise that such a strained 'oxaspiropentane' can be made. BF_4^- is chosen as the counterion since it is non-nucleophilic and will not open the epoxide or attack the carbonyl group.

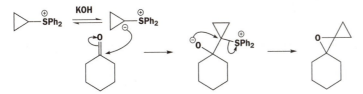

The epoxide rearrangement is triggered by acid. It opens to give the tertiary cation on the six-membered ring, as the cation on the three-membered ring would be very strained (60° ring angle to accommodate a 120° angle between bonds). The OH group pushes one of the C–C bonds in the three-membered ring across in a ring expansion.

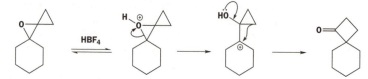

■ This chemistry was developed as part of a synthesis of aphidicolin, an antiviral natural product, by B. M. Trost *et al., J. Am. Chem. Soc.,* 1979, **101**, 1328.

No other derivatives of the tertiary cation, such as the allylic alcohol in the last part, can be formed this way because the rearrangement is too fast. The reaction with PhSe–SePh and NaBH$_4$ is a trick to get this allylic alcohol. The true reagent is PhSe$^-$, formed by reduction of the Se–Se bond. It is very nucleophilic and will attack even the tertiary epoxide to give a selenide. Oxidation to the selenoxide leads to a fast concerted *cis* elimination. The intermediate selenide is unstable as well as very smelly and must be oxidized immediately before it decomposes.

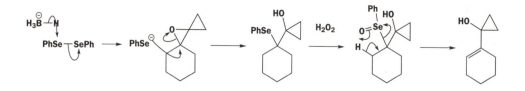

Problem 8

Give mechanisms for these reactions, explaining the role of sulfur.

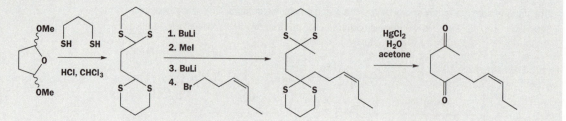

Purpose of the problem

Sulfur acetals come into their own as d^1 reagents (acyl anion equivalents).

Suggested solution

The first reaction is an acetal exchange controlled by entropy: three molecules go in and four come out (the product, two molecules of methanol, and one of water). We show just part of the mechanism.

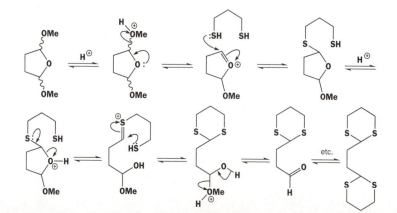

Now the two dithioacetals work as d¹ reagents (acyl anion equivalents, p. 1256) by deprotonation and alkylation. Hydrolysis needs mercury, Hg(II) as described on p. 1255.

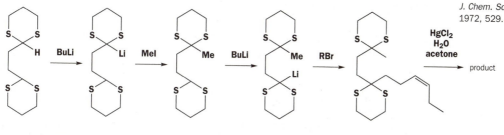

■ This chemistry was used to make the perfume *cis*-jasmone by R. A. Ellison and W. D. Woessner, *J. Chem. Soc., Chem. Commun.,* 1972, 529.

Problem 9

Suggest a mechanism for this formation of a nine-membered ring. Warning! The weak hindered base is not strong enough to form an enolate from the lactone.

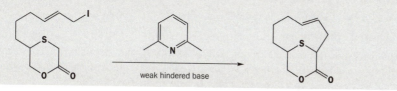

weak hindered base

Purpose of the problem

A slightly tricky sulfonium ylid rearrangement.

Suggested solution

The weak base can do nothing to our sulfur compound so it must do something by itself – alkylation at sulfur is the answer. Sulfur is a good soft nucleophile and allylic iodides are excellent electrophiles. Now the weak base can remove a proton because the result is an enolate and a sulfonium ylid as well.

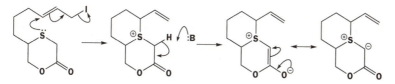

The ylid is perfectly arranged for a [2,3]-sigmatropic rearrangement forming a nine-membered ring with a *trans* (*E*-) alkene inside it.

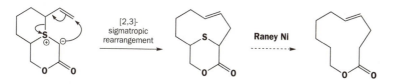

■ If the sulfur were removed by Raney nickel, we should have an eleven-membered ring and this was the idea of E. Vedejs *et al., J. Org. Chem.,* 1981, **46**, 5452.

Problem 10

Comment on the role of sulfur in the steps in this synthesis of the turmeric flavour compound *ar*-turmerone.

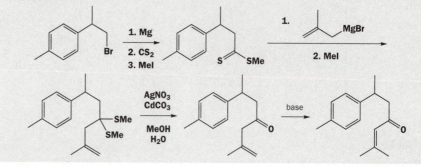

Purpose of the problem

So you can see for yourself the versatility of sulfur.

Suggested solution

The Grignard reagent from the starting material adds to CS_2 and the resulting anion is trapped with MeI to give a dithioester. This is attacked by a second Grignard reagent and the anion again trapped so that we have built up a dithioacetal by making two new C–C bonds.

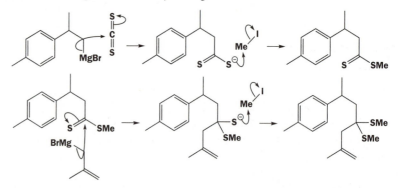

The dithioacetal is hydrolysed (mechanism like that on p. 1255), silver and cadmium being used instead of mercury (not much gain in environmental friendliness there!), and the double bond moves into conjugation with the resulting ketone to complete the synthesis of *ar*-turmerone carried out at Caen by A. Thuillier and group, *J. Org. Chem.*, 1979, **44**, 2807.

Problem 11

Explain how the presence of the sulfur-containing group allows this cyclization to occur regio- and stereoselectively.

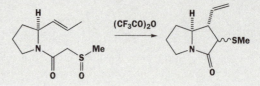

Purpose of the problem

Your first innocent encounter with the Pummerer rearrangement!

Suggested solution

Everything is set up for a Pummerer rearrangement – the sulfoxide, the acidic hydrogens on one side, the anhydride – so we must first react the oxygen of the sulfoxide with the anhydride and make the Pummerer cation.

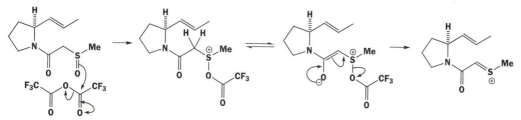

Now what will the cation do? It could capture the external nucleophile, trifluoroacetate, but that is very weak and there is another nucleophile close at hand: the alkene. We know the regioselectivity from the structure of the product, but it looks as though the alkene prefers to act in an *exo* manner, probably because of better overlap of the orbitals. The stereochemistry of the side chain comes from a transition state already having the side chain on the outside of the molecule as it folds and the lack of control in the SMe group is unimportant as it can be removed with Raney nickel anyway.

■ In Baldwin's rules terms, the reaction is 5-*exo-trig* in both the electrophile and the nucleophile.

■ This chemistry was part of an asymmetric alkaloid synthesis (the starting material was made from proline) by H. Ishibashi *et al.*, *J. Chem. Soc., Chem. Commun.*, 1986, 654.

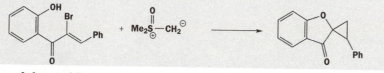

Problem 12

Problem 9 in Chapter 32 asked you to interpret the NMR spectrum of a cyclopropane (A). This compound was formed using a sulfur ylid. What is the mechanism of the reaction?

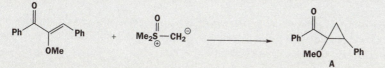

Attempts to repeat this synthesis on the bromo compound below led to a different product. What is different this time?

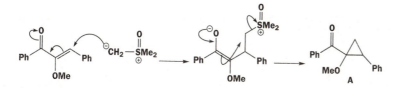

Purpose of the problem

Can you disentangle a curious variation in a simple mechanism?

Suggested solution

The first reaction is a straightforward cyclopropane formation with a sulfoxonium ylid and a conjugated ketone. The only strange feature, the OMe group, seems to make no difference.

■ J. A. Donnelly *et al.*, *J. Chem. Soc., Perkin Trans. 1*, 1979, 2629.

But evidently a bromine atom does make a difference. No doubt the reaction starts the same way and forms a cyclopropane. Under the reaction conditions the phenol will exist as an oxyanion and this can displace the bromine atom.

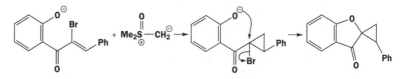

Problem 13

Epoxides may be transformed into allylic alcohols by the sequence shown here. Give mechanisms for the reactions and explain why the elimination of the selenium gives an allylic alcohol rather than an enol.

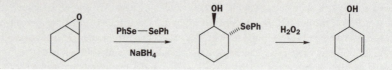

Purpose of the problem

Selenium compounds are reduced and oxidized and both processes are useful.

Suggested solution

Reduction of the Se–Se bond produces the excellent nucleophile PhSe- and this attacks the epoxide to give the adduct with *trans* stereochemistry.

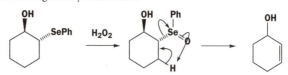

■ Except in the Heck reaction, pp. 1321–4.

Now oxidation to the selenoxide is followed by a *cis* elimination to give the allylic alcohol. There are *cis* hydrogen atoms on both sides of the selenoxide so the choice cannot be steric. Eliminations towards oxygen atoms are generally disfavoured.

■ This was the first useful piece of organoselenium chemistry and was found by K. B. Sharpless and R. F. Lauer, *J. Am. Chem. Soc.*, 1973, **95**, 2697.

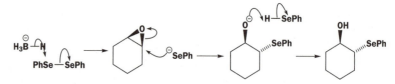

Problem 14

In a process resembling the Mitsunobu reaction (Chapter 17), alcohols and acids can be coupled to give esters, even macrocyclic lactones as shown below. In contrast to the Mitsunobu reaction, this reaction leads to retention of stereochemistry at the alcohol. Propose a mechanism that explains the stereochemistry. Why is sulfur necessary here?

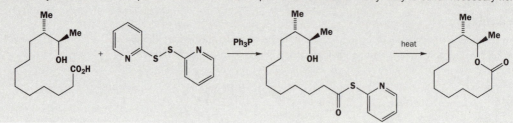

Purpose of the problem

Sulfur and phosphorus combine to make macrolactonization possible.

Suggested solution

There is no problem in getting lactones to form in smaller rings but the difficulty with medium and large rings is to get the carboxyl group and the hydroxyl group close together. You may not have seen the explanation we give below but you should have been able to deduce the roles of the phosphine and the disulfide in outline. In a Mitsunobu-style reaction the phosphine attacks the disulfide and prepares a reagent that acylates the thiolate anion. The resulting thiolester can trap a proton between the carbonyl oxygen and the nitrogen of the pyridine. This proton could come from the alcohol, and the hydrogen-bonded structure holds the alcohol and the acid close together. Because it is attached to the pyridine, the thiolate is a good leaving group and the macrolactone forms. As the alcohol is the nucleophile, there is no change in the stereochemistry.

Sulfur is necessary to provide a good nucleophile, a good leaving group from phosphorus, and a stable starting material that reacts in an ionic sense (an O–O analogue would react to form radicals).

■ This macrolactonization was devised by E. J. Corey and K. C. Nicolaou, *J. Am. Chem. Soc.*, 1974, **96**, 5614.

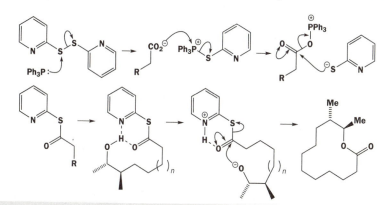

Problem 15

Suggest mechanisms for these reactions, explaining any selectivity.

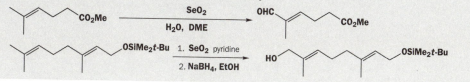

Purpose of the problem

Rehearsal of the important mechanism that leads to functionalization of allylic positions with selenium.

Suggested solution

The first step in both cases is an ene reaction with the Se=O bond (pp. 1270–1). The electrophilic selenium attacks the less substituted end (largest HOMO coefficient) of the alkene and a proton is removed from the methyl group *trans* to the main chain. Then a [2,3]-sigmatropic rearrangement puts the double bond back where it was (*trans* selectively) and functionalizes the old methyl group with an oxygen atom.

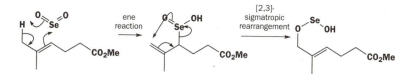

The selenium is lost from this unstable compound to give an aldehyde or, if the work-up i
reductive, as in the second case with NaBH$_4$, the alcohol. The second case also shows regioselectivit·
in that only the terminal alkene is attacked. If the other alkene were to react, a proton next t·
oxygen would have to be removed and this is unfavourable.

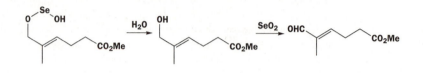

Suggested solutions for Chapter 47

47

Problem 1

The Hammett ρ value for the following reaction is –4.8. Explain this in terms of a mechanism. If the reaction were carried out in deuterated solvent, would the rate change and would there be any deuterium incorporation into the product? What is the silicon-containing product?

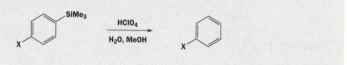

Purpose of the problem

Revision of electrophilic aromatic substitution (Chapter 22), the Hammett relationship from Chapter 41 (pp. 1090–100), and establishing the importance of the β-silyl cation.

Suggested solution

This is a 'proto-desilylation' and the large negative Hammett ρ value shows that electrons are being drained out of the ring in a big way in the rate-determining step. Protonation at the '*ipso*' position (where the silicon is) is the only way to get a β-silyl cation (pp. 1291–3) and attack of a nucleophile such as water removes the silicon.

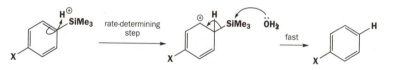

In a deuterated solvent, rate-determining proton transfer to carbon would be slower and the H marked in the product would instead be deuterium. The silicon-containing product is initially the silanol $Me_3Si–OH$ but this quickly gives the ether $Me_3Si–O–SiMe_3$.

■ This reaction was studied by C. Eaborn and J. A. Sperry, *J. Chem. Soc.*, 1961, 4921.

Problem 2

Identify the intermediates in this reaction sequence and draw mechanisms for the reactions, explaining the special role of the Me_3Si group.

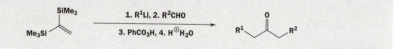

Purpose of the problem

Introducing the stabilization of anions by silicon, the Peterson reaction, and the chemistry of silyl epoxides.

Suggested solution

Addition of R^1Li to the double bond gives an anion stabilized by two silicon atoms. Peterson reaction gives a vinyl silane and epoxidation and hydrolysis (p. 1302) gives the carbonyl compound. There is no need to concern ourselves with stereochemistry as it all disappears in the final product. The silicon atom is essential in every step except the final tautomerism.

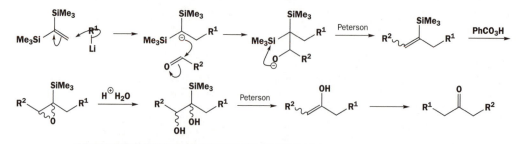

Problem 3

The synthesis of a compound used in a problem in Chapter 38 (fragmentation) is given below. Give mechanisms for the reactions explaining the role of silicon.

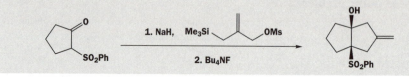

Purpose of the problem

Reminder of the anion-stabilizing ability of sulfones and the excellence of the mesylate leaving group (Chapter 46) plus the special role of fluoride as a nucleophile for silicon.

Suggested solution

■ This product was the starting material for Problem 5 in Chapter 38 and was used in a three-carbon ring expansion devised by B. M. Trost and J. E. Vincent, *J. Am. Chem. Soc.*, 1980, **102**, 5680.

Sodium hydride removes a proton from the sulfone to give an anion you can represent as an enolate. Displacement of mesylate gives an allyl silane converted into an allyl anion with fluoride. Addition to the ketone gives the 5/5 fused system, necessarily with *cis* stereochemistry.

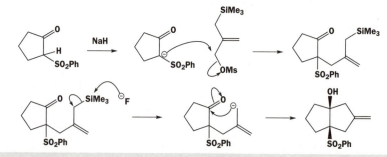

Problem 4

Give mechanisms for the following reactions, drawing structures for all the intermediates including stereochemistry. How would the reaction with Bu_3SnH have to be done?

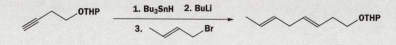

Purpose of the problem

Introduction to tin compounds as (a) reagents for radical additions, especially to alkynes, and (b) sources of stereochemically controlled vinyl anion reagents by tin-lithium exchange.

Suggested solution

The reaction of the alkyne with Bu_3SnH is a radical addition initially giving the *cis* (*Z*-) vinyl stannane but this equilibrates to the *trans* (*E*-) if a (slight) excess of Bu_3SnH is used so this is how the reaction must be done.

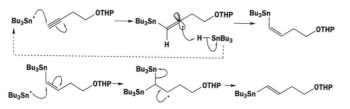

Exchange with lithium via an 'ate' complex gives a vinyl-lithium with retention of configuration and this is alkylated again with retention. These reactions go with retention because they are *electro*philic substitution – the electrophile attacks the filled σ orbital of the C–Sn or C–Li bond.

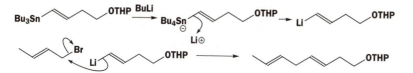

Problem 5

Explain the following reactions. In particular, explain the role of tin and why it is necessary and discuss the stereochemistry.

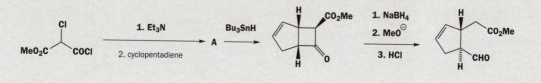

Purpose of the problem

More tin chemistry with revision of stereochemical control in cyclic compounds (Chapter 33), cycloaddition reactions (Chapter 35), and fragmentation (Chapter 38).

Suggested solution

The action of the tertiary amine on the acid chloride produces a ketene that does a [2 + 2] cycloaddition to cyclopentadiene (pp. 929–32). The ring junction in the adduct must be *cis* but the other centre need not be controlled as it is lost during the dehalogenation by the tin hydride.

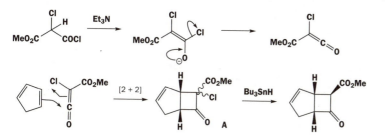

The tin hydride reaction is a standard dehalogenation using the radical to sustain a chain reaction (pp. 1040–1). The stereochemistry is that of the most stable product with the CO_2Me group on the outside of the folded molecule. In fact, this stereochemistry is also unimportant as it disappears in the next reaction.

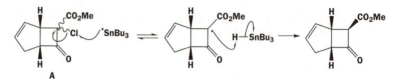

■ This chemistry was used by L. Ghosez and group in a synthesis of prostaglandins (*J. Am. Chem. Soc.*, 1981, **103**, 4616).

Reduction of the ketone gives an alcohol that fragments in a reverse aldol reaction to relieve the strain in the four-membered ring. The product initially retains the *cis* stereochemistry but equilibrates through the enolate (or enol) of the aldehyde to the more stable *trans* arrangement.

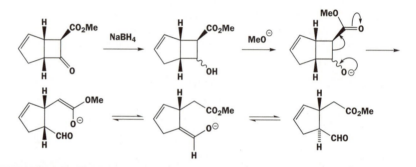

Problem 6

Explain the stereochemistry and mechanism of this hydroboration–carbonylation sequence.

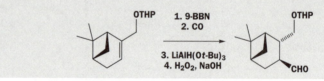

Purpose of the problem

Checking that you understand regio- and stereoselectivity in hydroboration.

Suggested solution

Hydroboration occurs on the top face, away from the *gem*-dimethyl group on the lower bridge. Carbon monoxide is one of those carbon nucleophiles that attacks boron and inserts with retention (p. 1283) into the C–B bond.

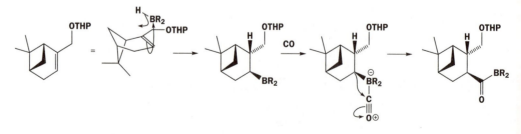

The acyl borane is unstable and easily reduced to the alcohol and that gives the aldehyde when the normal boron to oxygen migration is initiated with alkaline hydrogen peroxide. This reaction does not affect any of the chiral centres.

■ This sequence was used by M. P. L. Caton and group to make thromboxane analogues, *Tetrahedron Lett.*, 1979, 4497.

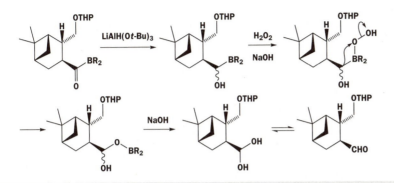

Problem 7

Revision content. Give mechanisms for these reactions, commenting on the role of silicon and the stereochemistry of the cyclization. The LiAlH₄ simply reduces the ketone to the corresponding alcohol. If you have trouble with the Hg(II)-catalysed step, there is help in Chapter 36.

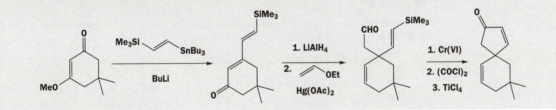

Purpose of the problem

Revision of Chapter 36 (sigmatropic rearrangements), some tin-lithium exchange chemistry, and an exercise in vinyl silane mechanisms.

Suggested solution

The drawing of the first two molecules gives you the clue that, after stereospecific tin-lithium exchange (pp. 1306–8), the vinyl-lithium attacks the carbonyl group direct. The ketone in the first product is formed by hydrolysis of the enol ether (pp. 540–1) in the work-up.

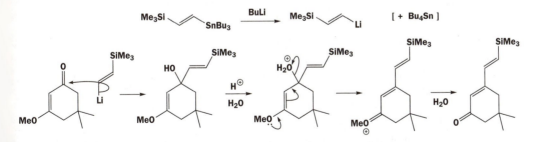

The next step is a Claisen rearrangement, a [3,3]-sigmatropic rearrangement of a vinyl allyl ether (pp. 944–6). The first two steps are just to set up that vital intermediate, using Hg(II) as a soft Lewis acid, ideal for carbon.

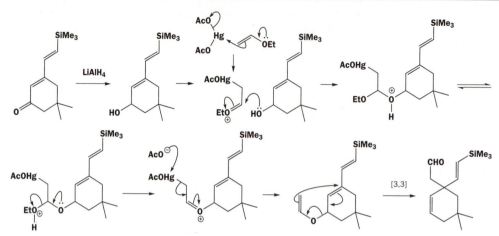

■ This chemistry formed the basis of a general approach to spirocyclic compounds by S. D. Burke *et al.*, *J. Org. Chem.*, 1981, **46**, 2400.

Now the aldehyde is oxidized to an acid and the acid chloride does an intramolecular, Lewis-acid-catalysed, aliphatic Friedel–Crafts reaction on the vinyl silane. Notice attack at the *ipso* carbon to give a stable β-silyl cation intermediate and that this vinyl silane has to react with inversion as a *trans*-alkene is impossible in the ring.

Problem 8

Give mechanisms for these reactions explaining (a) the regio- and stereoselectivity of the hydroboration and (b) why such an odd method was used to close the lactone ring.

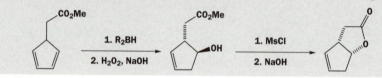

Purpose of the problem

Another check on your understanding of the important regio- and stereoselectivity in hydroboration.

Suggested solution

■ This was also important in the Diels–Alder reaction, pp. 919–21.

The most important interaction in hydroboration is between the HOMO of the alkene (here a diene) and the empty orbital (LUMO) on boron (p. 1279). Here the ends of the diene have the larger coefficients. The reaction occurs on the opposite face of the diene to the one occupied by the substituent, especially as it must occur next to that substituent. In the oxidation, the less substituted group migrates with retention.

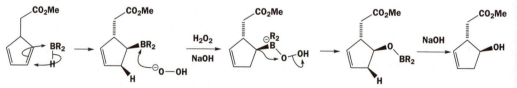

■ This chemistry was used in an asymmetric synthesis of prostaglandins at Hoffman-La Roche (J. J. Partridge *et al.*, *J. Am. Chem. Soc.*, 1973, **95**, 7171).

The closure of the lactone must be done with inversion of configuration at the alcohol centre. We should normally use a Mitsunobu reaction for this but mesylation is a good alternative. Mesylation occurs via the sulfene with retention and then displacement of this excellent leaving group requires inversion and the *cis* lactone is formed.

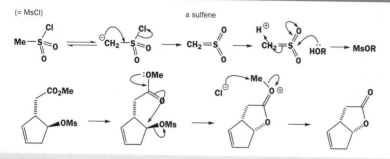

Problem 9

Give mechanisms for these reactions, explaining the role of silicon. Why is this type of lactone difficult to make by ordinary acid- or base-catalysed reactions?

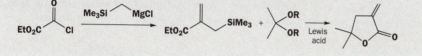

Purpose of the problem

Basic silicon chemistry: the Peterson reaction and allyl silanes as nucleophiles.

Suggested solution

The acylation of the Grignard reagent is followed by an second attack on the ketone as expected but the tertiary alcohol is a Peterson intermediate and eliminates to give the alkene (pp. 812–14 and 1296–7).

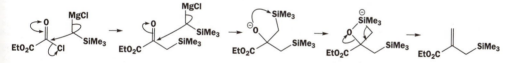

Now a Lewis-acid-catalysed reaction of the allyl silane via a β-silyl cation, gives the lactone. The double bond in these 'exo-methylene' lactones easily moves into the ring in acid or base so the mild conditions for allyl silane reactions are ideal.

■ This silicon chemistry was devised by A. Haider, *Synthesis*, 1985, 271.

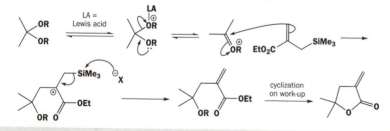

Problem 10

Revision of Chapters 38 and 46. How would you prepare the starting material for these reactions? Give mechanisms for the various steps. Why are these sequences useful?

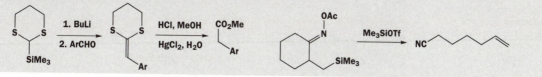

Purpose of the problem

More basic silicon chemistry: the Peterson reaction again with sulfur around and β-silyl cation doing something new. Revision of the Beckmann fragmentation.

Suggested solution

The first reaction is a straightforward Peterson reaction (pp. 812–14 and 1296–7), without even worry over stereochemistry.

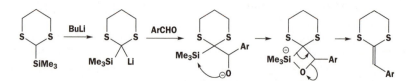

■ This chemistry played a vital part in a synthesis of steganone by E. R. Larson and R. A. Raphael, *Tetrahedron Lett.*, 1979, 5041.

The second step is a Hg(II)-catalysed hydrolysis of a dithiane. There are many variations on this mechanism – the only important thing is to show the Hg(II) making off with the sulfur.

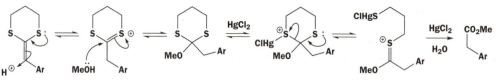

■ The 'silicon-directed Beckmann fragmentation' was discovered by K. Itoh and group, *Tetrahedron Lett.*, 1983, **24**, 4021.

The second reaction is a fragmentation and it shows silicon in a useful role of directing a reaction by stabilizing a β-cation. Without the Me₃Si group, the oxime would do the Beckmann rearrangement. Only with tertiary centres does a fragmentation occur (p. 1003). The Me₃Si group makes the cation as stable as a tertiary cation (at least) and fragmentation occurs. Notice the use of Me₃Si–OTf as a Lewis acid to set the fragmentation going.

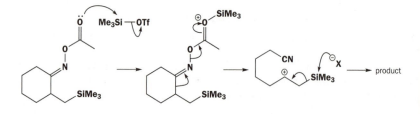

Problem 11

How would you carry out the first step in this sequence? Give a mechanism for the second step and suggest an explanation for the stereochemistry. You may find that a Newman projection (Chapters 32 and 33) helps.

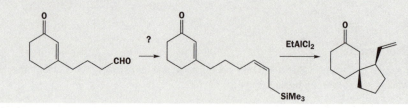

Purpose of the problem

How to make an allyl silane and one important reaction.

Suggested solution

The best route to the allyl silane is the Wittig reaction suggested in the chapter (pp. 1296–7). The ylid is not stabilized by extra conjugation so the Z-alkene is favoured.

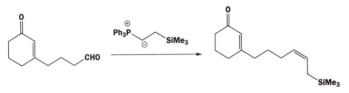

The reaction with EtAlCl₂ is a Lewis-acid-catalysed conjugate addition of the allyl silane on to the enone. Conjugate addition is preferred because the nucleophile (allyl silane) is tethered to the electrophile (enone) and the five-membered ring is easier to form than the alternative seven-membered ring.

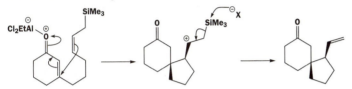

■ D. Schinzer, *Angew. Chem., Int. Ed. Engl.*, 1984, **23**, 308.

The stereochemistry comes from the way the molecule prefers to fold. The Newman projection below should be clear. The hydrogen atom on the allyl silane tucks under the six-membered ring and the double bond of the allyl silane projects out into space to give the stereochemistry found in the product. The ratio of this and the other diastereoisomer varies from 2:1 to 7.5:1 depending on the conditions so there isn't all that much in it.

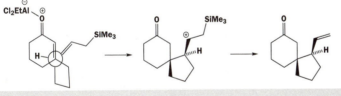

Problem 12

Revision of Chapter 36. Give a mechanism for this reaction and explain why it goes in this direction.

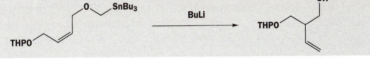

Purpose of the problem

Another reminder of tin-lithium exchange and revision of sigmatropic rearrangements from Chapter 36.

Suggested solution

The reaction must start with the usual tin-lithium exchange. You might have continued by numbering the carbons to see which atom has gone where. Here is the answer though you might reasonably have drawn the intermediates as anions rather than as covalent lithium compounds.

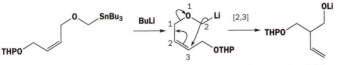

■ This rearrangement provided a starting material for the synthesis of the antibiotic talaromycin B (A. P. Kozikowski and J. G. Scripko, *J. Am. Chem. Soc.*, 1984, **106**, 353).

This is a [2,3]-sigmatropic rearrangement (pp. 951–2) and goes in this direction because an OLi derivative is more stable than a C–Li derivative.

Problem 13

The Nazarov cyclization (Chapter 36) normally gives a cyclopentenone with the alkene in the more substituted position. That can be altered by the following sequence. Give a mechanism for the reaction and explain why the silicon makes all the difference.

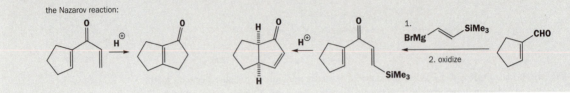

the Nazarov reaction:

Purpose of the problem

To show how silicon can alter the regioselectivity of another reaction (compare Problems 9 and 10).

Suggested solution

■ This 'silicon-directed Nazarov reaction' was discovered by T. K. Jones and S. E. Denmark, *Helv. Chim. Acta*, 1983, **66**, 2377.

The Grignard reagent adds direct to the enal as we should expect and oxidation gives the dienone. Now for the Nazarov cyclization itself (p. 962). We have drawn the cyclization in a conrotatory fashion to give H and SiMe₃ *cis*, not that this matters, and we have put the positive charge in the intermediate β to the SiMe₃ group, though that doesn't matter either as it is delocalized. The key point is that the SiMe₃ group is selectively removed in preference to a proton so the double bond in the product is where the SiMe₃ group was. The stereochemistry of the ring junction comes from the protonation of the enol and is under thermodynamic control.

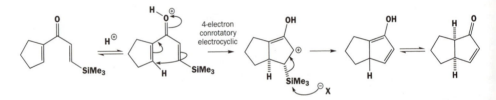

Problem 14

This is rather a long problem but it gives you the chance to see an advanced piece of chemistry involving several elements – P, Si, Sn, Mg, B, Ni, Cr, Os, and Li – and it revises material from Chapters 23, 33, and 45 at least. It starts with the synthesis of this phosphorus compound: what is the mechanism and selectivity?

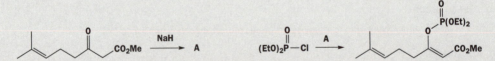

Next, reaction with a silicon-substituted Grignard reagent in the presence of Ni(II) gives an allyl silane. What kind of reaction is this, what was the role of phosphorus, and why was a metal other than sodium added? (You know nothing specific about Ni as yet but you should see the comparison with another metal. Consult Chapter 23 if you need help.)

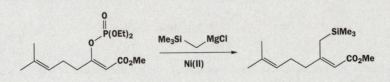

Asymmetric dihydroxylation (Chapter 45) is straightforward though you might like to comment on the chemoselectivity. The diol is converted into the epoxide and you should explain the regio- and chemoselectivity of this step. The next step is perhaps the most interesting: what is the mechanism of the cyclization, what is the role of silicon, and how is the stereochemistry controlled?

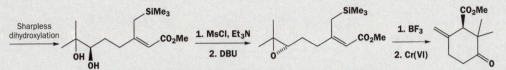

Reaction of this ketone with a stannyl-lithium reagent gives one diastereoisomer of a bridged lactone. Again, give a mechanism for this step and explain the stereochemistry. Make a good conformational drawing of the lactone.

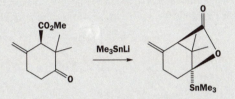

Treatment of the tin compound with MeLi and a complex aldehyde represented as RCHO gave an adduct that was used in the synthesis of some compounds related to Taxol™. What is the mechanism of the reaction, and why is tin necessary?

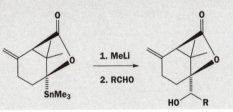

Purpose of the problem

To show how silicon and tin are used in serious chemical operations.

Suggested solution

This is too long a problem to argue our way through. We simply give reasonable solutions. The first part is the reaction of the hard oxygen atom of an enolate with a hard phosphorus electrophile.

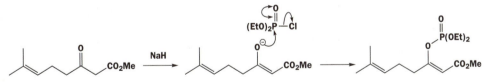

Now the organometallic compound does a conjugate substitution on the conjugated ester with the phosphate acting as a leaving group. The nickel evidently acts in a copper type of role promoting conjugate addition rather than direct attack at C=O or P=O.

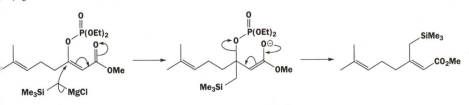

The dihydroxylation occurs on the more nucleophilic of the two alkenes. Mesylation is sele for the secondary alcohol and DBU removes the remaining proton and closes the epoxide inversion at the secondary centre.

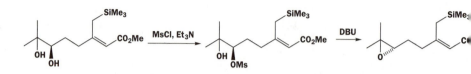

The Lewis acid (BF_3) opens the epoxide to give the tertiary cation, which cyclizes on to the silane to give a β-silyl cation that loses the Me_3Si group in the usual way. It is clear from the pro what the stereochemistry of this intermediate must be and it looks as though cyclization give more stable di-equatorial product (margin). Probably the cation folds up in this conformation. alcohol controls the new centre and is then removed by oxidation to a ketone.

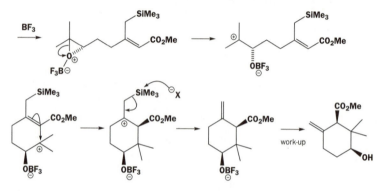

Addition of the tinlithium reagent to the ketone must occur from underneath (axial attac formation of the lactone is to succeed and these steps are best seen with conformational drawing is necessary to flip the six-membered ring to put the two reactive groups axial before cyclization occur.

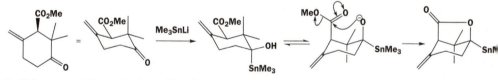

■ R. J. Armstrong and L. Weiler, *Can. J. Chem.*, 1983, **61**, 2530.

The mechanism of the final step is simple enough except that the organolithium species r remain tetrahedral throughout. It would not, of course, be possible to make such a specie deprotonation but it is possible with tin chemistry. The stereochemistry of the reaction has t retention as inversion is impossible.

Suggested solutions for Chapter 48

48

Problem 1

Suggest mechanisms for these reactions, explaining the role of palladium in the first step.

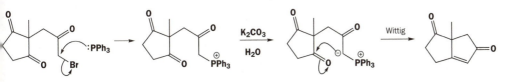

Purpose of the problem

Revision of enol ethers and their bromination (Chapter 21), the Wittig reaction (Chapters 14 and 31), and, of course, first steps in palladium chemistry.

Suggested solution

The first step is a reaction of an enol with an allylic acetate catalysed by palladium (0) via an η^3 allyl cation (pp. 1330–2). There is no regiochemistry to worry about as the diketone and the allylic acetate are both symmetrical.

■ You might have drawn the η^3 allyl complex in various satisfactory ways, some mentioned on p. 1331.

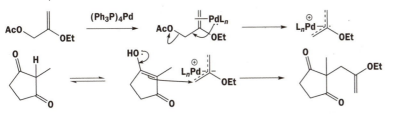

NBS in aqueous solution is a polar brominating reagent, ideal for reaction with an enol ether. The intermediate is hydrolysed to the ketone by the usual acetal style mechanism (pp. 540–2).

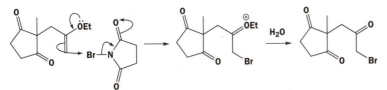

Finally, an intramolecular Wittig reaction. This is a slightly unusual way to do what amounts to an aldol reaction but the 5/5 fused enone system is strained and the Wittig reaction went under very mild conditions (K_2CO_3 in aqueous solution). The stereochemistry of the new double bond is the only one possible and Wittig reactions with stabilized ylids are under thermodynamic control so it will find a way.

■ This process is a general way to make 5/5 fused systems devised by B. M. Trost and D. P. Curran, *J. Am. Chem. Soc.*, 1980, **102**, 5699.

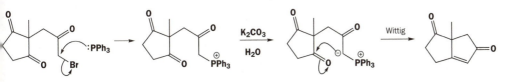

Problem 2

This Heck style reaction does not lead to regeneration of the alkene. Why not? What is the purpose of the formic acid (HCO_2H) in the reaction mixture?

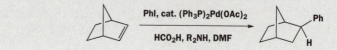

Purpose of the problem

Making sure you understand the basic Heck reaction.

Suggested solution

The reaction must start with the oxidative insertion of Pd(0) into the Ph–I bond. The reagent added is Pd(II) so one of the reduction methods listed in the box on p. 1322 must provide enough Pd(0) to start the reaction going. The oxidative insertion gives PhPdI and this does the Heck reaction on the alkene. Addition occurs from the less hindered top (*exo-*) face and the phenyl group is transferred to that same face.

■ A heterocyclic version of this reaction was part of a synthesis of the natural analgesic epibatidine by S. C. Clayton and A. C. Regan, *Tetrahedron Lett.*, 1993, **34**, 7493.

Normally now, the alkyl palladium(II) species would lose palladium by a β-elimination. This is impossible in this example as there is no hydrogen atom *cis* to the PdI group. Instead, an external reducing agent is needed and that is the role of the formate anion: it provides a hydride equivalent by 'transfer hydrogenation' when it loses CO_2.

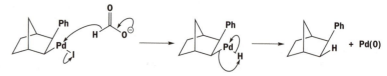

Problem 3

Cyclization of this unsaturated amine with catalytic Pd(II) under an atmosphere of oxygen gives a cyclic unsaturated amine in 95% yield. How does the reaction work? Why is the atmosphere of oxygen necessary? Explain the stereochemistry and regiochemistry of the reaction. How would you remove the CO_2Bn group from the product?

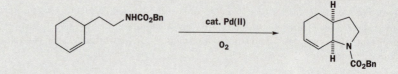

Purpose of the problem

Introducing you to 'aminopalladation' – like oxypalladation – nucleophilic attack on a palladium π-complex.

Suggested solution

The π-complex of the alkene and Pd(II) allows nucleophilic attack by the amide on the nearer end and in a *cis* fashion because the nucleophile is tethered to the alkene by only two carbon atoms. Nucleophilic attack and elimination of palladium(0) occur in the usual way. The removal of the CO$_2$Bn group would normally be by hydrogenolysis but in this case ester hydrolysis by, say, HBr treatment would be preferred to avoid reduction of the alkene. The free acid decarboxylates spontaneously.

■ This general synthesis of heterocycles was introduced by J.-E. Bäckvall and group, *Tetrahedron Lett.*, 1995, **36**, 7749.

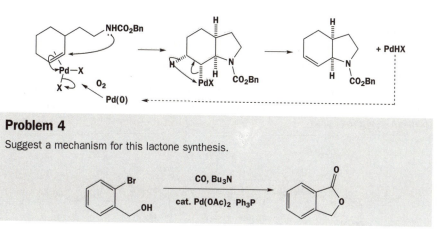

Problem 4

Suggest a mechanism for this lactone synthesis.

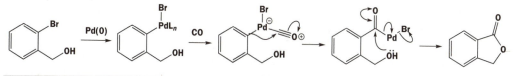

Purpose of the problem

Introducing you to carbonyl insertion into a palladium (II) σ-complex.

Suggested solution

Oxidative insertion into the arylbromide, carbonylation, and nucleophilic attack on the carbonyl group with elimination of Pd(0) completes the catalytic cycle. No doubt the palladium has a number (1–2) of phosphines complexed to it during the reaction and these keep the Pd(0) in solution between cycles.

■ M. Mori *et al.*, *Heterocycles*, 1979, **12**, 921.

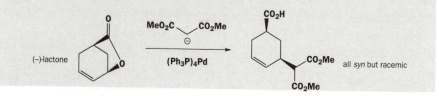

Problem 5

Explain why enantiomerically pure lactone gives all *syn* but racemic product in this palladium-catalysed reaction.

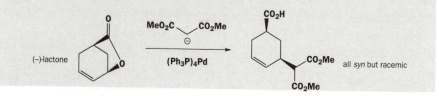

(−)-lactone   MeO$_2$C⎯CO$_2$Me (⊖)   (Ph$_3$P)$_4$Pd   CO$_2$H ... CO$_2$Me ... CO$_2$Me   all *syn* but racemic

Purpose of the problem

Helping you to understand the details of the palladium-catalysed allylation.

Suggested solution

Following the usual mechanism (pp. 1330–4), the palladium complexes to the face of the alkene opposite the bridge and the ester group leaves to give an allyl cation complex. This is attacked by the malonate anion from the opposite face to the palladium. So the overall result is retention, the *cis* starting material giving the *cis* product.

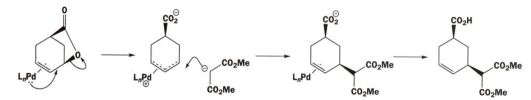

■ This investigation helped to establish the mechanism of these reactions: B. M. Trost and N. R. Schmuff, *Tetrahedron Lett.*, 1981, **22**, 2999.

The racemization comes from the structure of the allyl cation complex. It is, in fact, symmetrical with a plane of symmetry and attack occurs equally at the two ends of the allyl system giving the two enantiomers of the product.

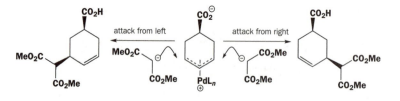

Problem 6

Revision of Chapter 47. The synthesis of a bridged tricyclic amine shown below starts with an enantiomerically pure allyl silane. Give mechanisms for the reactions, explaining how the stereochemistry is controlled in each step.

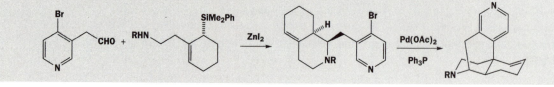

Purpose of the problem

Revision of allyl silane chemistry (pp. 1296–300) and the stereochemistry and mechanism of an intramolecular Heck reaction.

Suggested solution

The first reaction is a Lewis-acid-catalysed reaction between the amine and the aldehyde and addition of the allyl silane to the resulting iminium salt. Addition occurs at the other end of the allylic system and on the opposite face to the silicon. The most obvious way to put the reagents together looks excellent until we realize, from the Newman projection, that the two Hs in the product would then be *anti* whereas they are actually *syn*.

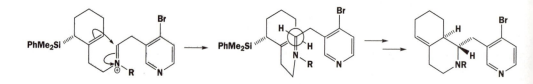

We can reverse this stereoselectivity by using the other isomer of the iminium salt. Evidently, 'R' is larger than CH$_2$. Addition now gives a β-silyl cation that loses silicon to give the *syn* isomer of the product.

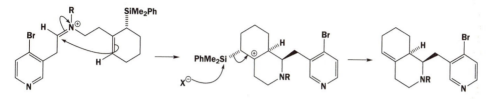

Now the Heck reaction. Palladium (0) inserts oxidatively into the pyridinebromine bond. Addition to only one end of the alkene is possible because of the tether made clear by a conformational drawing roughly in the shape of the product. There is only one *cis* H to the Pd so that must be lost.

■ This requires a preliminary reduction of Pd(OAc)$_2$ by one of the methods on p. 1322.
■ This chemistry led to a general synthesis of potential painkillers by L. E. Overman and group, *Tetrahedron Lett.*, 1997, **38**, 8439.

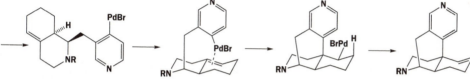

Problem 7

Revision of Chapter 44. Explain the reactions in this sequence commenting on the regioselectivity of the organometallic steps.

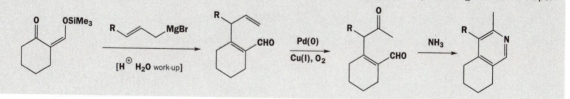

Purpose of the problem

Revision of allyl Grignard chemistry, the synthesis of pyridines (pp. 1191–5), and the mechanism of the Wacker oxidation.

Suggested solution

The allylic Grignard reagent does direct addition at its 'wrong' (or more remote) end, as you should expect. Hydrolysis of the silyl enol ether reveals an aldehyde.

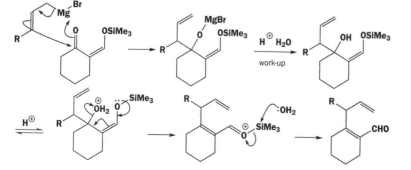

Now, the Wacker oxidation (p. 1337), by whatever detailed mechanism you prefer, must involve the addition of water to a Pd(II) π-complex of the alkene and β-elimination of palladium to give Pd (0), which is recycled by oxidation with oxygen mediated by copper.

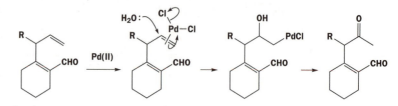

■ This general aromatic annelation sequence came from Hawaii (M. A. Tius, *Tetrahedron Lett.*, 1982, **23**, 2819).

Finally, the pyridine synthesis is simply a double enamine/imine formation between ammonia and the two carbonyl groups. Probably the aldehyde reacts first.

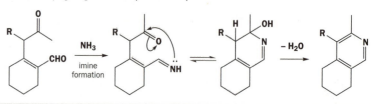

Problem 8

Give a mechanism for this carbonylation reaction. Comment on the stereochemistry and explain why the yield is higher if the reaction is carried out under a carbon monoxide atmosphere.

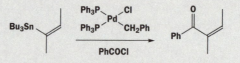

Hence explain this synthesis of part of the antifungal compound pyrenophorin.

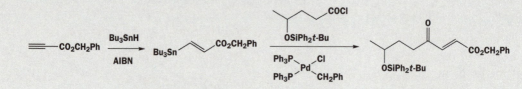

Purpose of the problem

Revision of Chapter 47 with practice at Stille coupling and carbonylation.

Suggested solution

The tin-palladium exchange (transmetallation) occurs (pp. 1326–7) with retention of configuration at the double bond. The exchange of the benzyl group for the benzoyl group is necessary just to get the reaction started.

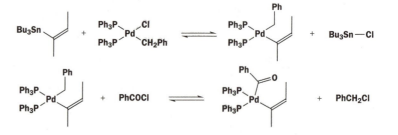

Now the coupling can take place on the palladium atom producing the product and Pd(0), which can insert oxidatively in the C–Cl bond. Transmetallation now sets up a sustainable cycle of reactions. It is better to have an atmosphere of carbon monoxide because the acyl palladium complex can give off CO and leave a PdPh σ-complex. The atmosphere of CO reverses this reaction.

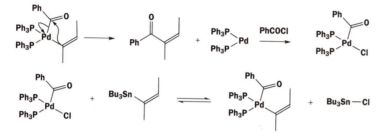

The second sequence starts with a radical hydrostannylation (pp. 1305–6) giving the *E*-vinyl stannane preferentially if a slight excess of Bu₃SnH is used.

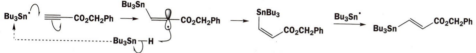

Now the coupling reaction with the acid chloride takes place as before though this time we have an aliphatic carbonyl complex. There is no problem with β-elimination as that would give a ketene. Again, the stereochemistry of the vinyl stannane is retained in the product.

■ This chemistry was developed by J. W. Labadie and J. K. Stille, *Tetrahedron Lett.*, 1983, **24**, 4283.

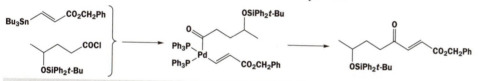

Problem 9

Explain the mechanism and stereochemistry of these reactions. The first is revision and the second is rather easy!

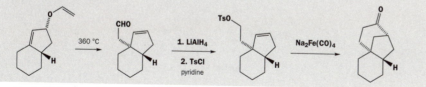

Purpose of the problem

Revision of sigmatropic rearrangements from Chapter 36 and just one example of Collman's ferrate reagent.

Suggested solution

The first step is a [3,3]-sigmatropic Claisen rearrangement with suprafacial transfer of the vinyl group across the allyl part of the molecule. Then there is a simple reduction and tosylation.

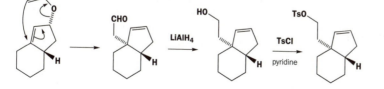

■ This is part of a synthesis of the antiviral compound aphidicolin by J. E. McMurry *et al.*, *J. Am. Chem. Soc.*, 1979, **101**, 1330 using chemistry developed by J. P. Collman, *Acc. Chem. Res.*, 1975, **8**, 342.

Now for the organometallic bit. As described in the chapter (p. 1318) the ferrate anion is very nucleophilic and displaces the tosylate from the primary carbon. Carbonyl insertion gives an iron acyl complex and the neutral iron atom forms a π-complex with the alkene. The carbonyl is transferred to the far end of the alkene as six-membered rings are preferred to five-membered ones and protonation of the C–Fe bond completes the reaction.

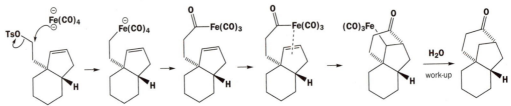

Problem 10

The synthesis of an antifungal drug was completed by this palladium-catalysed reaction. Give a mechanism and explain the regio- and stereoselectivity.

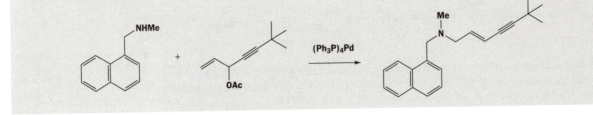

Purpose of the problem

A nice simple example for once! The idea is to give you confidence.

Suggested solution

The palladium forms the usual allyl cation complex and the nitrogen nucleophile attacks the less hindered end also retaining the conjugation. Attack at the triple bond would give an allene.

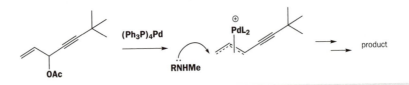

Problem 11

Some revision content. Work out the structures of the compounds in this sequence and suggest mechanisms for the reactions, explaining any selectivity.

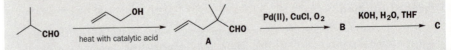

B has IR: 1730, 1710 cm^{-1}; δ_H (p.p.m.) 9.4 (1H, s), 2.6 (2H, s), 2.0 (3H, s), and 1.0 (6H, s).
C has IR: 1710 cm^{-1}; δ_H (p.p.m.) 7.3 (1H, d, J 5.5 Hz), 6.8 (1H, d, J 5.5 Hz), 2.1 (2H, s), and 1.15 (6H, s).

Purpose of the problem

A nice simple example again: a Wacker oxidation and an intramolecular aldol reaction (Chapter 27).

Suggested solution

B is clearly an aldehyde and a ketone with nothing but singlets in the NMR. On the other hand, C has a *cis*-disubstituted alkene with a small *J* value and is a cyclopentenone. Here are their structures.

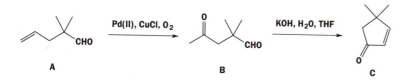

Problem 12

Revision of Chapter 36. What would be the starting materials for the synthesis of these cyclopentenones by the Nazarov reaction and by the Pauson–Khand reaction? Which do you prefer in each case?

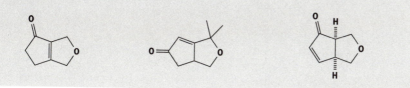

Purpose of the problem

Looking at organometallic chemistry from the other side: how is it used?

Suggested solution

The simplest disconnections for the two reactions are remarkably similar. Both make one bond next to the ketone and the bond on the far side of the ring. In addition, the Pauson–Khand reaction requires disconnection on the other side of the ketone as CO is one of the reagents. The position of the final alkene in the Nazarov product is not determined by the position of the double bond in the starting materials (pp. 962–3).

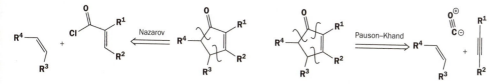

　　Our preferred solution for the first compound is a Nazarov reaction as the starting material is symmetrical and it is not necessary to use a vinyl silane, though this would work well too. The starting alkene is easily made from *cis*-butenediol. The double bond in the product will much prefer to go inside the ring where it is more highly substituted.

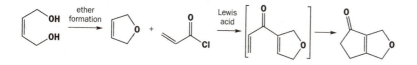

■ The best yield (75%) comes on heating with silica as discovered by a joint Russian/American programme (W. A. Smit, R. Caple, and co-workers, *Tetrahedron Lett.*, 1986, **27**, 1245.

We prefer Pauson–Khand reactions for the other two compounds as they have both been made that way. The first needs a simple allyl propargyl ether readily formed from the adduct of acetylene and acetone and allyl bromide. The cobalt carbonyl complex gives a good yield of the cyclopentenone.

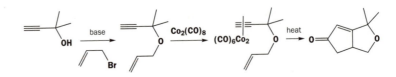

■ The reaction goes very well under an atmosphere of acetylene and carbon monoxide giving 85% yield of the bicyclic enone used in a synthesis of the antibiotic cyclosarkomycin by D. C. Billington, *Tetrahedron Lett.*, 1983, **24**, 2905.

The third compound was made by a Pauson–Khand reaction using the same starting material as the first. The only difference between these two target molecules is the position of the double bond. In the Nazarov reaction, it goes into the thermodynamically most favourable position but in the Pauson–Khand reaction it goes where the alkyne was. So we simply react the cyclic ether with acetylene cobalt carbonyl complex. The *cis* stereochemistry is inevitable.

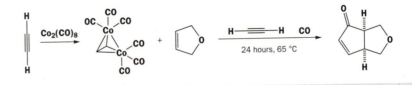

Problem 13

A variation on the Vollhardt co-trimerization allows the synthesis of substituted pyridines. Draw the structures of the intermediates in this sequence. In the presence of an excess of the cyanoacetate a second product is formed. Account for this too.

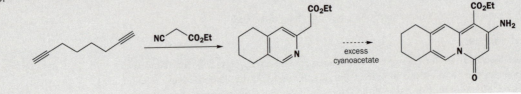

Purpose of the problem

Practice at unravelling the Vollhardt co-trimerization.

Suggested solution

Starting the Vollhardt reaction in the usual way (p. 1340) we get a cobalt heterocycle that can do a Diels-Alder reaction on to the nitrile of the ester. Loss of cobalt (the cyclopentadiene is omitted from this step for clarity) gives a pyridine.

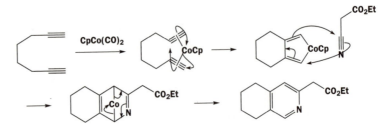

Now a second molecule of cyanoacetate condenses with the pyridine. The nitrogen of the pyridine attacks the ester and the enol(ate) of the pyridine ester attacks the new nitrile. Take your choice on the order of events.

■ This work is taken from K. P. C. Vollhardt's review in *Angew. Chem., Int. Ed. Engl.*, 1984, **23**, 539 where he prefers the alternative mechanism via a seven-membered cobalt heterocycle (p. 1340).

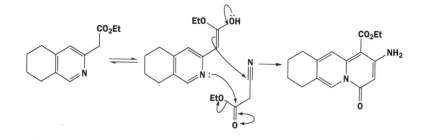

Problem 14

The synthesis of the Bristol–Myers Squibb anti-migraine drug Avitriptan (a 5-HT1D receptor antagonist) involves this palladium-catalysed indole synthesis. Suggest a mechanism and comment on the regioselectivity of the alkyne attachment.

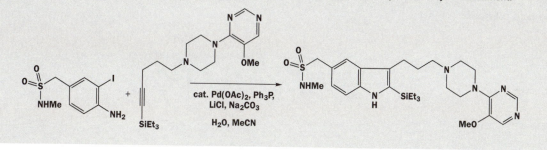

Purpose of the problem

A new reaction for you to try – a palladium-catalysed indole synthesis.

Suggested solution

Though palladium (II) is added to the solution, the aryl iodide tells you that this is an oxidative insertion of Pd(0) produced by one of the methods on p. 1322. The resulting Pd(II) species complexes to the alkyne and the amine can now attack it. This gives a heterocycle with Pd(II) in the ring. Coupling of the two organic fragments extrudes Pd(0) to start a new cycle and gives the indole. The nitrogen atoms attacks the more hindered end of the alkyne so that the palladium atom can occupy the less hindered end.

■ This is the Larock indole synthesis (R. C. Larock and E. K. Yum, *J. Am. Chem. Soc.*, 1991, **113**, 6689) and is used in production of Avitriptan (P. R. Brodfuehrer *et al.*, *J. Org. Chem.*, 1997, **62**, 9192).

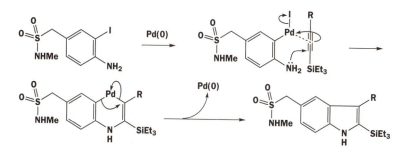

Problem 15

A synthesis of the natural product γ-lycorane starts with a palladium-catalysed reaction. What sort of a reaction is this, and how does it work?

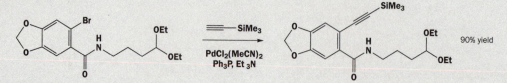

The next two steps are a bit of revision: draw mechanisms for them and comment on the survival of the Me₃Si group.

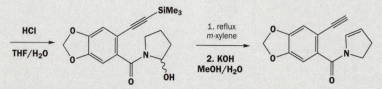

Now the key step – and you should recognize this easily. What is happening here? Though the product is a mixture of isomers, this does not matter. Why not?

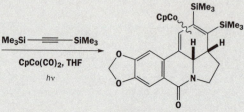

65% yield, 3:2 Co up : Co down

Finally, this mixture must be converted into γ-lycorane: suggest how this might be done.

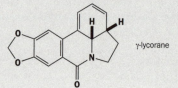

γ-lycorane

Purpose of the problem

Looking at organometallic chemistry from the other side: how is it used?

Suggested solution

The first reaction is a Sonogashira coupling between an aryl bromide and an alkyne. There is no need to have anything but hydrogen on the alkyne for this reaction in contrast to the Stille reaction (R₃Sn) or Suzuki reaction (boronic acid).

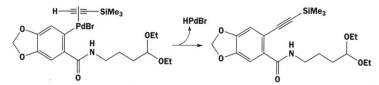

Then the acetal is hydrolysed and the resulting aldehyde immediately cyclizes on to the nitrogen atom of the amide. This is a very unusual reaction but the formation of a five-membered ring is very favourable. The water might have hydrolysed the C–Si bond but, as we see in the next reaction, that requires stronger nucleophiles.

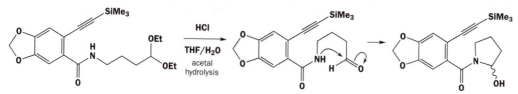

Heating dehydrates the hemiacetal-like compound: the electrons on nitrogen push the OH group out. Treatment with KOH in water does then cleave the C–Si bond by nucleophilic attack at silicon. KOH would not be strong enough to remove a proton from an alkyne but attack on silicon is more favourable because of the strong O–Si bond.

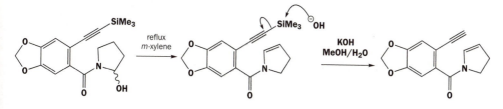

Now we have a rather unusual Vollhardt co-trimerization. We can start by making the cobalt heterocycle but we cannot then do the Diels–Alder mechanism. Instead the cobalt must combine the reagents to form a seven-membered ring from which it withdraws as it couples its two ligands. The hydrogens in the product are *cis* because they were *cis* in the enamine but there is no stereoselectivity in which side of the new ring is occupied by the cobalt. Fortunately, this doesn't matter as both diastereoisomers give the same product after the cobalt is removed.

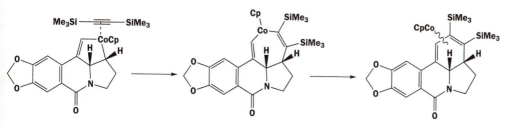

To complete the synthesis we must remove the cobalt by oxidation, Fe(III) is a popular choice. Then the two Me₃Si groups must be removed. You might have suggested acid (CF₃CO₂H does the job) or fluoride (Bu₄NF, 'TBAF', is better).

■ This lycorane synthesis comes from the laboratories of Vollhardt himself at Berkeley (D. B. Grotjahn and K. P. C. Vollhardt, *Synthesis*, 1993, 579).

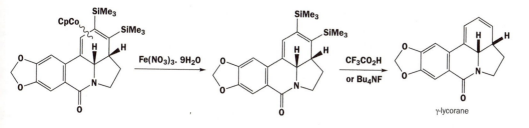

Suggested solutions for Chapter 49 **49**

Problem 1

Do you consider that thymine and caffeine are aromatic compounds? Explain.

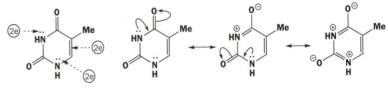

thymine

caffeine

Purpose of the problem

Revision of aromaticity (Chapters 7 and 22) and exploration of the structure of nucleic acid bases.

Suggested solution

Thymine, a pyrimidine, has an alkene and lone pair electrons on two nitrogen atoms, making six in all for an aromatic structure. You might have shown this by drawing delocalized structures.

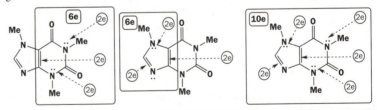

Caffeine, a purine, is slightly more complicated as it has two rings. You might have said that each is aromatic with six electrons, counting all the nitrogen lone pairs except that on the pyridine-like nitrogen in the five-membered ring, or you might have said that the whole system is aromatic with ten electrons. You might also have drawn delocalized structures (not shown here) like those drawn for thymine.

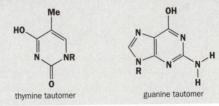

Problem 2

It is important that we draw certain of the purine and pyrimidine bases in their preferred tautomeric forms. The correct pairings are given early in the chapter. What alternative pairings would be possible with these (minor) tautomers of thymine and guanine? Suggest reasons (referring to Chapter 43 if necessary) why the major tautomers are preferred.

thymine tautomer

guanine tautomer

urpose of the problem

hecking on your understandzing of the binding of the two strands in DNA.

uggested solution

he true pairs have hydrogen bonds between NHs and carbonyl groups or pyridine-like nitrogen
_oms. In each case we must have a purine-pyrimidine pair. Two purines would be pushed too close
•gether inside the DNA double helix to form hydrogen bonds and two pyrimidines would be too
_r apart.

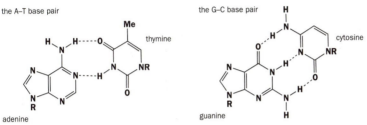

There are no OHs in the major tautomers but there are in the minor tautomers. So a group, say
_=O, that accepts a hydrogen bond in the main tautomer becomes a hydrogen bond donor in the
ninor tautomer. You might have chosen a number of different possibilities; here are just two. We
_ave chosen these because they show the GT pairs not observed in natural DNA.

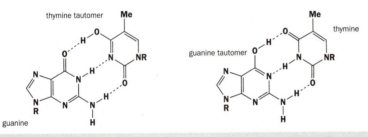

Problem 3

Dialkyl phosphates are generally hydrolysed quite slowly at near-neutral pHs but this example
hydrolyses much more rapidly. What is the mechanism and what relevance has it to RNA
chemistry?

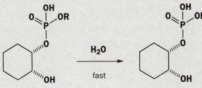

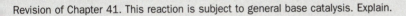

Revision of Chapter 41. This reaction is subject to general base catalysis. Explain.

Purpose of the problem

Exploring the similarities of phosphate ester hydrolysis to ordinary ester hydrolysis.

Suggested solution

The completely uncatalysed reaction involves nucleophilic attack by the OH group on the P=O
bond to form a five-covalent intermediate much like the tetrahedral intermediate in carboxylate
ester hydrolysis. There is a difference. Phosphorus prefers five-membered rings to any other

■ We omit fast proton transfers in this summary mechanism.

size and the cyclic ester is formed rapidly with loss of ROH. Attack of water on the cyclic ester also fast: in the complete reaction the phosphate may have migrated from one OH group to th other.

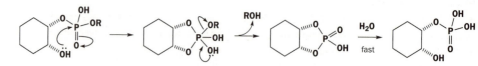

The reaction is subject to general base catalysis because the OH group is a poor nucleophile an removal of its proton during the ring closure accelerates the reaction. We now have a mechanisı very similar to that on p. 1352 for the hydrolysis of RNA. In full this is general base catalysis o nucleophilic catalysis of phosphate ester hydrolysis.

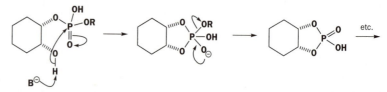

Problem 4

Primary amines are not usually made by displacement reactions on halides with ammonia. Why not? The natural amino acids can be made by this means in quite good yield. Here is an example.

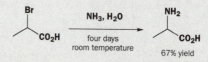

Why does this example work? Comment on the state of the reagents and products under the reaction conditions. What is the product and how does it differ from the natural amino acid?

Purpose of the problem

Exploration of simple amino acid chemistry.

Suggested solution

The problem with displacement of alkyl halides with ammonia is not that the reaction doesn't worł but that it works too well. The primary amine RNH_2 is more nucleophilic than ammonia and react again and so on until a useless mixture of primary, secondary, and tertiary amines is formed.

$$X{-}R \quad :NH_3 \longrightarrow R{-}\overset{\oplus}{N}H_3 \rightleftharpoons R{-}\ddot{N}H_2 \quad R{-}X \longrightarrow R_2NH \longrightarrow R_3N$$

The synthesis of amino acids by this method is rather different. A large excess of ammonia means that there is always more ammonia than primary amine in solution and the amino acid exists mostly as the zwitterion in water so that the primary amino group is no longer nucleophilic.

■ It is actually isolated by evaporation and crystallization from aqueous methanol (B. S. Furniss et al., Vogel's textbook of organic chemistry (5th edn), Longmans, Harlow, 1989, pp. 751–2).

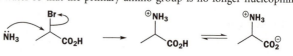

The reagents and product will be mostly as carboxylate anions in this solution but the product will be mostly zwitterion. The product is the amino acid alanine and it differs from natural alanine by being racemic.

Purpose of the problem

Some slightly more complicated amino acid chemistry including stereochemistry and the SH group.

Problem 5

Human hair is a good source of cystine, the disulfide dimer of cysteine. The hair is boiled with aqueous HCl and HCO_2H for a day, the solution concentrated, and a large amount of sodium acetate added. About 5% of the hair by weight crystallizes out as pure cystine $[\alpha]_D$ –216. How does the process work? Why is such a high proportion of hair cystine? Why is no cysteine isolated by this process? What is the stereochemistry of cystine? Make a good drawing of cystine to show its symmetry. How would you convert the cystine to cysteine?

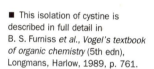

(S)–cysteine

Suggested solution

Prolonged boiling with HCl hydrolyses the peptide linkages (shown as thick bonds) and breaks the hair down into the constituent amino acids. These crystallize as the zwitterions at about neutral pHs and the mixture of NaOAc and HCl provides a buffer solution. Hair is much cross-linked by disulfide bridges and these do not break down under the hydrolysis conditions so intact cystine emerges.

■ This isolation of cystine is described in full detail in B. S. Furniss *et al.*, *Vogel's textbook of organic chemistry* (5th edn), Longmans, Harlow, 1989, p. 761.

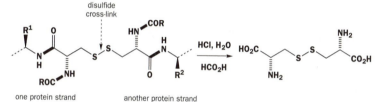

No cysteine is isolated because: (1) most of it is present as cystine in the hair; (2) any cysteine released in the hydrolysis will be oxidized to cystine by air. The stereochemistry of cysteine is preserved in cystine, which has C_2 symmetry and can be best represented by diagrams that show its symmetry – two are suggested. It is not important whether you draw the zwitterion or the neutral structure. Reduction of the S–S bond by, say, $NaBH_4$ is needed to convert cystine into cysteine.

Problem 6

A simple preparation of a dipeptide is given below. Explain the reactions, drawing mechanisms for the interesting steps. Which steps are protection, activation, coupling, and deprotection? Explain the reasons for protection and the nature of the activation. Why is the glycine added to the coupling step as its hydrochloride? What reagent(s) would you use for the final deprotection step?

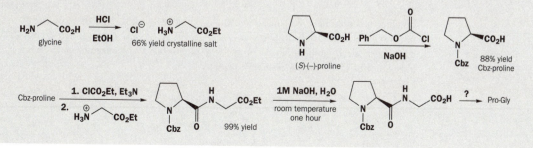

Purpose of the problem

Following the details of a simple dipeptide formation.

Suggested solution

The glycine ester needs to be prepared as its salt to prevent the free amino group forming an amide with the ester. Protection of the amino group of proline as its Cbz derivative (p. 653) is a standard method and relies on the greater reactivity of the amino group (greater than that of the carboxylate anion) towards carbonyl substitution reactions. The chloroformate gives a mixed anhydride that reacts cleanly with glycine ethyl ester in the presence of Et₃N to give the protected dipeptide. Nucleophilic attack occurs at the more electrophilic acyl group rather than at the more delocalized carbonate group.

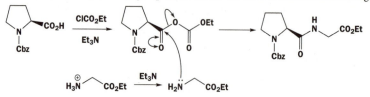

■ Full details are in B. S. Furniss et al., *Vogel's textbook of organic chemistry* (5th edn), Longmans, Harlow, 1989, pp. 761–2.

Finally, the protecting groups are removed: the ester by hydrolysis and the Cbz group by hydrogenation (p. 653). The designation of the steps is straightforward.

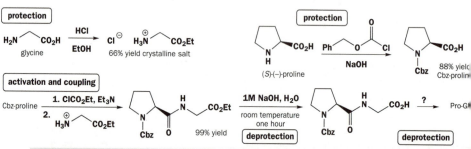

Problem 7

Suggest how glutathione might detoxify these dangerous chemicals in living things. Why are they still toxic in spite of this protection?

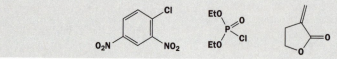

Purpose of the problem

Sulfur chemistry in action: can you make sensible suggestions based on what you know? Revision of Chapters 10 (conjugate addition), 12 (substitution at carbonyl groups), and 23 (electrophilic alkenes).

Suggested solution

The full structure of glutathione (p. 1356) is a tripeptide but the reactive group is a thiol (SH) on a cysteine in the middle. We shall represent glutathione as GSH. The first compound reacts by nucleophilic aromatic substitution (pp. 590–5) aided by the nitro groups. After the reaction, the dangerously electrophilic dinitrochlorobenzene cannot react with enzymes or DNA but is carried away attached to a short water-soluble peptide.

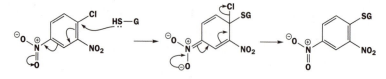

The second compound is a nerve gas style of phosphorus compound that interferes with acetyl choline metabolism and could destroy nerve function. Thiols are excellent nucleophiles for phosphorus and glutathione forms a thiol ester with the phosphate. The thiol is a better nucleophile than the OH group on the enzyme and, in any case, gets there first.

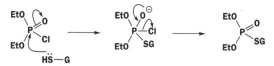

The third compound, an *exo*-methylene lactone, is carcinogenic and might react by conjugate addition with proteins and nucleic acids if glutathione does not react more quickly. Fortunately, thiols are excellent at conjugate addition.

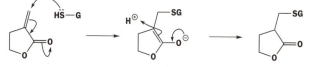

In all these cases, detoxification by glutathione uses the very reaction that makes the toxin dangerous in the first place. If glutathione is not around in sufficient quantity, or if an enzyme is just too quick at reaction with the toxin, glutathione is not good enough to prevent damage.

Problem 8

Alanine can be resolved by the following method, using a pig kidney acylase. Draw a mechanism for the acylation step. Which isomer of alanine acylates faster? In the enzyme-catalysed reaction, which isomer of the amide hydrolyses faster? In the separation, why is the mixture heated in acid solution, and what is filtered off? How does the separation of the free alanine by dissolution in ethanol work? If the acylation is carried out carelessly, particularly if the heating is too long or too strong, a

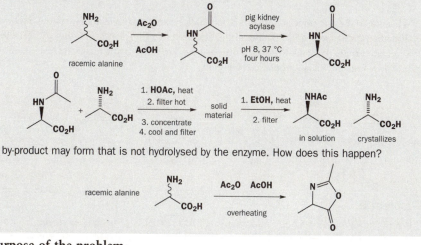

by-product may form that is not hydrolysed by the enzyme. How does this happen?

Purpose of the problem

Rehearsal of some basic amino acid and enzyme chemistry plus revision of stereochemistry (Chapter 16) and asymmetric synthesis (Chapter 45).

Suggested solution

The acylation takes place by the normal mechanism for the formation of amides from anhydrides, that is, by nucleophilic attack on the carbonyl group and loss of the most stable anion (acetate) from the tetrahedral intermediate. The two isomers of alanine are enantiomers. Enantiomers *must* react at identical rates with an achiral reagent like acetic anhydride.

■ Do you feel that was a catch question? It was in a way but it is very important that you cling on to the fact that the chemistry of enantiomers is the same except with chiral reagents.

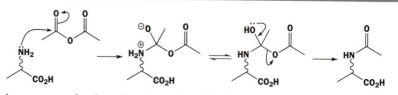

In the enzyme-catalysed reaction, the acylase hydrolyses the amide group of one enantiomer while leaving the other unchanged. This time the two enantiomers do *not* react at the same rate because the reagent (or catalyst if you prefer) is chiral. Not surprisingly, the enzyme attacks the amide of natural alanine and leaves the unnatural enantiomer alone.

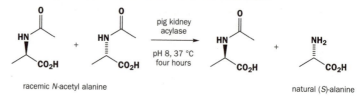

racemic *N*-acetyl alanine   natural (*S*)-alanine

The purification and separation first requires removal of the enzyme. This is, of course, soluble in the aqueous buffer solution at pH 8, but acidification and heating denatures the protein (rather like heating egg white) and destroys its structure. The solid material filtered off is the denatured enzyme. The separation in ethanol works because the very polar amino acid is soluble only in water but the more 'organic' amide is soluble in ethanol. The amide has an extra organic group and only one very polar group (CO$_2$H). In addition, it does not form a zwitterion.

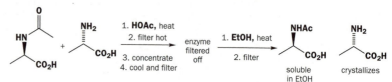

soluble   crystallizes
in EtOH

■ The practical details of all this chemistry are in L. F. Fieser, *Organic experiments* (2nd edn), D. C. Heath, Lexington Mass., 1968, pp. 139–42.

Overheating in the acid solution causes cyclization of the amide oxygen atom on to the carboxylic acid. This reaction happens only because of the formation of a five-membered ring, an 'azlactone'. These compounds are particularly dreaded by amino acid chemists because they racemize easily by enolization and the enol is an aromatic compound.

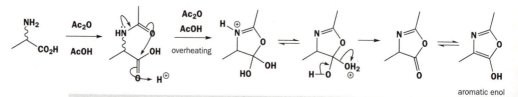

aromatic enol

Problem 9

A patent discloses this method of making the anti-AIDS drug d4T. The first few stages involve differentiating the three hydroxyl groups of 5-methyluridine as shown below. Explain the reactions, especially the stereochemistry at the position of the bromine atom.

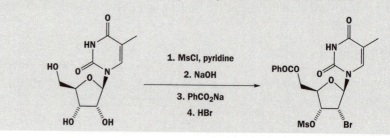

1. MsCl, pyridine
2. NaOH
3. PhCO$_2$Na
4. HBr

Suggest how the synthesis might be completed.

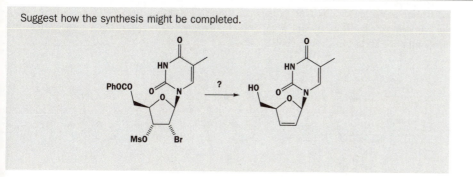

Purpose of the problem

A chance for you to explore nucleoside chemistry, particularly the remarkable control the heterocyclic base can exert over the stereochemistry of the sugar.

Suggested solution

There is remarkable regio- and stereochemical control in this sequence. How are the three OH groups converted into three different functional groups with retention of configuration? The first step must be the formation of the trimesylate. Then treatment with base brings the pyrimidine into play and allows replacement of one mesylate by participation through a five-membered ring.

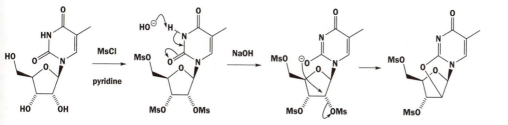

Now the weakly nucleophilic benzoate can replace only the primary mesylate. The participation process is then brought to completion with HBr. Opening the ring gives a bromide with double inversion, that is, retention.

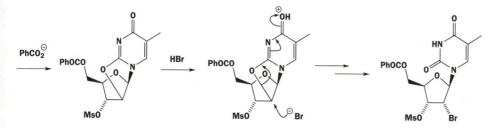

To complete the synthesis of the drug, some sort of elimination is needed, removing both Br and OMs in a *syn* fashion. You might have chosen a number of ways to do this probably involving

■ The synthesis of d4T is disclosed by the Bristol-Myers Squibb company in 1997 by US Patent 5, 672, 698. Some details are in B. Chen *et al.*, *Tetrahedron Lett.*, 1995, **36**, 7957.

metallation of the bromide and loss of mesylate. It turns out that the two-electron donor zinc does this job well. Finally, the protecting group must be removed. There are many ways to do this and butylamine was in fact used.

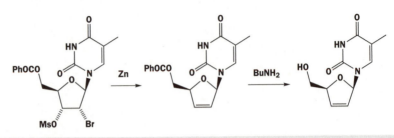

Problem 10

Mannose usually exists as the pyranoside shown below. This is in equilibrium with the furanoside. What is the conformation of the pyranoside and what is the stereochemistry of the furanoside? What other stereochemical change will occur more quickly than this isomerization?

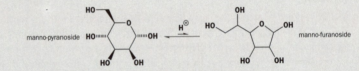

Treatment of mannose with acetone and HCl gives the acetal shown. Explain the selectivity.

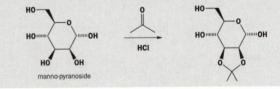

Purpose of the problem

Checking on your understanding of the stereochemistry of sugars and their selective protection.

Suggested solution

It's quite tricky to unravel the sugar and do it up again but that's what we have to do. It's easy to draw the conformation of the pyranoside and to answer the question about the 'other stereochemical change'. This is the hemiacetal equilibration at the anomeric centre.

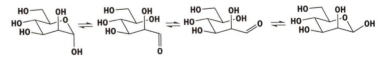

Now for the unravelling. The same intermediate (open-chain aldehyde) is involved but the hemiacetal is formed with a different secondary OH group. There are many ways to work this out. The front OHs stay the same, of course; you just have to twist the back round.

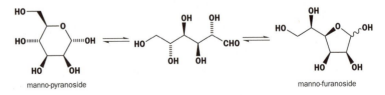

Isopropylidene acetals, formed with acetone, prefer a *cis* ring fusion even with a six-membered sugar (pp. 1361–3). The one formed is the only possible such acetal. Acetal formation is under thermodynamic control.

Purpose of the problem

Problem 11

How are glycosides formed from phenols (in Nature or in the laboratory)? Why is the stereochemistry of the glycoside not related to that of the original sugar?

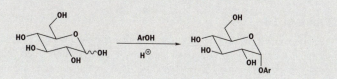

Revision of the mechanism of acetal formation (Chapter 14) and of the anomeric effect (Chapter 42).

Suggested solution

The hemiacetal gives a locally planar oxonium ion that can add the phenol from the top or bottom face. The bottom face is preferred in this instance as axial C–O bonds are more stable in acetals because of the anomeric effect (p. 1130) and acetal formation is under thermodynamic control.

Purpose of the problem

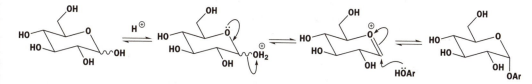

Problem 12

Draw all the keto and enol forms of ascorbic acid (vitamin C). Why is the one shown the most stable?

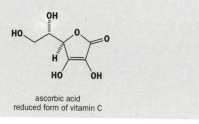

ascorbic acid
reduced form of vitamin C

Revision of enols (Chapter 21) and an assessment of stability by conjugation.

Suggested solution

There can be one or two keto (or ester) forms with a carbonyl group at one end or the other (but not both) of the ene-diol.

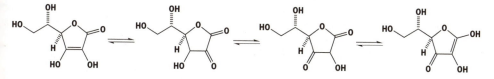

Two forms have greater conjugation than the others and the favoured form preserves the ester rather than a ketone and so has extra conjugation.

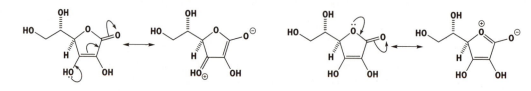

Problem 13

'Caustic soda' (NaOH) was used to clean ovens and clear blocked drains. Many commercial products for these jobs with fancy names still contain NaOH. Even concentrated sodium carbonate (Na_2CO_3) does quite a good job. How do these cleaners work? Why is NaOH so dangerous to humans, particularly if it gets in the eye?

Purpose of the problem

Relating the structures and chemistry of fats to everyday things as well as everyday chemical reactions.

Suggested solution

Fats are triesters of glycerol (triglycerides, pp. 1374–7). If the fatty acid is saturated (R in the diagram = n-alkyl), the fats are usually solids and can block drains. Hydrolysis with NaOH and water, or even, less efficiently but more safely, Na_2CO_3 and water gives liquid glycerol and solid but water-soluble sodium salts of the fatty acids.

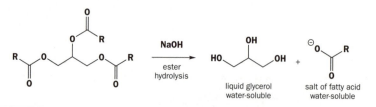

Problem 14

Bacterial cell walls contain the unnatural amino acid D-alanine. If you wanted to prepare a sample of D-Ala, how would you go about it? (*Hint.* There is not enough in bacteria to make that a worthwhile source, but have you done Problem 8 yet?)

Purpose of the problem

Combining biological chemistry with revision of asymmetric synthesis (Chapter 45).

Suggested solution

As the chiral pool strategy is evidently no good, we are left with resolution or asymmetric synthesis. We could try one of the asymmetric catalytic reductions in Chapter 45, though the best of them works only for aromatic amino acids (p. 1236), or we could resolve. The method of Problem 8 – a kinetic resolution with an enzyme – certainly works and the unwanted L-amino acid can be racemized to get a greater yield. Resolution by chiral HPLC is effective with various commercially available columns. The answers to Problems 1 and 7 in Chapter 45 offer other suggestions. The synthesis of unnatural D-amino acids is now big business and inventing a good new method might pay dividends.

Suggested solutions for Chapter 50

50

Problem 1

On standing in alkali in the laboratory, prephenic acid rearranges to 4-hydroxyphenyl-lactic acid with specific incorporation of deuterium label as shown. Suggest a mechanism, being careful to draw realistic conformations.

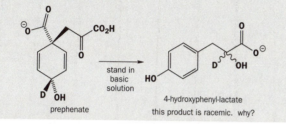

prephenate

4-hydroxyphenyl-lactate
this product is racemic. why?

Purpose of the problem

Revision of the use of isotopic labelling in determining mechanisms from Chapter 41 (pp. 1086–8) and exploration of the different chemistry of prephenic acid in Nature (p. 1403) and in the laboratory.

Suggested solution

Base obviously removes the protons from both carboxylic acids but the reaction really starts with the removal of the proton from the hydroxyl group. This oxyanion promotes a deuterium (hydride if not labelled) shift to the ketone. The groups look a long way away but the diene ring adopts a boat conformation that brings them close together. Decarboxylation to give an aromatic ring follows.

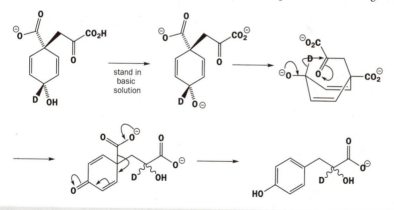

Problem 2

Write a full reaction scheme for the conversion of ammonia and pyruvate to alanine in living things. You will need to refer to the section of the chapter on pyridoxal to be able to give a complete answer.

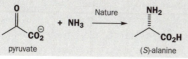

pyruvate

(S)-alanine

Purpose of the problem

Making sure you understand the way pyridoxal works (pp. 1384–8).

Suggested solution

The full structures of pyridoxamine and pyridoxal are on p. 1384. The incorporation of ammonia into α-keto-glutarate and the formation of glutamic acid by NADPH reduction of the imine is on p. 1386. The transamination from glutamic acid to pyridoxamine is on p. 1385. We start from pyridoxamine, whose structure we abbreviate, and pyruvate. Imine formation (full mechanism on pp. 348–50) followed by proton removal and replacement gives a new imine whose hydrolysis (full mechanism on pp. 350–1) gives alanine and pyridoxal. The alanine is a single enantiomer because enzyme-directed protonation occurs on one face of the imine. Pyridoxal is recycled by transamination with glutamic acid.

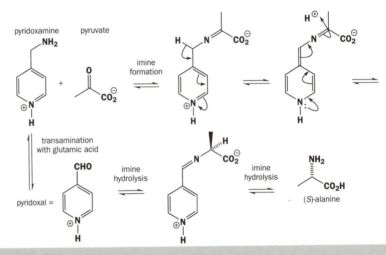

Problem 3

Give a mechanism for this reaction. You will find the Stetter catalyst described in the chapter. How is this sequence biomimetic?

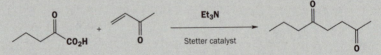

What starting material would be required for formation of the natural product *cis*-jasmone by an intramolecular aldol reaction (Chapter 27)? How would you make this compound using a Stetter reaction?

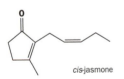

cis-jasmone

Purpose of the problem

Exploring the mechanism of a d^1 reagent inspired by Nature and encounter with biomimetic reactions.

Suggested solution

The full structures of thiamine and of Stetter's catalyst appear in the chapter (pp. 1392–7). We use an abbreviated structure for the mechanism of conjugate addition. The essential features of this catalyst are that it is biomimetic and a d^1 reagent that does conjugate additions. The first stages are the formation of the ylid, direct attack on the most reactive carbonyl group, the ketone of the α-keto-acid, and loss of CO_2 to give the key intermediate.

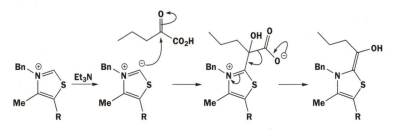

This intermediate is an enol but it is also an enamine and a vinyl sulfide. Nitrogen and sulfur together are more electron-donating than OH by itself so it does the conjugate addition from the 'wrong' end of the enol. Proton transfer and reversal of the first addition give the product and the catalyst ready for another cycle. The sequence is not only biomimetic in using a thiamine mimic. It also uses Nature's familiar trick: decarboxylation rather than removal of a proton from carbon.

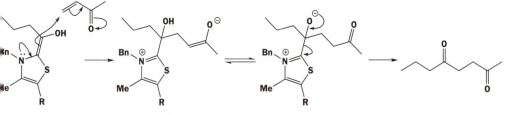

Now to make *cis*-jasmone. The compound is an enone so an aldol reaction is called for and the starting material is a 1,4-diketone. Then we can use the Stetter reaction.

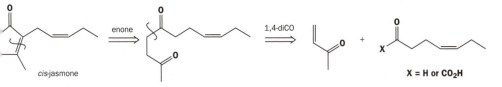

The required aldehyde or keto-acid can be made in a number of ways with the *cis*-alkene probably best derived from an alkyne. Here are two possibilities.

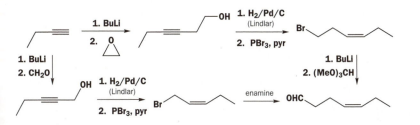

■ H. Stetter and H. Kuhlmann, *Synthesis*, 1975, 379.

The published synthesis uses Stetter's method with the aldehyde rather than the keto-acid as the source of the d^1 reagent.

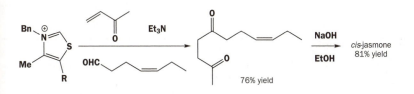

Problem 4

The amino acid cyanoalanine is found in leguminous plants (*Lathyrus*) but not in proteins. It is made in the plant from cysteine and cyanide by a two-step process catalysed by pyridoxal phosphate. Suggest a detailed mechanism.

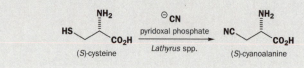

Purpose of the problem

Exploration of a new reaction in pyridoxal chemistry using pyridoxal itself rather th‍ pyridoxamine.

Suggested solution

The reaction starts the same way as all the others and what looks at first like an S_N2 displacement tur‍ out to be an elimination followed by a conjugate addition. Any attempt at an S_N2 displacement wou‍ just remove the proton from sulfur. We use the shorthand pyridoxal structure again.

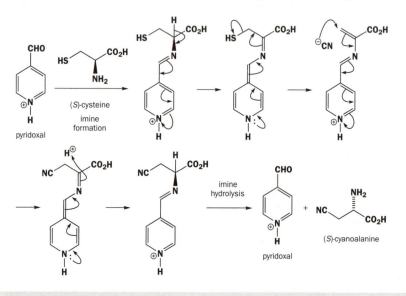

Problem 5

This chemical reaction might be said to be similar to a reaction in the shikimic acid pathway. Compare the two mechanisms and suggest how the model might be made closer and more interesting.

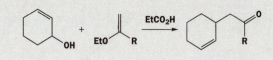

Purpose of the problem

Anchoring the key reaction of the shikimic acid pathway with chemistry from Chapter 36.

Suggested solution

The 'reaction in the shikimic acid pathway' is, of course, the [3,3]-sigmatropic shift in which chorismic acid rearranges to prephenic acid on the way to aromatic rings (p. 1403). The simpler reaction given here is one of the family of reactions from Chapter 36 (pp. 944–6) using an allylic alcohol and an enol derivative of a carbonyl compound. In this case we have the enol ether of a ketone. We must combine these to make an allyl vinyl ether for rearrangement.

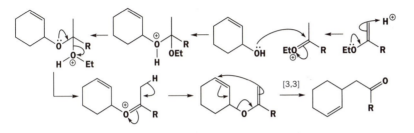

The biomimicry is quite good in that the rearrangement takes place around a six-membered ring and a ketone is formed in the right place in the side chain. It would be better if a keto-acid were formed and if there were some stereochemistry elsewhere in the six-membered ring so that we could see that it is a suprafacial migration. There is also a second alkene in chorismic acid. However, this reaction gives us the reassurance that [3,3]-sigmatropic rearrangements of the type involved in the shikimic acid pathway do indeed take place.

Problem 6

Stereospecific deuteration of the substrate for enolase, the enzyme that makes phosphoenol pyruvate, gives the results shown below. What does this tell us definitely about the reaction and what might it suggest about the mechanism?

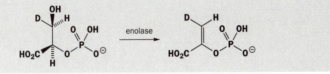

Purpose of the problem

Showing that biological mechanisms too can be explored by isotopic labelling and stereochemistry.

Suggested solution

The reaction is stereospecifically *anti*. This would fit an E2 mechanism but such a reaction could not occur with hydroxide as a leaving group. The OH group would first have to be turned into a good leaving group by a reaction such as phosphorylation. A more likely mechanism would be an E1cB via the enol of the acid but that would not be stereospecific if it were a chemical reaction. When the molecule is bound to the enzyme it might not be able to rotate and an E1cB mechanism might be possible. Studying stereochemistry by isotopic labelling is a standard technique for exploring enzymatic mechanisms.

■ M. Cohn *et al.*, *J. Am. Chem. Soc.*, 1970, **92**, 4095.

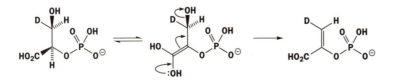

Problem 7

This rearrangement was studied as a biomimetic version of the NIH shift. Write a mechanism for the reaction. Do you consider it a good model reaction? If not, how might it be made better?

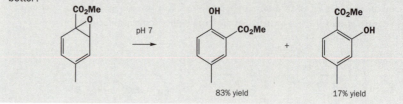

83% yield   17% yield

Purpose of the problem

Exploration of an enzymic mechanism by model reactions.

Suggested solution

The epoxide opens to give mainly a cation stabilized by the alkenes but not conjugated to the destabilizing ester group. The CO_2Et group then migrates. The minor product comes from opening the epoxide the other way round. The methyl group at the bottom of the ring is important as a marker to show definitely that the CO_2Et group has migrated.

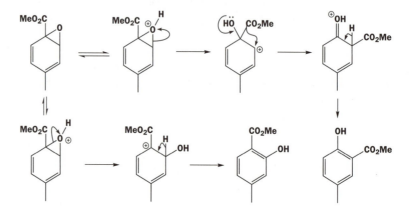

This is quite a good model reaction in that it shows that monoepoxides of benzenes can undergo a rearrangement quite like the NIH shift (pp. 1407–10) but in the real thing a hydrogen atom migrates with its electrons rather than the very different CO_2Et group. However, it is very difficult to make epoxides of aromatic rings without some substituent other than hydrogen and this is one limitation of model studies.

Problem 8

The following experiments relate to the chemical and biological behaviour of NADH. Explain what they tell us.

(a) This FAD analogue can be reduced *in vitro* with NADH in D_2O with deuterium incorporation in the product as shown.

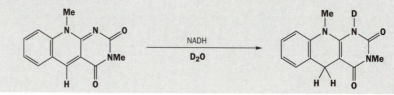

(b) NADH does not reduce benzaldehyde *in vitro* but it does reduce this compound.

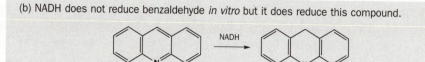

Purpose of the problem

Exploration of the mechanism of coenzyme reduction by model reactions using deuterium labelling and structural variation. Both these reactions were carried out in aqueous (D_2O in the first case) buffer solutions *without* enzyme.

Suggested solution

The full structure of NADH is on p. 1382 and that of FAD on p. 1407. We shall use abbreviated structures. The FAD analogue in part (a) of the question is very close to the real thing. In the reduction two atoms of hydrogen are added, one from NADH and one from the solution. The labelling experiment tells us which is which: hydride is transferred from NADH to the carbon atom of the FAD analogue and the nitrogen atom takes a proton from the solution. The same mechanism probably occurs in the real thing.

■ The original workers comment specifically on the analogy to FAD (M. Brüstlein and T. C. Bruice, *J. Am. Chem. Soc.*, 1972, **94**, 6548).

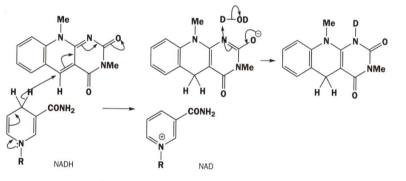

Reaction (b) was intended to model a simpler reduction by NADH. In the laboratory it is very difficult to find any simple carbonyl compounds that are reduced by NADH in the absence of the enzyme. Evidently, the mild general acid catalysis at the carbonyl group is needed for the reaction to occur. This model works because (1) the iminium salt is more reactive than a simple aldehyde and (2) the product contains two benzene rings.

■ 'The reduction of *N*-methylacridinium is considerably more rapid than other nonenzymic transhydrogenation reactions.' (D. S. Sigman and group, *J. Am. Chem. Soc.*, 1973, **95**, 6855)

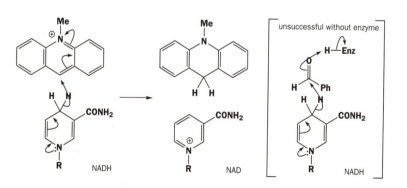

Problem 9

Oxidation of this simple thiol ester gives a five-membered cyclic disulfide. The reaction is proposed as a model for the behaviour of lipoic acid in living things. Draw a mechanism for the reaction and make the comparison.

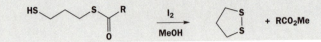

Purpose of the problem

Model reactions involving thiol oxidation to lead you into lipoic acid chemistry.

Suggested solution

Sulfur is a good soft nucleophile and the more reactive (not conjugated) SH group reacts first. The cyclization gives a sulfonium ion that is deacylated by methanol acting as a hard nucleophile and attacking the carbonyl group.

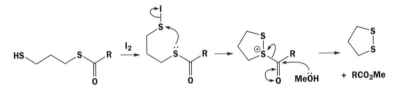

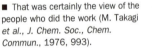

■ That was certainly the view of the people who did the work (M. Takagi et al., J. Chem. Soc., Chem. Commun., 1976, 993).

The chemistry of lipoic acid is described in the chapter (pp. 1394–5). A thiol ester of exactly the kind described in the problem is indeed involved and it is deacylated, though by CoASH rather than an alcohol, and the dithiol is indeed recycled as the disulfide. The oxidation to the cyclic disulfide is carried out in Nature by FAD. In the chapter we suggest that the oxidation of the dithiol to disulfide is a separate step from the transesterification. You may well think that the evidence in this problem suggests that the oxidation may take place first.

Problem 10

This curious compound is chiral – indeed it has been prepared as the (–) enantiomer. Explain the nature of the chirality.

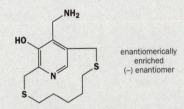

enantiomerically enriched (–) enantiomer

This compound has been used as a chemical model for pyridoxamine. For example, it transaminates phenylpyruvate under the conditions shown here. Comment on the analogy and the role of Zn(II). In what ways is the model compound worse and in what ways better than pyridoxamine itself?

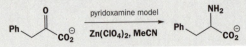

pyridoxamine model

urpose of the problem

evision of chirality without a chiral centre from Chapter 16 (pp. 398–9) and further exploration of ridoxal reactions.

uggested solution

hough the molecule is drawn in a planar fashion, with the sulfur-containing side chain lying in the ane of the pyridine ring, such a conformation is not possible as the chain is too short. In fact, the ain must twist across the top or bottom faces of the pyridine ring and cannot cross to the other de because the marked substituents – including the hydrogen atom – get in the way. It is not nown which enantiomer has the (–) rotation.

■ In spite of the low ees, this is an interesting example (H. Kuzuhara et al., Tetrahedron Lett., 1978, 3563).

The molecule is a chiral analogue of pyridoxamine and transamination occurs chemically by a echanism similar to the biological one described in the chapter (pp. 1384–6). The zinc holds the olecule in a fixed conformation during reaction. The key step is the protonation of the enamine as at produces the new chiral centre. If the chain is across the top of the ring, protonation occurs referentially from underneath. Hydrolysis gives the new amino acid (Phe) and the pyridoxal nalogue, which can be recycled by reductive amination via the oxime.

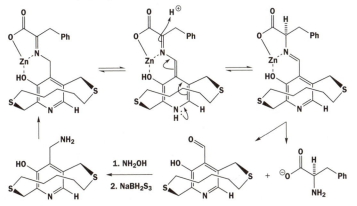

You may feel that the chain is too far away from the site of protonation to have much effect and ou'd be right. The best they could do was about 25% ee. In the biological system, the enzyme holds ne molecule in a fixed conformation and delivers the proton from one face.

Problem 11

Enzymes such as aldolase, thought to operate by the formation of an imine and/or an enamine with a lysine in the enzyme, can be studied by adding $NaBH_4$ to a mixture of enzyme and substrate. For example, treatment of the enzyme with the aldehyde shown below and $NaBH_4$ gives a permanently inhibited enzyme that on hydrolysis reveals a modified amino acid in place of one of the lysines. What is the structure of the modified amino acid, and why is this particular aldehyde chosen?

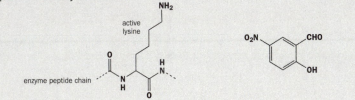

Purpose of the problem

Exploration of a method of finding the active site of an enzyme by irreversible inhibition.

Suggested solution

■ This was part of a programme of investigations into enzyme mechanisms by F. H. Westheimer and group, *J. Am. Chem. Soc.*, 1971, **93**, 7266, 7270. They called the *p*-nitrophenol a 'reporter group'.

The reaction used to inhibit the enzyme is our old friend, reductive amination. The amino group o the lysine is the most nucleophilic in the enzyme because it is involved in the active site and it form an imine with the aldehyde. Reduction fixes the aldehyde to the enzyme as a secondary amin Hydrolysis of the protein reveals a new amino acid in place of that lysine. This particular aldehy was used because it contains a UV-active (actually coloured) *p*-nitrophenol so that the modifie amino acid can be found among the many natural amino acids.

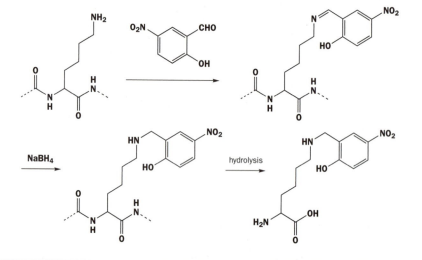

Problem 12

This question is about the hydrolysis of esters by 'serine' enzymes. First, interpret these results: The hydrolysis of this ester is very much faster than that of ethyl benzoate itself. It is catalysed by imidazole and then there is a primary isotope effect (Chapter 41) $k_{(OH)}/k_{(OD)} = 3.5$. What is the mechanism? What is the role of the histidine?

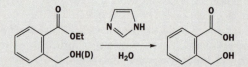

The serine enzymes have a serine residue vital for catalysis. The serine OH group is known to act as a nucleophilic catalyst. Draw out the mechanism for the hydrolysis of *p*-nitrophenyl acetate.

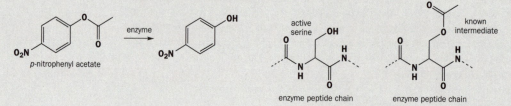

The enzyme also has a histidine residue vital for catalysis. Use your mechanism from the first part of the question to say how the histidine residue might help. The histidine residue is known to help both the formation and the hydrolysis of the intermediate. The enzyme hydrolyses both *p*-nitrophenyl acetate and *p*-nitrophenyl thiolacetate at the same rate. Which is the rate-determining step?

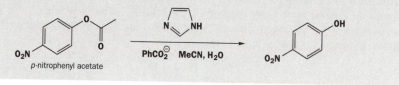

p-nitrophenyl acetate   p-nitrophenyl thiolacetate

Finally, an aspartic acid residue is necessary for full catalysis and this residue is thought to use its CO_2^- group as a general base. A chemical model shows that the hydrolysis of *p*-nitrophenyl acetate in aqueous acetonitrile containing sodium benzoate and imidazole follows the rate law: rate = k[*p*-nitrophenyl acetate] [benzoate] [imidazole]. Suggest a mechanism for the chemical reaction.

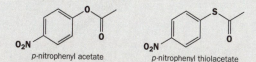

p-nitrophenyl acetate

Purpose of the problem

Exploration of a method of finding the active site of an enzyme by kinetics and by model studies. Revision of chapter 41.

Suggested solution

The normal primary deuterium isotope effect in the first part shows that an OH bond is being broken in the rate-determining step. Imidazole is too weak a base to remove the OH proton completely so its role must be as general base catalyst. Attack on the carbonyl is the slow step with faster breakdown of the tetrahedral intermediate and hydrolysis of the lactone. Lactones are hydrolysed faster than esters because they lack anomeric stabilization (p. 1134). The role of the OH group is intramolecular nucleophilic catalyst.

■ This investigation was specifically aimed at understanding the mechanism of action of serine enzymes (T. H. Fife and B. M. Benjamin, *J. Am. Chem. Soc.*, 1973, **95**, 2059).

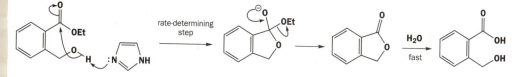

In outline, the mechanism of the enzyme-catalysed hydrolysis must use the serine OH as a nucleophilic catalyst to form the known intermediate and then hydrolyse that intermediate with water. The intermediate is an ordinary ester rather than a lactone so its hydrolysis will need help from the enzyme too.

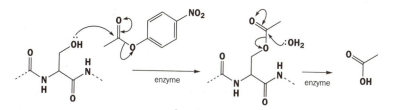

With the involvement of the imidazole group of histidine as a general base, as we deduced in the first part of the question, we can fill in more details of this outline mechanism. If the ester and the

■ The thiol ester experiments were performed by A. Frankfater and F. J. Kézdy, *J. Am. Chem. Soc.*, 1971, **93**, 4039.

thiol ester hydrolyse at the same rate, the rate-determining step must be the hydrolysis of the intermediate as this is the same in both cases.

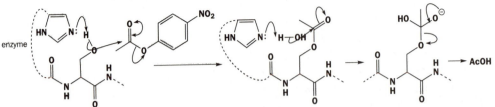

■ The 'proton relay' involvement of the carboxylate group was discovered by P. Haake and co-workers, *J. Am. Chem. Soc.*, 1971, **93**, 4938.

This mechanism is still not complete as we need extra acid or basic groups to put protons on to the tetrahedral intermediates as they form. From the requirement for the carboxylate anion, it looks as though the removal of the other proton from the imidazole of histidine helps. The complete mechanism is very complicated to draw and we shall draw just the rate-determining step of the chemical mechanism to show the main features.

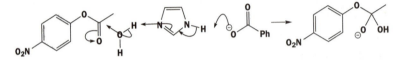

Problem 13

Give mechanisms for the biological formation of biopterin hydroperoxide and its reaction with phenylalanine. The reactions were discussed in the chapter but no details were given.

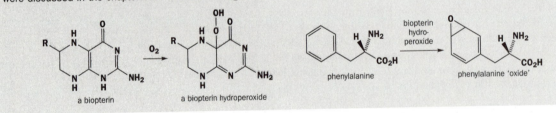

a biopterin   a biopterin hydroperoxide   phenylalanine   phenylalanine 'oxide'

Purpose of the problem

Development of ideas from the chapter (p. 1409).

Suggested solution

The reactions are likely to resemble those of FADH$_2$ with oxygen and benzene oxide formation (p. 1408). Don't forget that ordinary oxygen is a triplet (diradical) good at abstracting hydrogen atoms. The intermediate radical is tertiary and is stabilized by the carbonyl group, a nitrogen atom, and imine conjugation so it is long-lived enough to combine with a hydroperoxide radical.

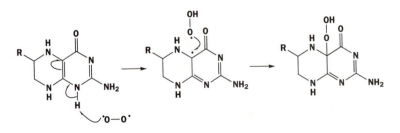

The biopterin hydroperoxide donates an electrophilic oxygen atom to the benzene ring, rather in the way of *m*-CPBA, though that reagent would not do the trick. The enzyme is essential. The other product is an oxidized biopterin and is recycled by reduction by NADPH back to the biopterin.

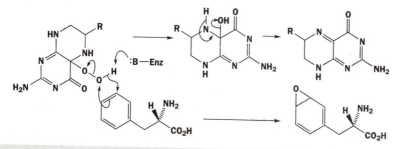

Problem 14

Revision of Chapter 48. How many electrons are there on the iron atom in the oxyhaemoglobin structure shown in the chapter? Does it matter if you consider the complex to be of Fe(II) or Fe(III)? Why are zinc porphyrins perfectly stable *without* extra ligands (L in diagram)?

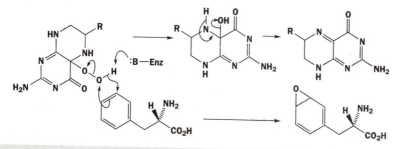

Purpose of the problem

Reminder of electron counting in organometallic compounds. Minor exploration of porphyrins.

Suggested solution

The iron porphyrin in the chapter (p. 1406) is an Fe(II) complex with 6 electrons from the iron and two each from the Fe–N bonds (see tables on pp. 1312–14). This makes 14 electrons in all so coordination to two more ligands would give the very favourable 18 electrons required for a stable complex. If you consider it to be an Fe(0) complex, you count 8 for the metal, one each from the two 'real' Fe–N bonds, and two each from the other two nitrogen atoms giving the same number. If you merely 'consider' it to be an Fe(III) complex you again get the same answer, but if you actually remove an electron, making it an Fe(III) cation, then there really is one electron fewer. Only one ligand makes a sixteen-electron complex able to react with oxygen.

The zinc porphyrin is a Zn(II) species (two 'real' Zn–N bonds and two dotted bonds) so it has 10 electrons. If we add eight for the four nitrogen ligands, we get 18 electrons. These compounds are stable and several substituted versions are sold by chemical companies. They will accept one extra ligand, for example, a pyridine, but this then pulls the zinc atom out of the plane of the porphyrin.

Suggested solutions for Chapter 51

51

Problem 1

Assign each of these natural products to a general class (such as amino acid metabolite, terpene, polyketide) explaining what makes you choose that class. Then assign them to a more specific part of the general class (for example, tetraketide, sesquiterpene).

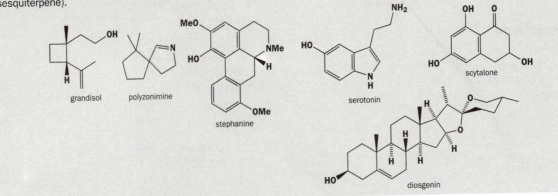

grandisol   polyzonimine   stephanine   serotonin   scytalone   diosgenin

Purpose of the problem

Practice at the pattern recognition needed to classify natural products.

Suggested solution

■ They are also an insect pheromone (grandisol) and defence substance (polyzonimine), stephanine is an aporphine alkaloid, scytalone a fungal metabolite, and diosgenin is a rare vegetable saponin from *Dioscorea*, the Mexican yam. It is useful as animal steroids can be made from it (p. 999).

Grandisol and polyzonimine have ten carbon atoms each with branched chains having methyl groups at the branchpoints. They are terpenes, specifically, monoterpenes. You might also have said that polyzonimine is an alkaloid as it is a basic nitrogen compound. Stephanine is definitely an alkaloid and belongs to the benzyl isoquinoline class. Serotonin is an amino acid metabolite derived from tryptophan. Scytalone has the characteristic unbranched chain and alternating oxygen substitution of a polyketide, an aromatic pentaketide in fact. Finally, diosgenine is steroid-like or, if you prefer, a triterpene metabolite.

Problem 2

Some compounds can arise from different sources in different organisms. 2,5-Dihydroxybenzoic acid comes from shikimic acid (Chapter 50) in *Primula acaulis* but from acetate in *Penicillium* species. Outline details.

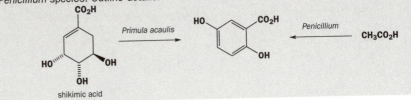

shikimic acid

Purpose of the problem

Applying known biosynthetic pathways to simple natural products. Showing that oxidations occur readily in Nature.

Suggested solution

There are various possibilities, depending on the line you choose for the oxidation of the aromatic ring. It is known that the pathway from shikimic acid involves phenylalanine and perhaps elimination of ammonia (p. 1404) but the stage at which you put the OH groups on the ring is uncertain.

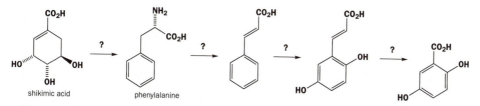

The most obvious pathway from acetate involves polyketide assembly, some sort of cyclization, oxidation of the methyl group, and oxidation of the ring. Again, you may have chosen a different order of events from ours or even a different pathway.

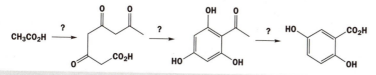

Problem 3

The piperidine alkaloid pelletierine was mentioned in the chapter but full details of its biosynthesis were not given. There follows an outline of the intermediates and reagents used. Fill in the details. Pyridoxal chemistry is discussed in Chapter 50.

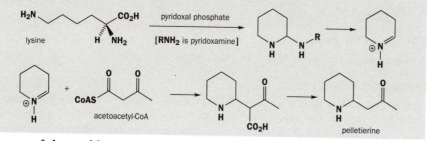

Purpose of the problem

A more thorough exploration of the biosynthesis of piperidine alkaloids.

Suggested solution

The piperidine alkaloids were mentioned in the chapter (p. 1418) where the pathway was described as 'similar to that of the pyrrolidine alkaloids'. We can follow the same reactions starting from lysine instead of ornithine. The first stage produces the cyclic iminium salt from lysine by decarboxylation with pyridoxal.

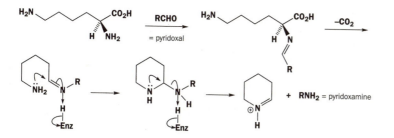

■ The saturated six-membered cyclic amine is piperidine.

Now the enol of acetyl CoA adds to the iminium salt to complete the skeleton of the piperidine alkaloids. Hydrolysis and decarboxylation by the usual cyclic mechanism (p. 678) gives pelletierine.

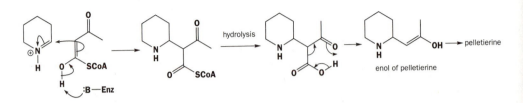

Problem 4

The rather similar alkaloids anabasine and anatabine come from different biosynthetic pathways. Labelling experiments outlined below show the origin of one carbon atom from lysine and others from nicotinic acid. Suggest detailed pathways. (*Hint.* Nicotinic acid and the intermediate you have been using in Problem 3 in the biosynthesis of the piperidine alkaloid are both electrophilic at position 2. You also need an intermediate derived from nicotinic acid which is nucleophilic at position 3. The biosynthesis involves reduction.)

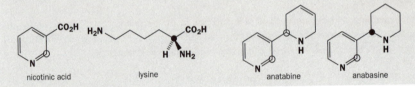

nicotinic acid	lysine	anatabine	anabasine

Purpose of the problem

A more thorough exploration of the biosynthesis of piperidine alkaloids.

Suggested solution

■ Anatabine and anabasine are both alkaloids of *Nicotiana* (tobacco) species. Their biosynthesis was elucidated by E. Leete, *J. Chem. Soc., Chem. Commun.*, 1975, 9.

The labelling and the hints given in the problem suggest an outline biosynthesis in which two molecules of nicotinic acid, one made nucleophilic by reduction, combine to give anatabine whilst the iminium salt we made in the last problem is attacked by the same nucleophilic derivative of nicotinic acid to give anabasine.

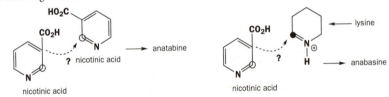

The most obvious derivative of nicotinic acid that is nucleophilic in the 3-position would be an enamine, easily made by reduction. Let us try that first with the cyclic iminium salt to see if we can get anabasine.

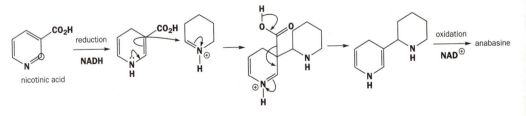

That worked rather well! Though reduction is needed to get the enamine, the final product must be oxidized back to a pyridine again so there is no overall oxidation or reduction. Let's try adding the same enamine to nicotinic acid as an electrophile. Now we need decarboxylation of both rings, oxidation of the left-hand ring, and reduction of the right-hand ring, all easily achieved with imines or enamines.

■ The same reduction of pyridinium salts to tetrahydropyridines with a 3,4 double bond occurs in the laboratory with NaBH$_4$.

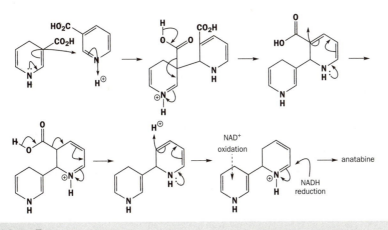

Problem 5

The three steps in the biosynthesis of papaverine set out below involve pyridoxal (or pyridoxamine). Write detailed mechanisms.

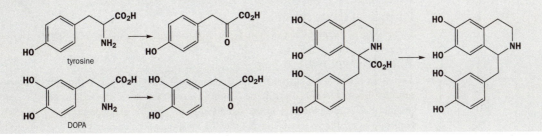

Purpose of the problem

Revision of pyridoxal mechanisms from Chapter 50 and extension to the biosynthesis of benzyl isoquinoline alkaloids.

Suggested solution

The basic transamination mechanism (pp. 1384–7) can be drawn for both tyrosine and DOPA by writing 'Ar' for either benzene ring.

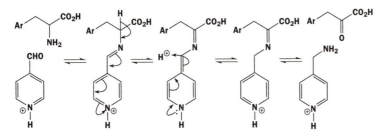

The decarboxylation of the complete benzyl isoquinoline system is slightly more complicated as the amine cannot form a simple imine, only an iminium salt. That is quite enough.

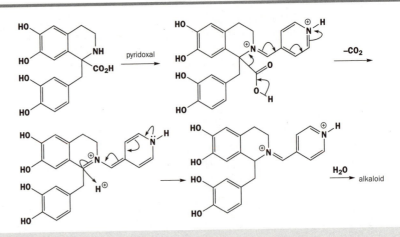

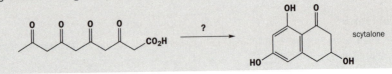

Problem 6

Concentrate now on the biosynthesis of scytalone in the first problem. You should have identified it as a pentaketide. Now consider how many different ways the pentaketide chain might be folded to give scytalone.

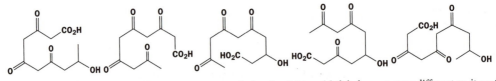

Purpose of the problem

An exercise in the folding of the polyketide chain to make a simple pentaketide.

Suggested solution

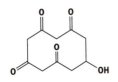

This is not as obvious as it seems. At first, the simplest thing to do is to restore the carbonyl groups from the benzene ring and break the central bond so that we see the fifth carbonyl group clearly (margin). We might then consider that the remaining OH group must have been formed by reduction of a carbonyl group and that the ring must have been closed by some sort of Claisen ester condensation. We can put the CO_2H group at any of the five sites so that gives us five possibilities.

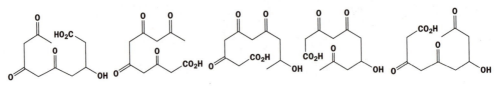

However, scytalone is not symmetrical – the right and left halves are very different so it matters which way round we wind the chain. This gives us another five possibilities.

■ The folding of the polyketide chain into scytalone was studied by double ^{13}C labelling by U. Sankawa et al., *Tetrahedron Lett.*, 1977, 483; 487.

Finally, and most easily overlooked, we don't have to wrap the chain round the outside of the two rings – we can twist it through the middle so that the central bond is not made in the cyclization. This gives us two more possibilities. There are 12 ways to fold the skeleton so that the results could be distinguished by the labelling pattern (particularly double labelling, p. 1426).

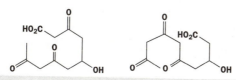

Problem 7

This question concerns the biosynthesis of stephanine, another compound mentioned in Problem 1. You should have deduced that it is a benzylisoquinoline alkaloid. Now suggest a biosynthesis from orientaline.

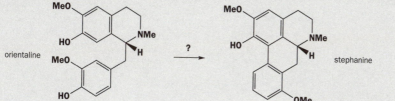

Purpose of the problem

Practice at recognizing the patterns of phenol coupling and what they mean.

Suggested solution

Stephanine looks like an *ortho/meta* coupling but we know it can't be – it must result from a rearrangement step. If you follow the obvious pathway, you should find something like what we have drawn. You may have done the migration before the reduction, and that's fine. It is actually thought to be a dienol-benzene rearrangement rather than a dienone-phenol. You may also be worried by our redrawing the *spiro* compound. This is just to make the migration easier to see and the two rings are at right angles to each other so you can draw them either way.

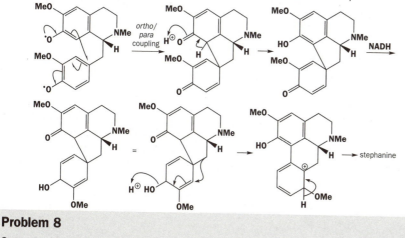

Problem 8

Suggest a biosynthesis of olivetol.

Purpose of the problem

Practice at recognizing the patterns of polyketides and the need for reduction and decarboxylation.

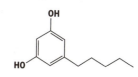

olivetol

Suggested solution

Olivetol has the characteristic 1,3-diOH pattern of a polyketide around the benzene ring but it has only 11 carbon atoms so CO_2 was probably lost somewhere and the long side chain has certainly been reduced. The acyl-polymalonate pathway (pp. 1425–36) needs to be used. We start from acetate and follow the sequence on pp. 1426–7 to get saturated butyroyl-SCoA. Then one acylation with reduction and further acylations without reduction take us to a compound that can be decarboxylated and cyclized to olivetol. The decarboxylation and cyclization might be concerted or either might occur before the other.

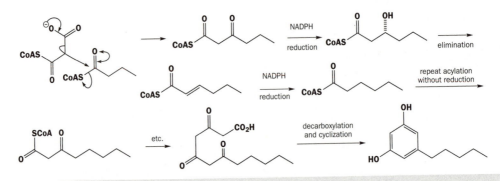

Problem 9

Tetrahydrocannabinol, the major psychoactive compound in marijuana, is derived in the *Cannabis* plant from olivetol and geranyl pyrophosphate. Details of the pathway are unknown. Make some suggestions and outline a labelling experiment to establish whether your suggestions are correct.

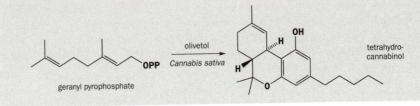

geranyl pyrophosphate   olivetol / *Cannabis sativa*   tetrahydro-cannabinol

Purpose of the problem

Your chance to speculate freely on the biosynthesis of a natural product from 'mixed metabolism'.

Suggested solution

The structure of olivetol is given in the last problem and it is immediately obvious which parts of tetrahydrocannabinol (THC) come from which starting material. The only question is: how do they join up? There is also the question of the geometry of the alkene. It is *E* in geranyl pyrophosphate and that cannot cyclize. However, we know at least one way (p. 1440) – allylic rearrangement – of solving that problem.

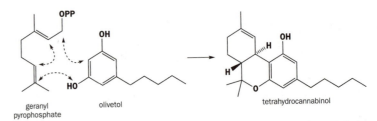

geranyl pyrophosphate   olivetol   tetrahydrocannabinol

There are many possible approaches all involving *O*-alkylation at one end, *C*-alkylation at the other, and cyclization. Here is one scheme; you may have thought of others as good. The first step is electrophilic aromatic substitution (Chapter 22) on a reactive benzene ring (two OH groups) with a reactive allylic pyrophosphate.

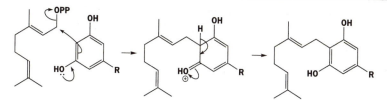

Next, some sort of allylic oxidation at the position between the alkene and the benzene ring. This would be with some oxygenation reagent such as FAD peroxide. The position is very reactive and would easily form radicals (Chapter 39).

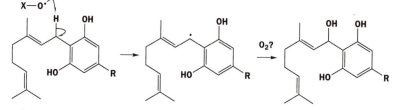

Now the stage is set for the allylic rearrangement, presumably of the pyrophosphate, and cyclization either in two steps or in the style of the polyolefin cyclizations in the chapter (pp. 1444–7). In either case we should expect a *trans* ring junction to be preferred.

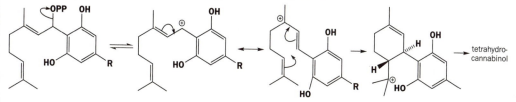

The experiments you would want to do would include studying the incorporation of labelled (even doubly labelled) geranyl pyrophosphate and olivetol, synthesizing some of your proposed intermediates in labelled form, and seeing if they were incorporated into THC in the plant. However, attempts at these experiments have not been very successful because of poor incorporation of labelled compounds into the plant.

Problem 10

Both humulene, mentioned in the chapter, and caryophyllene are made in nature from farnesyl pyrophosphate in different plants. Suggest detailed pathways. How do the enzymes control which product is formed?

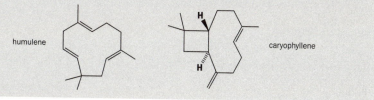

Purpose of the problem

Some serious terpene biosynthesis for you to unravel.

Suggested solution

These closely related compounds are clearly sesquiterpenes from the number of carbon atoms (15) and the pattern of their methyl groups. They can both be derived from the same intermediate by cyclization of farnesyl pyrophosphate without the need to isomerize an alkene. The eleven-membered ring can take three *E*-alkenes. Humulene is formed simply by losing a proton.

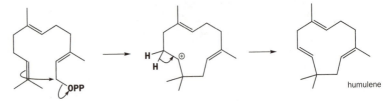

Caryophyllene needs a second cyclization to give a four-membered ring – the stereochemistry is already there as the molecule folds – and a proton loss. The enzymes control which product is formed by providing basic groups to remove protons from specific sites in the molecules and by folding the chain so that the reaction centres are close to each other.

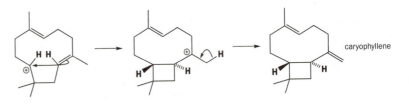

Problem 11

Abietic acid is formed in nature from mevalonate via the intermediates shown. Give some more details of the cyclization and rearrangement steps and compare this route with the biosynthesis of the steroids.

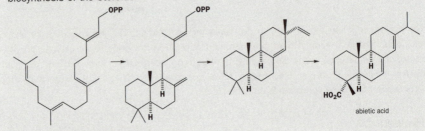

Purpose of the problem

An exercise in terpene synthesis by polyolefin cyclization.

Suggested solution

The initial cyclization must occur by protonation of the terminal alkene. It is easier to see what happens if we move immediately to conformational drawings and write 'R' for the side chain that does nothing in this first cyclization. The stereochemistry of the A/B ring junction is stereospecifically produced from the *E* geometry of the alkene and that of the B/C ring junction from the chair-like folding of the chain.

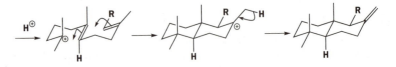

It is easy to draw a mechanism for the second cyclization as the nucleophilic end of the *exo*-methylene alkene attacks the allylic pyrophosphate to form a new six-membered ring.

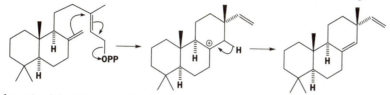

Conformational drawings reveal that the product has the larger group equatorial but it is not so easy this time to draw good conformations for the cyclization itself. We need to get the p orbitals of the alkenes pointing at each other. The best we can do is to form the new six-membered ring initially in a twist-boat conformation and flip it into a chair afterwards. You may well have done better.

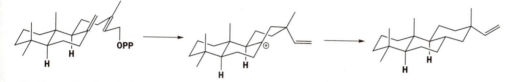

Finally, a migrationelimination sequence starting with protonation of the vinyl group leads to abietic acid itself. The methyl migration and proton loss in this step and the two previous cyclizations are, in their own small way, quite like the sequence of reactions in steroid biosynthesis (pp. 1445–7) except that the sequences are initiated by protonation rather than epoxidation and they are a series of small steps rather than one majestic sequence. The conformational control by folding of the chain is very similar.

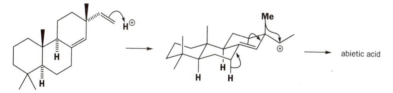

Problem 12

Borneol, camphene, and α-pinene are made in nature from geranyl pyrophosphate. The biosynthesis of α-pinene and the related camphor are described in the chapter. In the laboratory bornyl chloride and camphene can be made from α-pinene by the reactions described below. Give mechanisms for these reactions and say whether you consider them to be biomimetic.

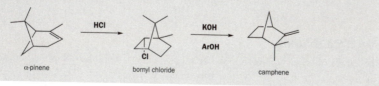

α-pinene   bornyl chloride   camphene

Purpose of the problem

Some model reactions based on monoterpene biosynthesis.

Suggested solution

Protonation of α-pinene, migration of the bridge, and attack by chloride give bornyl chloride. Or do they? Two questions must be answered. Why does that bridge (CMe$_2$) migrate and not the other

■ This reaction was briefly discussed on p. 983 but its stereoselectivity was not then revealed.

(CH$_2$) and why is only one diastereoisomer of the chloride formed? It is reasonable for the more highly substituted carbon atom to migrate (pp. 993–7). However we know that the resulting cation, if formed, would give the other diastereoisomer of the chloride (frame) so chloride addition must be concerted with migration and *anti* to the migrating group.

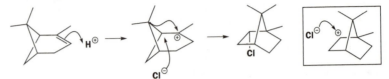

■ A similar rearrangement is discussed on p. 981.

The second reaction involves formation of the cation (the one that is not an intermediate in the first reaction), migration, and elimination. It is difficult at first to see the structure of the resulting cation so we have first drawn it deliberately badly but in the shape of the starting material. These reactions are very biomimetic as almost identical rearrangements occur in the biosynthesis of camphor and its relatives (p. 1440).

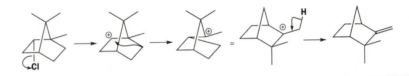

Problem 13

Suggest a biosynthetic route to the monoterpene chrysanthemic acid that uses a reaction similar to the formation of squalene in steroid biosynthesis.

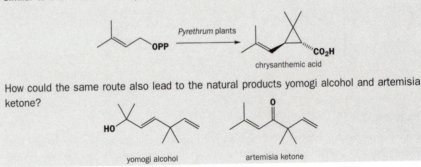

chrysanthemic acid

How could the same route also lead to the natural products yomogi alcohol and artemisia ketone?

yomogi alcohol   artemisia ketone

Purpose of the problem

To show that the complicated but important formation of squalene has less important but more easily understood analogies.

Suggested solution

The details of squalene biosynthesis are on pp. 1442–4. The biosynthesis of chrysanthemic acid requires only the first few steps of this sequence. The greatest simplification is that 'R' in the squalene biosynthesis is now just a methyl group. The stereochemistry in the three-membered ring comes from the stereoselectivity of the enzyme-controlled ring closure reaction. Hydrolysis of the pyrophosphate and oxidation of the alcohol are needed to get the acid.

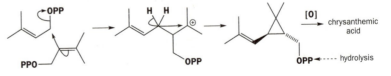

The other two natural products seem to have the wrong skeleton for monoterpenes and that might have given you the clue that they are derived from the three-membered ring before it has lost the pyrophosphate group. This chemistry resembles a simpler version of the chemistry of presqualene pyrophosphate (pp. 1443–4). The intermediate allyl cation can add water from either end.

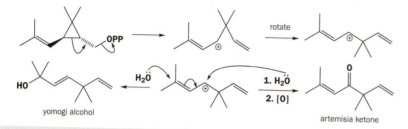

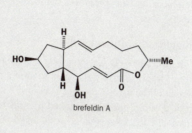

yomogi alcohol

artemisia ketone

Problem 14

In the chapter we suggested that you could detect an acetate starter unit and seven malonate additional units in the skeleton of brefeldin. Give the mechanism of the addition of the first malonyl CoA unit to acetate. Draw out the structure of the complete acyl polymalonate chain and state clearly what must happen to each section of it (reduction, elimination, etc.) to get brefeldin A.

brefeldin A

Purpose of the problem

Reminding you of the different and varied possible fates of each malonate unit added to the polyketide chain.

Suggested solution

The details of the polyketide pathway are on pp. 1425–30. The first addition is a decarboxylation of the malonate and a nucleophilic substitution on acetyl CoA in one step (in frame). After the addition of each unit, reduction to an alcohol (stereoselective), elimination to the unsaturated acid, and reduction to the saturated acid may occur.

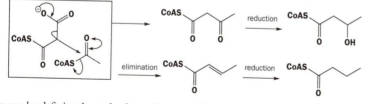

We can be definite about the fates of some of the units of brefeldin, as the diagram shows. The units forming the five-membered ring are less certain. If the ring is formed by oxidation, as in prostaglandin synthesis (pp. 1431–2), elimination is needed before oxidation.

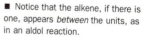

■ Notice that the alkene, if there is one, appears *between* the units, as in an aldol reaction.

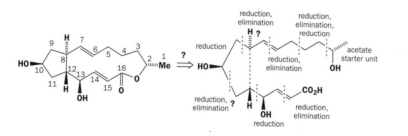

Problem 15

This chemical experiment aims to imitate the biosynthesis of terpenes. A mixture of products results. Draw a mechanism for the reaction. To what extent is it biomimetic, and what can the natural system do better?

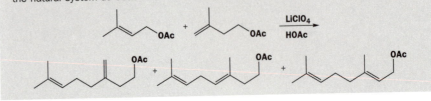

Purpose of the problem

Reminder of the weaknesses inherent in, and the reassurance possible from, biomimetic experiments.

Suggested solution

■ These experiments still give us confidence that the rather remarkable mechanisms proposed in biosynthesis are feasible (M. Julia et al., *J. Chem. Res (S)*, 1978, 268, 269).

The relatively weak leaving group (acetate) is lost from the allylic reagent with Lewis acid catalysis from LiClO$_4$ to give a stable allyl cation. This couples with the isopentenyl acetate in a way very similar to the natural process. However, what happens to the resulting cation is not well controlled. Loss of one of the three marked protons from the cation gives each of the three products. In the enzymatic reaction, proton loss would be concerted with C–C bond formation and a basic group, such as the imidazole of histidine or a carboxylate anion, would be at the right position to remove one of the protons selectively.

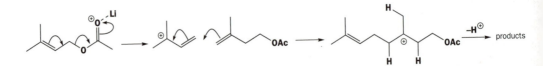

Suggested solutions for Chapter 52

52

Problem 1

The monomer bisphenol A is made by the following reaction. Suggest a detailed mechanism.

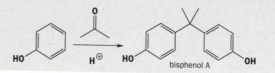

Purpose of the problem

Revision of electrophilic aromatic substitution (Chapter 22) and formation of a difunctional spacer, first step in polymerization.

Suggested solution

Phenol is a nucleophilic aromatic compound (p. 557) and reacts even with acetone in acidic solution. The second reaction with a tertiary benzylic alcohol, is a more conventional Friedel–Crafts alkylation. Phenol is *ortho*, *para*-directing and the *para* product is favoured in both cases for steric reasons.

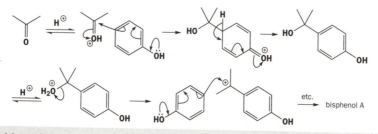

Problem 2

An alternative synthesis of 18-crown-6 to the one given in the chapter is outlined below. How would you describe the product in polymer terms? What is the monomer? How would you make 15-crown-5?

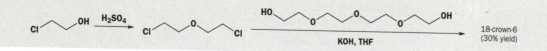

Purpose of the problem

Getting used to the ideas and terminology of polymerization.

Suggested solution

The product is a cyclic oligomer or, more exactly, a hexamer and the diol intermediate a tetramer. The monomer is ethylene glycol (1,2-dihydroxyethane), though that is not actually used in the synthesis. Using chloroethanol in acid to start with allows dimer formation without danger of polymerization as only the protonated alcohol can be displaced by the alcohol as nucleophile.

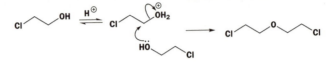

For the second reaction, basic conditions are used so that chloride can be displaced by the oxyanions. The leaving groups (Cl) are in one molecule and the nucleophiles in the other so that the first reaction must be intermolecular. Thereafter, cyclization is preferred to a second intermolecular reaction. The complexation of potassium in the reacting species helps the cyclization. Even so, the yield is only moderate.

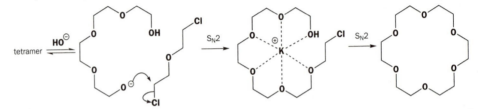

To make 15-crown-5 we should use the same bischloroether but use the diol trimer as the nucleophile. The cyclization will not be so favourable but should be all right.

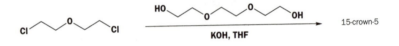

Problem 3

Melamine is formed by the trimerization of cyanamide and a hint was given in the chapter as to the mechanism of this process. Expand that hint into a full mechanism.

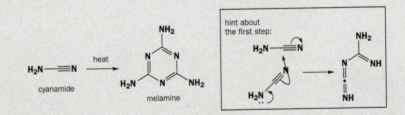

Melamine is polymerized with formaldehyde to make formica. Draw a mechanism for the first step in this process.

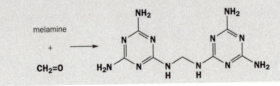

Purpose of the problem

An exercise in trimerization and simple linear polymerization.

uggested solution

sing the hint, we have the dimerization step and then we have a number of options to get to the yclic trimer. As long as you use amines as nucleophiles and imines as electrophiles, the precise rder of events doesn't matter very much. Cyclization to form a six-membered ring will beat ddition of a fourth cyanamide molecule and thermodynamically the trimerization is very avourable as melamine is aromatic.

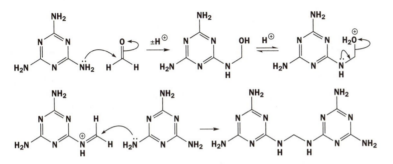

The reaction with formaldehyde is essentially aminal formation. The chemistry of the olymerization and cross-linking of this mixture is in the chapter (p. 1468).

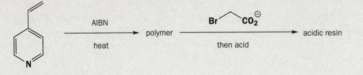

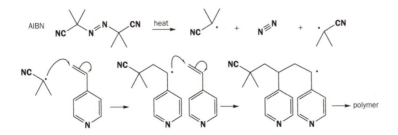

Problem 4

An acidic resin can be made by the polymerization of 4-vinylpyridine initiated by AIBN and heat followed by treatment of the polymer with bromoacetate. Explain what is happening and give a representative part structure of the acidic resin.

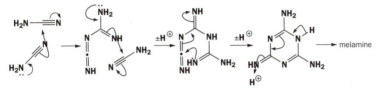

Purpose of the problem

Exploring radical polymerization and functionalization of polymers.

Suggested solution

AIBN forms radicals (p. 1041) that add to the vinyl group and initiate a radical polymerization. The polymer is rather like polystyrene except for the nitrogen atom in the pyridine ring.

Bromoacetate is an excellent electrophile for S_N2 reactions so some of the pyridine rings in the polymer displace bromide and form a neutral functionalized polymer. Protonation gives the free acid

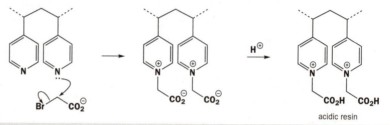

acidic resin

Problem 5

An artificial rubber may be made by cationic polymerization of isobutene using acid initiation with BF_3 and water. What is the mechanism of the polymerization, and what is the structure of the polymer?

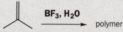

This rubber is too weak to be used commercially and 5–10% isoprene is incorporated into the polymerizing mixture to give a different polymer that can be cross-linked by heating with sulfur (or other radical generators). Draw representative structures for sections of the new polymer and show how it can be cross-linked with sulfur.

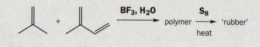

Purpose of the problem

Exploring cationic polymerization with an introduction to strengthening by cross-linking.

Suggested solution

The stable *t*-butyl cation initiates the polymerization. As each monomer is added, a new tertiary cation is formed and the polymer grows with alternate CH_2 and CMe_2 groups.

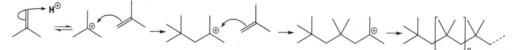

When the diene is incorporated, some methyl groups will be replaced by vinyl side chains. These can be cross-linked with sulfur in the manner of rubber. Rubber is, of course, polyisoprene.

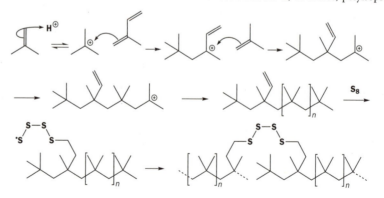

Problem 6

When sodium metal is dissolved in a solution of naphthalene in THF, a green solution of a radical anion is produced. What is its structure?

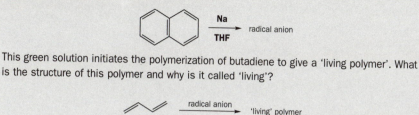

This green solution initiates the polymerization of butadiene to give a 'living polymer'. What is the structure of this polymer and why is it called 'living'?

Purpose of the problem

Introduction to electron transfer polymerization.

Suggested solution

Naphthalene acts as a stable source of single electrons. It accepts an electron from sodium atoms to give a radical anion that can be drawn in various ways and provides the green solution. We have drawn it like a Birch reduction intermediate (pp. 628–9).

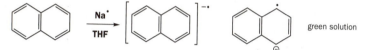

Transfer of this electron to the diene gives a delocalized allylic radical anion, which can be drawn in various ways and which attacks another diene. This process can also be drawn in various ways: as a radical dimerization, for example.

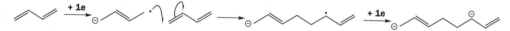

This dimer is symmetrical with an allylic anion at each end (we have drawn them differently but they are delocalized). Each can add to further butadienes in anionic polymerization until the butadiene is used up. We have shown the arrows for the next two additions.

The process ends when the butadiene is all gone and at that moment the polymer has an active allylic anion at each end. It is therefore able to polymerize further – it is a 'living polymer' – and the process is often brought to an end by adding some styrene to cap the polymer and then some water to protonate the anions produced.

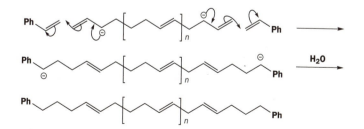

Problem 7

We introduced the idea of a spacer between a benzene ring (in a polystyrene resin) and a functional group in the chapter. If a polymer is being designed to do Wittig reactions, why would it be better to have a Ph_2P group joined directly to the benzene ring than to have a CH_2 spacer between them?

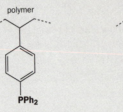

useful for Wittig reactions   useless for Wittig reactions

If you need a hint, draw out the reagents you would add to the polymer to do a Wittig reaction and work out what you would get in each case.

Purpose of the problem

Introduction to polymer-supported reagents and your chance to avoid a trap.

Suggested solution

The first polymer, with the Ph_2P group joined directly to the polymer, does a straightforward Wittig reaction without mishap. After the reaction, the P=O groups in the polymer must be reduced back to Ph_2P groups before the next application.

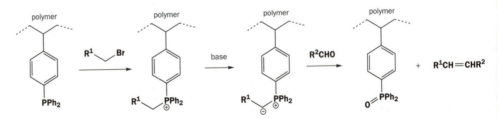

The other polymer starts in the same way, but treatment with base removes the wrong proton as the CH_2 group between the polymer and the PPh_2 group is benzylic and has more acidic protons than those on the other CH_2 group unless R^1 is a very anion-stabilizing group like CO_2Et. Worse, the product is left attached to the polymer while the phosphorus atom is lost and the polymer cannot be used again.

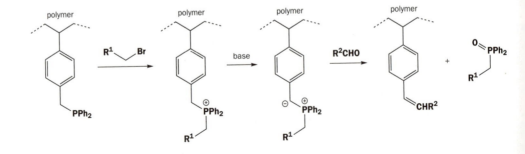

Problem 8

A useful reagent for the oxidation of alcohols is 'PCC' (pyridinium chlorochromate). Design a polymeric (or at least polymer-bound) reagent that should show similar reactivity. What would be the advantage of the polymer-bound reagent over normal PCC?

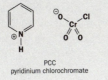

PCC
pyridinium chlorochromate

Purpose of the problem

Extension of your understanding of polymer-bound reagents into invention.

Suggested solution

The most obvious thing would be to use the pyridine polymer from Problem 4 in this chapter and to react it with CrO_3 and HCl, the way to make normal PCC. Some of the pyridine rings in the polymer will react. You might object that the oxidizing agent, the Cr(VI)-containing anion, is not bound to the polymer, but it is. In solution, cations and anions float around independently of each other but on the polymer the two are held together by electrostatic attraction. There is no need of a covalent bond.

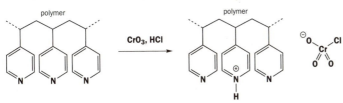

The advantage of the polymer-bound reagent would be easy isolation of the product by simply filtering off the reagent. Removing soluble Cr(III) salts from organic products can be a nuisance. The spent polymer, now containing Cr(III), could be reactivated by a more powerful oxidizing agent.

Problem 9

A polymer that might bind specifically to metal ions and be able to extract them from solution would be based on a crown ether. How would you make a polymer such as this?

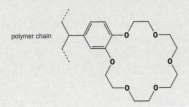

polymer chain

Purpose of the problem

A more difficult creative exercise in the synthesis of a functionalized polymer.

Suggested solution

Since the required polymer is a functionalized polystyrene, the most sensible approach would be to co-polymerize styrene and some 3,4-dihydroxystyrene, perhaps protected as an acetal or a silyl derivative. A proportion of the benzene rings in the polystyrene would have the correct functionalization and the crown ether could be built on to them by passing a large excess of a suitable reagent, such as one of those we discussed in Problem 2 of this chapter or in the main text (p. 1456), deprotecting as required. A potassium salt would be used as a base in the final cyclization to take advantage of complexation by the crown ether. The various methods of polymerizing styrene (radical, anionic, etc.) are described in the chapter (pp. 1459–62).

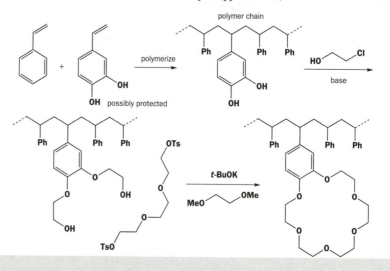

Problem 10

What is a 'block co-polymer'? What polymer would be produced by this sequence of reactions? What special physical properties would it have?

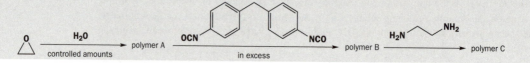

Purpose of the problem

Moving on to block co-polymerization as a development of the chemistry in the previous two problems.

Suggested solution

The polymerization of ethylene oxide cannot be controlled if there is enough water. It can be controlled only by limiting the amount of water when good yields of reasonable polymer lengths are formed rapidly, each containing one molecule of water (p. 1457). Polymer A is polyethylene glycol.

Each terminal OH group can react with an isocyanate to make a urethane at each end of the old polymer (A) with a free isocyanate on the extreme end of the new polymer, polymer B. Polymer B is sometimes called a 'pre-polymer' since it is the next step that links these units together in the final polymer.

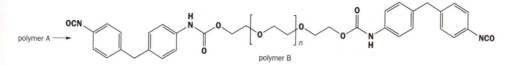

polymer A

polymer B

Now the diamine reacts with the free isocyanates by forming a urea at each end of the chain to link them together. It is possible to show only a small fragment of this massive polymer but that is enough to see the block of polyethylene glycol, the block of (one) diarylmethane, and the double urea linkage beyond. There is more detail of this sort of reaction in the chapter (p. 1458).

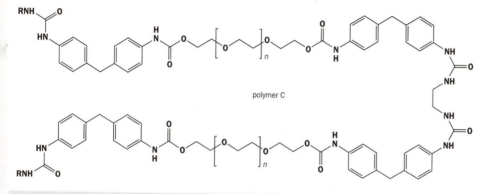

polymer C

Problem 11

Why does polymerization occur only at relatively low temperatures, often below 200°C? What occurs at higher temperatures? Formaldehyde polymerizes only below about 100°C but ethylene still polymerizes up to about 500°C. Why the difference?

Purpose of the problem

Going back to the fundamentals of polymerization, particularly the thermodynamics.

Suggested solution

Polymerization is favourable thermodynamically if the conversion of many π bonds into σ bonds is sufficiently favourable to make up for the inevitable negative entropy of the reaction. Polymerization always involves many molecules giving one so the entropy is always large and negative. The vital equation is (p. 313)

$$-RT \ln K = \Delta H^\circ - T\Delta S^\circ$$

Thus there must be a compensating gain in ΔH°, that is, in the stability of the bonds in the polymer compared with those in the monomer, to offset the loss of entropy. The relative contribution made by entropy increases as the temperature goes up so polymerization tends to be favoured at lower temperatures. At higher temperatures depolymerization tends to occur.

■ Depolymerization doesn't always occur as the polymer may decompose in some other way.

Formaldehyde contains a very strong C=O double bond while polyformaldehyde has C–O single bonds instead. The single bond is only slightly stronger than two double bonds (in other carbonyl compounds such as ketones, C=O is more than twice as strong as C–O). Low temperatures are needed for entropy not to play too big a part in the thermodynamics. Ethylene (ethene), on the other hand, has a weak C=C double bond while the polymer has C–C single bonds. These are much stronger than half a C=C so the gain in enthalpy during polymerization compensates for the loss of entropy at much higher temperatures.

Problem 12

Polyvinyl chloride (PVC) is used for rigid structures like window frames and gutters with only small amounts of additives such as pigments. If PVC is used for flexible things like plastic bags, about 20–30% of dialkyl phthalates such as the compound below are incorporated during polymerization. Why is this?

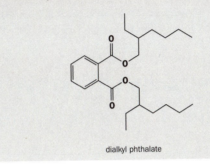

dialkyl phthalate

Purpose of the problem

One small problem on the composition of plastics rather than just polymers.

Suggested solution

PVC is a regular polymer with chains that fit together well to form a rigid solid ideal for window frames and gutters but it is inevitably unable to bend far without breaking. Dialkyl phthalates are plasticizers that fit around the rigid chains rather in the way that a solvent fits round solute molecules. The difference here is that the polymer molecules are larger than the 'solvent'. You could also think of the plasticizer as a lubricant to allow the polymer chains to move across each other to make the whole structure flexible. Plasticizers are used in substantial quantities (20–30% suggested in the question). In fact, dialkyl phthalates are made for the plastics industry in amounts greater than those for almost any polymer.